Series on Grey System

Series Editors

Sifeng Liu, Institute for Grey Systems Studies, Nanjing University of Aeronautics and Astronautics, Nanjing, Jiangsu, China

Yingjie Yang, Institute of Artificial Intelligence, De Montfort University, Leicester, UK

Jeffrey Yi-Lin Forrest, Department of Mathematics, Slippery Rock University, PA, PA, USA

This proposal aims to publish a series on grey system and various applications in the fields of natural sciences, social sciences and engineering.

This series is devoted to the international advancement of the theory and application of grey system. It seeks to foster professional exchanges between scientists and practitioners who are interested in the models, methods and applications of grey system. Through the pioneering work completed over 40 years, grey data analysis methods have become powerful tools in addressing system with poor information.

Books published with this series will explore the models and applications of grey system, in order to tackle poor information more effectively and efficiently. The series aims to provide state-of-the-art information and case studies on new developments and trends in grey system research and its potential application to solve practical problems.

Coverage includes, but is not limited to:

- Foundations of grey systems theory
- Grey sequence operators
- Grey relational analysis models
- Grey clustering evaluations models
- Techniques for grey system forecasting
- Grey models for decision-making
- Combined grey models
- Grey input-output models
- Techniques for grey control
- Various applications of grey system models in the fields of natural sciences, social sciences and engineering.

Sifeng Liu · Yingjie Yang · Jeffrey Yi-Lin Forrest

Grey Systems Analysis

Methods, Models and Applications

 Springer

Sifeng Liu
Institute for Grey Systems Studies
Nanjing University of Aeronautics
and Astronautics
Nanjing, Jiangsu, China

Faculty of Computing
Engineering and Media
School of Computer Science
and Informatics, Institute of Artificial
Intelligence
De Montfort University
Leicester, Leicestershire, UK

Jeffrey Yi-Lin Forrest
Institute for Grey Systems Studies
Nanjing University of Aeronautics
and Astronautics
Nanjing, Jiangsu, China

International Institute for General System
Studies
Slippery Rock University
Slippery Rock, PA, USA

Yingjie Yang
Institute for Grey Systems Studies,
Nanjing University of Aeronautics
and Astronautics,
Nanjing, Jiangsu, China

Faculty of Computing, Engineering
and Media, School of Computer Science
and Informatics, Institute of Artificial
Intelligence
De Montfort University
Leicester, Leicestershire, UK

ISSN 2731-4936 ISSN 2731-4944 (electronic)
Series on Grey System
ISBN 978-981-19-6162-5 ISBN 978-981-19-6160-1 (eBook)
https://doi.org/10.1007/978-981-19-6160-1

This Springer imprint is published by the registered company Springer Nature Singapore Pte Ltd.
The registered company address is: 152 Beach Road, #21-01/04 Gateway East, Singapore 189721,
Singapore

This work was made possible due to projects supported by the national major talent programme of China, the Marie Curie International Incoming Fellowship of the European Union, the National Natural Science Foundation of China, the Leverhulme Trust International Network, and the joint projects supported by the NSFC and the RS in the UK.

Series Preface

This series will publish the books on grey system theory and various applications in the fields of natural sciences, social sciences and engineering.

It is devoted to the international advancement of the theory and application of grey system theory and seeks to foster professional exchanges between scientists and practitioners who are interested in the models, methods and applications of grey system theory. Through the pioneering work completed over 40 years, grey system analysis methods have become powerful tools in addressing system with poor information.

Books published with this series will explore the models and applications of grey system theory, in order to tackle poor information more effectively and efficiently. The series aims to provide state-of-the-art information and case studies on new developments and trends in grey system research and its potential application to solve practical problems.

In the era of big data, the grey system theory based on poor information data mining has sprung up. It has become an effective tool for people to extract valuable information from massive data. In the past 40 years, grey system method and model have been widely used in many fields, such as social science, natural science and engineering technology, which has led to innovation and progress in various fields. More and more people interested in grey system theory and a lot of new results have been obtained in recent years. In particular, successful applications in many fields have won the attention of the international world of learning.

Scholars from more than 100 countries and regions in the world have published more than 300,000 documents of grey system research and applications.

On the 7th of September 2019, Angela Dorothea Merkel, then German Chancellor, praised grey system theory in her speech at Huazhong University of Science and Technology. She said that the work of Prof. Deng Julong, the founder of grey system theory, and Prof. Sifeng Liu, the editor of this series, "have made a profound impact on the world".

The coverage of this series includes, but is not limited to:

- Foundations of grey systems theory
- Grey sequence operators
- Grey relational analysis models
- Grey clustering evaluations models
- Techniques for grey system forecasting
- Grey models for decision-making
- Combined grey models
- Grey input–output models
- Techniques for grey control
- Various applications of grey system models in the fields of natural sciences, social sciences and engineering.

If you are interested in the series on grey systems, please contact with Ms. Emily Zhang at emily.zhang@springernature.com or Prof. Sifeng Liu at sfliu@nuaa.edu. cn.

Nanjing, China   Prof. Sifeng Liu, Ph.D.
Editor of the Book Series on Grey
System, Director of Institute for Grey
Systems Studies, NUAA, President
of International Association of Grey
System and Uncertain Analysis

Foreword by Dr. James M. Tien

It gives me great pleasure to be introducing this 8th edition of *Grey System Theory and Its Applications* by Prof. Sifeng Liu. The theory of grey systems was first introduced in 1982 by J. L. Deng (1933–2013) at Huazhong University of Science and Technology; it established a relatively new approach for addressing poorly defined problems with a high level of greyness or uncertainty. The theory enables one to model, analyse, monitor and control such partially defined systems by generating, excavating and extracting useful information from what is available. It built on the work of Dr. Lotfi A. Zadeh, who introduced the concept of fuzzy sets in the 1960s that in turn led to breakthroughs in neural networks and soft computing.

Grey System Theory actually combines two critical and overarching areas. The first concerns systems which attempt to synthesize the various components or subsystems into an overall functioning system or system of systems. Systems theory attempts to make transparent the deep connections and interactions among objects and events, all leading to the enrichment and progress of science and technology. Many of the historically difficult, hard-to-solve problems in the different scientific fields have been successfully resolved through the application of systems theory and its allied methodologies, including information theory, cybernetics, combinatorics, genetics, etc. The second concerns the greyness or uncertainty level that is implicit in all natural or man-made systems. Indeed, most modelling techniques assume the existence of uncertainty or stochasticity, as defined by either empirical evidence or assumed distributions, including fuzzy sets.

Grey System Theory, then, provides a realistic approach to modelling, analysing, monitoring and controlling systems. Professor Sifeng Liu has greatly extended, if not expanded, his earlier efforts. In the 1980s, he put forward a series of new models and concepts, including sequence operator, absolute degree of grey incidence, grey cluster evaluation model with fixed weight, and positioned coefficient of grey matrix. In the 1990s, he proposed a buffer operator and its axiom, generalized degree of grey incidence, grey number and measurement of its information content, drifting and positioning solution, the grey econometrics model GM(1,1), the grey Cobb–Douglas model, etc. More recently, he proposed the concept of general grey numbers, the grey algebraic system based on a kernel and degree of greyness, and different variations of the model GM(1,1).

The widespread recognition and application of grey system theory reflect its growing acceptance. A number of universities from around the world have adopted Prof. Sifeng Liu's monographs, both in Chinese and English, as their textbooks. In 2002, he won the World Organization of Systems and Cybernetics (WOSC) Prize. In 2008, as a preeminent Chinese scholar, he was elected an Honorary WOSC Fellow. In 2013, after a strict review by the European Commission, he was selected to be a Marie Curie International Fellow, thus honouring him as the first such Fellow with grey systems expertise.

As a systems scientist and engineer, I am honoured to write this preface for the 8th edition of *Grey System Theory and Its Applications*. I look forward to its widespread dissemination and its promulgation of grey system applications in science and engineering.

James M. Tien, Ph.D., D.Eng. (h.c.),
NAE
Distinguished Professor and Dean
Emeritus, College of Engineering
University of Miami
Coral Gables, FL, USA

Note Professor James M. Tien prepared this note for 8th edition of *Grey System Theory and Its Applications* (in Chinese) by the same authors, published in 2016. With his permission, it is printed here as a foreword for this current book.

Foreword by Dr. Keith William Hipel

Grey Systems: Theory and Applications
Written by Sifeng Liu and Jeffrey Yi-Lin Forrest
Springer-Verlag: Berlin, Heidelberg
2010, 379 pages, ISBN 978-3-642-16158-2 (cloth)
DOI: 10.1007/978-3-642-16158-2

Professors Sifeng Liu and Yi-Lin have written another pioneering book on the important topic of grey systems. In 2006, the same authors wrote the well-received book entitled *Grey Information: Theory and Practical Applications* which was also published by Springer-Verlag. I am pleased to say that their second book on Grey Systems constitutes a significant expansion and improvement of their previous fine book. Accordingly, if you already possess a copy of the 2006 book, you can make a worthwhile academic investment by obtaining a copy of their recent book in order to be cognizant of the latest ideas and advancements in the crucial field of grey systems.

The question that naturally arises is why grey systems are of such great import at this point in history. The answer is quite straightforward: many challenging problems facing society consist of interconnected complex systems of systems exhibiting high uncertainty and having few measurements. For example, in order to effectively combat climate change, one must understand as much as possible the complex interactions among natural systems such as atmospheric, oceanic, geological, and hydrological systems, with societal systems including energy production, industrial, agricultural and city systems. The deep uncertainty involved with these interconnected systems of systems and their potential emergent behaviour, coupled with a dearth of observations, mean that formal tools for handling this uncertainty are in high demand. Fortunately, an arsenal of mathematically based methodologies and techniques have been developed over the years: a rich variety of probabilistic-based tools, fuzzy sets founded by Lotfi Zadeh, rough sets started by Z. Pawlak, information-gap modelling perfected by Yakov Ben-Haim, uncertainty theory developed by Baoding Liu, and grey systems established by Julong Deng in 1982. The foregoing and other approaches to describing uncertainty are based upon different axioms and are thereby highly complementary for tackling a wide variety of uncertain situations.

Grey systems are purposefully designed for modelling uncertain systems, or systems of systems, problems having small samples and low-quality information. Grey systems are capable of dealing with partially known information through generating, excavating and extracting useful information from what is available. How this is accomplished is explained in depth in the timely grey systems book of Profs. Liu and Lin.

In their contemporary textbook, Liu and Lin systematically present the theory and practice of grey systems. In fact, the excellent ideas and applications contained in their book are based upon the authors' many years of developing theoretical concepts, applying their methods to real-world applications, testing and refining their new techniques with actual data, carrying out stimulating research with their students and colleagues, teaching their students about their exciting work and delivering research papers at international conferences around the globe. Their comprehensive book contains the latest theoretical and applied advances created by the authors and other scholars around the world in order to place the readers at the forefront of international research in grey systems.

The main body of their book contains ten well-explained and interconnected chapters: Introduction to Grey Systems Theory, Basic Building Blocks, Grey Incidence and Evaluation, Grey Systems Modelling, Discrete Grey Prediction Models, Combined Grey Models, Grey Models for Decision Making, Grey Game Models, Grey Control Systems and Introduction to Grey Systems Modelling Software. Moreover, the book includes a computer software package developed for grey systems modelling to permit both researchers and practitioners to use the new methodologies. Their book concludes with three appendices. The first appendix compares grey systems theory and interval analysis while revealing the fact that interval analysis is a part of grey mathematics. The second presents an array of different approaches to studying uncertainties. Finally, the last appendix shows how uncertainties occur using a general systems approach.

The book contains a wealth of mathematical results, techniques and algorithms which are presented by the authors for the first time. These contributions include an axiomatic system of buffer operators and a series of weakening and strengthening operators; axioms for measuring the greyness of grey numbers; general grey incidences (grey absolute incidence, grey relative incidence, grey comprehensive incidence, grey analogy incidence and grey nearness incidence); discrete grey models; fixed weight grey cluster evaluation; and grey evaluation methods based on triangular whitenization weight functions, multi-attribute intelligent grey target decision models, applicable range of the $G(1,1)$, grey econometrics (G-E), grey Cobb–Douglas (G-C-D), grey input–output (G-I-O) and grey game models (G-G).

In their well-written book, Drs. Liu and Lin do a thorough job in their presentation of many difficult technical concepts. The authors are able to convince the readers of their book regarding the power and usefulness of their new theory by presenting many interesting examples of practical applications to real-life problems. The challenging practical problems addressed in their book include urban economic planning, downtown traffic design, natural disaster prediction, relative strength evaluation of a state, investment projection of a company and employee performance evaluation.

The depth and scope of the advancements in grey systems covered in this book, in conjunction with clarity of explanation, make this seminal book attractive to researchers, students, teachers and practitioners working in many different fields. These areas of endeavour include image processing, video processing, multimedia security, computer vision, machinery, control, agriculture, water resources, medicine, astronomy, earth science, economics and management. I personally found grey systems useful for accurately forecasting wastewater time series for which there is a scarcity of data. I intend to keep a copy of this valuable book easily accessible in my university office and purchase more copies of the book for use by my students.

Keith William Hipel, Ph.D., P.Eng., FIEEE,
FINCOSE, FCAE, FEIC, FRSC, FAWRA
University Professor of Systems Design Engineering
University of Waterloo
Waterloo, ON, Canada
e-mail: kwhipel@uwaterloo.ca
Website: http://www.systems.uwaterloo/Faculty/Hipel
Senior Fellow
Centre for International Governance Innovation
Waterloo, ON, Canada

Note Professor Keith William Hipel prepared this note for one of the earlier book by the same authors, published in 2010. It is published in *Grey Systems: Theory and Application*, 2011, Vol. 1, No. 3. With his permission, it is printed here as a foreword for this current book.

Foreword by Dr. Hermann Haken

With human knowledge maturing and scientific exploration deepening and largely expanding in the course of time, mankind finally realizes the fundamental fact that due to both internal and external disturbances and limitations of human and technical sensing organs, all information received or collected contains some kind of uncertainty. Accompanying the progress of science and technology and the aforementioned realization, our understanding about various kinds of uncertainties has gradually been deepened. Attesting to this end, in the second half of the twentieth century, the continual appearance of several influential and different types of theories and methods on unascertained systems and information has become a major aspect of the modern world of learning. Each of these new theories was initiated and followed-up by some of the best minds of our modern time.

In their recent book, entitled "*Grey Information: Theory and Practical Applications*", published in its traditionally excellent way by Springer, Profs. Sifeng Liu and Yi-Lin presented in a systematic fashion the theory of grey system, which was first proposed by J. L. Deng in early 1980s and enthusiastically supported by hundreds of scientists and practitioners in the following years. Based on the hard work of these scholars in the past (nearly) thirty years, scholars from many countries currently are studying and working on the theory and various applications of this fruitful scientific endeavour. With this book published by such a prestigious leading publisher of the world, it can be expected that more scientific workers from different parts of the world will soon join hands and together make grey system and information a powerful theory capable of bringing forward practically beneficial impacts to the advancement of the human society.

This book focuses on the study of such unascertained systems that are known with small samples or "poor information". Different of all other relevant theories on uncertainties, this work introduces a system of many methods on how to deal with grey information. Starting off with a brief historical introduction, this book carries the reader through all the basics of the theory. And, each important method studied is accompanied with a real-life project the authors were involved in during their professional careers.

Many of the methods and techniques the reader will learn in this book were originally introduced by the authors. They show how from our knowledge based on partially and poorly known information can be obtained to accurate descriptions and effective controls of the systems of interest. Because this book shows how the theory of grey system and information was established and how each method could be practically applied, this book can easily be used as a reference by scholars who are interested in either theoretical exploration or practical applications or both. I recommend this book highly to anyone who has either a desire or a need to learn.

Stuttgart, Germany   Prof. Dr. Dr. h.c. mult. Hermann Haken
July 2007   Founder of Synergetics

Note Professor Hermann Haken prepared this note for one of the earlier book by the same authors, published in 2006. It is published in *Grey Systems: Theory and Application*, 2011, Vol. 1, No. 1.

Foreword by Dr. Robert Vallée

I am much interested and impressed by Dr. Sifeng Liu and Dr. Yi-Lin's recently published monograph on grey information, dealing with the theory and practical applications.

This book encompasses many aspects of mathematics under the aegis of uncertain information. I am greatly in favour of this attitude, concerning the uncertainty of information, which has been mine since a long time ago. Also, this book focuses on practice and aims at explorations of new knowledge. It is a comprehensive, all-in-one exposition, detailing not only with the theoretical foundation but also real-life applications. Because of this characteristic of quality and usefulness, Liu and Lin's book possesses the value of the widest possible range of reference by the workers and practitioners from all corners of natural and social sciences and technology.

In this book, Liu and Lin present the theory of grey information and systems starting on such background information as the relevant history, an attempt to establish an unified information theory, the basics of grey elements, and reaching all the most advanced topics of the theory. Complemented by many first-hand and practical project successes, the authors developed an organic theory and methodology of grey information and grey system, dealing with errors. In fact, there is much more to tell about error than about truth. Error (inexactitude) can be met everywhere and truth (exactitude) nowhere. But inexactitude contains a part of the truth. Greyness is the field we live in. Extremes, as whiteness and blackness, are inaccessible, but very useful, ideal concepts.

With the publication of such a book that contains not only a theory, aspects of magnificent real-life implications and explorations of new research, but also the history, the theorization of various difficult concepts, and directions for future works, there is no doubt that Drs. Liu and Lin have made a remarkable contribution to the development and applications of systems science.

Paris, France   Prof. Robert Vallée
June 2007   President of the World Organisation
                                        of Systems and Cybernetics, Université
                                                                      Paris-Nord

Note This note is a book review written by Prof. Robert Vallée for one of the earlier book by the same authors, published in 2006. It is published in *Kybernetes: The International Journal of Cybernetics, Systems and Management Science*, 2008, Vol. 37, No. 1.

Preface

In this book, we answer the calls of the readers of our previous publications and systematically present the main advances in grey system theory and applications. By following our readers' feedback and suggestions, this volume introduces the most recent research results and updates on what is presented in our earlier books. In particular, the following content, which represents the authors' recent research, is highlighted in the book: general grey numbers and their operations, negative grey relational analysis models and grey relational analysis models based on similarity and closeness, three-dimensional grey relational analysis models, grey clustering evaluation models based on mixed possibility functions, original difference grey model (ODGM), even difference grey model (EDGM), discrete grey model (DGM), fractional grey models, self-memory grey models, multi-attribute intelligent grey target decision models, weight vector group with kernel and weighted comprehensive clustering coefficient vector. We also attach a software designed for grey system modelling, which was developed by Bo Zeng using Visual C#, the widely employed C/S software tool. This user-friendly software allows users to conveniently input and/or upload data and clearly distinguish module functions. Also, the software has the ability to present users with operational details, as well as periodic and partial results. Additionally, users can adjust the levels of computational accuracy based on their practical needs.

During the writing of this book, we prioritized theoretical simplicity and clarity to make it easy for the reader to follow the main arguments made. With a good number of practical applications, we intended to illustrate the methodology of grey system theory and modelling techniques so that we could emphasize the practical applicability of grey system thinking. We drew on the most recent research developments from various research groups around the world and tried to present the most complete picture of this new area of scientific endeavour in a concise manner.

The overall planning and organization of topics contained in this book were carried out by Sifeng Liu (Nanjing, China), who also authored Chaps. 1, 2, 4, 6, 10 and 12. Yingjie Yang (Leichester,UK) produced Chaps. 3, and 11, Jeffrey Forrest (Slippery Rock, USA) composed Chaps. 7 and 8, Naiming Xie (Nanjing, China) wrote Chap. 9, and the Appendix and the attached computer software were developed by Zeng Bo

(Chongqing, China). Zhigeng Fang, Yaoguo Dang, Lirong Jian and Chunhua Su and colleagues also worked with the authors to refine some of the book's content. Sifeng Liu was responsible for unifying the terms used throughout the book and for finalizing the manuscript.

Finally, we would like to encourage you to communicate with us and send us any comments you might have about this book. It is only by working together, as a team, that we can grow and mature as researchers. Sifeng Liu can be reached at sfliu@nuaa.edu.cn, Yingjie Yang can be reached at yyang@dmu.ac.uk, and Jeffrey Forrest at jeffrey.forrest@sru.edu or jeffrey.forrest@iigss.net.

Nanjing, China/Leicester, UK   Sifeng Liu
June 2022

Acknowledgements

The work presented in this book is supported by a project of the national major talent programme of China (YQR20024), a project of Marie Curie International Incoming Fellowship of the European Union, (629051). It is also supported by a project of the Leverhulme Trust International Network (IN-2014-020), the joint projects of both the NSFC and the RS in the UK (71811530338, 71111130211), and projects linked to the National Natural Science Foundation of China (72071111, 71671091, 71671090, 91324003, 90924022, 70473037, 70971064). The projects of Intelligence Introduction base of the Ministry of Science and Technology of China (G2021181014L, G20190010178). A project of National Society Open Cooperation Demonstration Program of China Association for Science and Technology (2021KFHZ001). At the same time, the authors would like to acknowledge the partial support of the Fundamental Research Funds for the Central Universities of China (NC2019003, NP2015208). The Foundation for National Outstanding Teaching Group of China (No. 10td128), the Foundation for Doctoral Programs (grant No. 20020287001, 200802870020, 20093218120032) and the Key Project of Natural Science Foundation of Jiangsu Province (Grant No. BK2003211). The Science Foundation for the Excellent and Creative group in Science and Technology (on Grey System) of Nanjing University of Aeronautics and Astronautics and Jiangsu Province (No.Y0553-091), and the Foundation for the Excellent Young Scientists in Henan Province (No. 964040600) have also provided their support.

Over the years, our research has been highly commended by many first-class scholars, such as Julong Deng, the founder of grey system theory, L. A. Zadeh, the founder of fuzzy mathematics, Hermann Haken, the founder of synergetics, Robert Vallee, former president of the World Organization of Cybernetics and Systems, James Tien, former vice-president of IEEE and member of the American Academy of Engineering, K. W. Hipel, former president of The Academy of Science of the Royal Society of Canada and, Jifa Gu, former president of the International Federation for Systems Research. And praised or cited positively by more than 100 academicians from academies of science and engineering all over the world, such as E. K. Zavadskas, A. Bernard, Guozhi Xu, Qun Lin, Da Chen, Chunsheng Zhao, Haiyan Hu, Suzi Yang, Youlun Xiong, Zhongtuo Wang, Shanlin Yang, Xiaohong Chen and others. Xuesen Qian, the first winner of the National Highest Science and Technology Award of China, Jinpeng Huai, minister of the Ministry of Education of China, Angela Dorothea Merkel, former German Chancellor, have also praised us for the achievements in this area of research. Indeed, such positive feedback on our work has given us the impetus to develop this book.

Finally, many colleagues and administrators have supported us in the process of writing this book, as the authors have consulted many scholars and a wide range of relevant published literature throughout the development of this project. In particular, they are deeply in debt to Dr. Caroline Moraes, who proofread all chapters of the book, and Ms. Emily Zhang and colleagues of Springer Nature for their great help. The authors would like to express their appreciation to them all.

Contents

Chapter 1
Introduction

1.1 The Scientific Background of the Birth of Grey System Theory

On the basis of dividing the spectrum of scientific and technological endeavors into fine sections, the overall development of modern science has shown the tendency of synthesis at a high level. This higher level synthesis has led to the appearance of various studies of systems science with their specific methodological and episte-mological significance. Systems science reveals deep and intrinsic interconnections between objects and events, and has greatly enriched the overall progress of science and technology. Many of the historically difficult problems in different scientific fields have been resolved successfully along with the appearance of systems science and its specific branches. Furthermore, because of the emergence of various new areas in systems science, our understanding of nature and the laws that govern objective evolutions has been gradually deepened. At the end of the 1940s, there appeared systems theory, information theory, and cybernetics. Toward the end of 1960s and the start of 1970s, there appeared the theory of dissipative structures, synergetics, catastrophe, and fractal theory. Then, in the mid to late 1970s, new transfield and interfiled theories of systems science such as the hypercycle theory and dynamical systems theory emerged.

In the process of system research, due to the existence of internal and external disturbances and the limitation of human cognitive ability, the information obtained by people often has some uncertainty. With the development of science and technology and the progress of human society, people have gradually deepened their understanding of various system uncertainties, and the research on uncertain systems is also deepening day by day. Since the 1960s, a variety of uncertain system theories and methods have been proposed one after another. Among them, Fuzzy mathematics founded by Professor L. A. Zadeh in the 1960s (Zadeh, 1965), grey system theory advanced by Professor Deng Julong in the 1980s (Deng Julong, 1982), rough sets theory devoloped by Professor Z. Pawlak in the 1980s (Pawlak, 1991), etc., are all important achievements in the study of uncertain systems with extensive international

S. Liu, *Grey Systems Analysis*, Series on Grey System,
https://doi.org/10.1007/978-981-19-6160-1_1

influence. These uncertain theories discussed the theories and methods of describing and processing various kinds of uncertain information from different perspectives and aspects.

Grey system theory takes the "small data and poor information" uncertain system with "some information known and some information unknown" as the research object. It mainly extracts valuable information through the mining of "some" known information, and realizes the correct description of the system operation behavior and evolution law, so that people can use mathematical models to analyze and assess the "small data and poor information" uncertain system, then realize high-precision prediction, scientific decision-making and optimal control of the "small data and poor information" uncertain system. The uncertainty system of "small data and poor information" in the real world provides rich research resources and broad development space for grey system theory.

1.2   The Founder of Grey System Theory

The birth of grey system theory is an outcome of Professor Julong Deng who has been working with perseverance for decades.

Prof. Deng was born in Lianyuan County, Hunan Province of China in 1933. He got his degree in electrical machinery from Huazhong Institute of Technology and then joined the same institute in 1955 as a teaching assistant. Prof. Deng used to keep an eye on new ideas related to his field which led to his later investigation into multi-variable system control problems. In the 1960s, he put forward a new method—"control by removing redundant". His paper entitled "multivariable linear system shunt calibration device of a comprehensive approach" was published in 1965 (Deng, 1965). By the early 1970s, the method of "control by remove redundant" has been widely recognized as a representative methodology in cybernetics.

In 1965, Prof. L. A. Zadeh proposed Fuzzy Sets (Zadeh, 1965). Prof. Deng was involved in research of fuzzy mathematics. He published some papers in fuzzy mathematics. And served as a member of editorial board for several journals on fuzzy mathematics. In the late 1970s, Prof. Deng devoted himself to the study of "prediction and control problems of economic system". In dealing with systems where "some information is known, and some information is unknown", the main challenge is to develop an effective method to represent such systems. Despite the difficulties, Professor Deng and his colleagues have made significant progress in their explorations. In 1982, his pioneering paper titled "The Control Problems of Grey Systems" published by Systems and Control Letters (Deng, 1982). The publication of this seminal article indicated that grey system theory, a new branch of research, came into being.

Since the birth of Grey System Theory, it has received significant attention from academic communities and industries both in China and overseas, especially in real world applications.

So far, Prof. Deng's works has been cited over 50 thousand times. Prof. Deng won the award of founder of Grey System Theory at the 2007 IEEE International Conference on Grey Systems and Intelligent Services which held in Nanjing. In 2011, he was elected as the honor fellow of the World Organisation of Systems and Cybernetics at the joint conference of the 15th WOSC International Congress on Cybernetics and Systems and 2011 IEEE International Conference on Grey Systems and Intelligent Services.

1.3   Development of Grey Systems Theory

1.3.1   Building a Basic Team

In the early 1990s, Professor Julong Deng began to recruit and train doctoral students in the field of grey system theory in the discipline of system engineering of Huazhong University of Science and Technology. He has recruited and trained 10 doctoral students, most of them are young scholars who have been engaged in grey system theory research for many years before entered Prof. Deng's group. These scholars naturally become the first generation of grey system theory. They actively participate in the research of grey system theory, consciously assume the responsibility of developing and disseminating grey system theory, and unswervingly take the research and inheritance of grey system theory as their lifelong career.

In 2000, as the first distinguished professor introduced by Nanjing University of Aeronautics and Astronautics (NUAA), one of Prof. Deng's Ph.D. students, Professor Sifeng Liu joined this university with aerospace characteristics. In the same year, with Professor Sifeng Liu as the chief discipline leader, NUAA submitted an application to the Academic Degrees Committee of the State Council of China for the establishment of a doctoral degree authorization point in management science and engineering, which was successfully approved. Therefore, grey system theory has naturally become the characteristic and leading direction of the doctoral program of management science and engineering of NUAA. At the same time, as the founding director, Professor Liu established the Institute for Grey System Studies at NUAA. IGSS-NUAA has also become the center of grey system scholars. A group of outstanding young scholars gathered in IGSS-NUAA through talent introduction, entering the station to carry out post-doctoral research and pursuing doctoral degree, forming a highland of grey system research. IGSS-NUAA has 12 doctoral tutors (including 6 full-time doctoral tutors). Over the past 20 years, it has recruited and trained more than 200 doctoral students, post-doctors and visiting scholars at home and abroad in the field of grey system theory.

Many other universities are recruiting and funding doctoral and postdoctoral researchers in grey system theory and its application. Examples include Southeast University, Wuhan University of Technology, Fuzhou University, Shantou University, De Montfort University, Bucharest Economics University, Poznań University of

Technology, Bogazici University, Cape Town University, Central Florida University, Nebraska-Lincoln University, University of Waterloo, Pablo de Olavide University, Kanagawa University, National Cheng Kung University, etc.

Hundreds of doctoral graduates constitute the basic team of grey system theory research. Each Ph.D. graduates in grey system theory become a seed which take root in the new institution, then enlarge and spread one's power and influence gradually.

1.3.2 Establishment of Academic Organizations

In 1987, Wuhan Grey System Society, with members from provinces, cities and autonomous regions all over the country of China, was approved by Wuhan Association for Science and Technology.

In 2005, the Grey System Society of China, CSOOPEM, was approved by China Association for Science and Technology, and the Ministry of Civil Affairs of China. At the beginning of 2008, the Technical Committee of IEEE SMC on Grey Systems was established. In 2012, the first Workshop of European grey system research collaboration network was held by De Montfort University, and delegates from twelve member states of the European Union attended the event. In 2013, Professor Sifeng Liu was selected for a Marie Curie International Incoming Fellowship (FP7-PEOPLE- IIF-GA-2013–629,051) of the 7th Research Framework Program of the European Union. Furthermore, in 2014 an international network project entitled "Grey Systems and Its Applications" (IN-2014–020) was funded by The Leverhulme Trust. Supported by this project, a series of grey system theory cooperative research and academic exchange activities have been held in Europe, North America and China. In 2015, the International Association of Grey System and Uncertain Analysis (GSUA) was established.

Since 1984, 36 domestic and 16 international conferences on grey system theory and its applications have been held. Such conferences have been supported by IEEE, WOSC, GSUA, China Association for Science and Technology, China Center of Advanced Science and Technology, The Leverhulme Trust, Institute for Grey System Studies at Nanjing University of Aeronautics and Astronautics, De Montfort University, Stockholm University, Huawei Technology of Thailand, Wuhan University of Technology, Pudong Educational Society of Shanghai. A large number of young scholars has attracted to such events.

Many special sessions and tracks on grey system theory have been organized at significant international conferences such as International Conference on Uncertain System Modeling, International Conference on System Forecast and Control, International Conference on General System Studies, International Congress of World Organization of Systems and Cybernetics, IEEE International Conference on Systems, Man and Cybernetics, etc.. The topicality of grey systems theory and its popularity in such high-profile international conferences have certainly played an active role in furthering understanding of, and promoting, this theory among peers in the world of systems science.

1.3.3   Journals and Book Series on Grey System Theory

In 1989, The Journal of Grey System was launched by Research Information Ltd in the UK. In 2007, The Journal of Grey System is indexed in SCIE (Science Citation Index Expanded) and belongs to the categories of "Mathematics" and "Mathematics, Interdisciplinary Applications" in SCIE Currently, Journal of Grey System belong to JCR Q2 with an impact factor of 1.912. This publication is indexed by Mathematical Review of the United States and other important indexing agencies from around the world. In 2011, Emerald launched a new journal named Grey Systems: Theory and Application, edited by the faculty of the Institute for Grey System Studies at Nanjing University of Aeronautics and Astronautics. In 2019, Grey Systems Theory and Application is indexed in SCIE (Science Citation Index Expanded) and belongs to the categories of "Mathematics"and "Mathematics, Interdisciplinary Applications" in SCIE. At present, Emerald/ Grey Systems Theory and Application (GS) belong to JCR Q1 with an impact factor of 3.321. This journal is indexed by EBSCO, Scopus, Summon, ReadCube Discover and other important indexing agencies from around the world. There are currently over one thousand different professional journals in the world that have published papers in grey systems theory, many of which are top journals in a variety of fields. As of this writing, many journals and publishers such as the journal of the Association for Computing Machinery (USA), Communications in Fuzzy Mathematics (Taiwan, China), Kybernetes: The International Journal of Systems & Cybernetics, Transaction of Nanjing University of Aeronautics and Astronautics, China Ocean Press, Chinese Agricultural Science Press, Henan University Press, Huazhong University of Science and Technology Press Co. Ltd, IEEE Press, Springer-Verlag have respectively published special issues or proceedings on grey system theory (Liu and Lin, 2010; Liu et al., 2022).

Numerous publishing agencies such as Science Press, Defense Industries Press, Huazhong University of Science and Technology Press Co. Ltd, Jiangsu Science and Technology Press, Shandong People's Press, Science and Technology Literature Press of China, China Science and Technology Book Press of Taiwan, Gaoli Books Limited Company of Taiwan, ASE Press of Romania, Japan Polytechnic Press, IIGSS Academic Press, CRC of Taylor & Francis Group, Springer-Verlag, Springer-Verlag London Ltd, and John Wiley & Sons, Inc. have published hundreds of academic works on grey systems, in many different languages including Chinese, Traditional Chinese, English, Japanese, Korean, Romanian, and Persian.

Series of grey systems both in Chinese and English are published by Science Press and Springer-Nature Group respectively. Series of grey systems in Chinese was launched by Science Press in 2014. So far, 32 books have been published. Series of grey systems in English was launched by Springer-Nature Group in 2021. The three books that passed the review in the first phase have completed the signing process of publishing contracts now.

1.3.4 Grey System Theory Curriculums

Numerous universities around the world have set up grey system theory curriculums. For example, in Nanjing University of Aeronautics and Astronautics (NUAA), the curriculums of the grey system theory are found not only in Ph.D. and Master's programs, but also in undergraduate programs of many disciplines across the university, as an elective module. Prof. Liu Sifeng and his team at IGSS- NUAA did a lot of work to popularize and inherit the Grey System Theory. As a result, this course has been selected as the National Excellence Course beginning in 2008, the National Excellence Resource Sharing Course since 2013, the National Excellence Online Open Course starting in 2018, and the National first class courses of online and offline since 2020. Furthermore, Professor Liu Sifeng's team worked with a number of professors from universities in Europe, the United States and Canada, including Keith William Hipel, former president of the Royal Canadian Academy of Sciences, Professor Yingjie Yang, the executive president of the GSUA, to complete the online course in English, Grey Data Analysis, which became a free open learning resource for all grey system hobbyists since 2021.

1.3.5 Researchers of Grey System Theory Are All Over the World

Many scholars from USA, UK, Germany, France, Italy, Korea, Canada, Romania, Poland, Turkey, South Africa, Iran, India and Pakistan, etc. have joined IGSS-NUAA as visiting professor, research fellow or for joint project research. In recent years, some young scholars from different countries joining IGSS-NUAA as Ph.D. or Master students supported by Chinese government scholarship. It is helpful to promotion the popularization and international communication of grey system theory.

According to the retrieval results by the database of web of science, scholars from more than 100 countries and regions in the world have carried out research on grey system theory and applications and published relevant academic papers. Hundreds of thousands of master's and doctoral students around the world applying grey system thinking and methods to carry out scientific research and complete their dissertations.

Many prominent scholars have commended grey system research. Such scholars include Professor Qian Xuesen, famous scientist and winner of the national highest science award, China, Professor Lotfi A. Zadeh (USA), the founder of fuzzy mathematics, Professor Herman Haken (Germany), the founder of synergetics, Professor James M. Tien (USA), former vice-president of IEEE and member of the National Academy of Engineering, Professor Robert Vallee (France), former president of World Organization of Systems and Cybernetics, Professor Alex Andrew (UK), former secretary General of the World Organization of Systems and Cybernetics, and Keith William Hipel, former president of the Canadian Royal Academy of Sciences, as well as many Academicians of the Chinese Academy of Sciences and the Chinese

Academy of Engineering, including Professor Yang Shuzi, Professor Xiong Youlun, Professor Lin Qun, Professor Chen Da, Professor Zhao Chunsheng, Professor Hu Haiyan, Professor Xu Guozhi, Professor Wang Zhongtuo, Professor Yang Shanlin, Professor Chen Xiaohong, and Professor Shan Zhongde, et al.

It attracts not only the affirmation and support from international leading scholars, but also many early career researcher from different disciplines of social sciences, natural sciences and engineering technology as well. Successful applications have been found in more than 100 countries and regions. It has been established as a new scientific branch in data analytics and uncertainty modelling (Liu et al., 2022).

On 7th September, 2019, during the visit to China, Angela Dorothea Merkel-then, German Chancellor praised Chinese original grey system theory. She said that the work of professor Deng Julong, the founder of grey system theory, and professor Liu Sifeng and three other Alumni of HUST "profoundly affecting the world."

1.3.6 Papers of Grey Systems Theory Are Growing Rapidly

The rapid development of grey system theory benefits from the strong promotion of practical application needs.

In the information age, people in various fields begin to deeply realize that data analysis method has become an indispensable skill for everyone. Just like the gold buried in the sand sea, the laws and characteristics that people want to understand and control are deeply covered up by the chaotic and complicated data information with extremely low information density and value. There is an urgent need for effective scientific methods. To meet the needs of the times, grey system theory came into being.

Just like any new thing, the growth process of a new theory is naturally full of hardships and twists and turns. When the grey system theory came out, it was inevitably criticized and questioned by some people. The desire for poor information data analysis methods in human social practice has formed a strong driving force, so that the grey system theory can still attract the positive attention of a large number of people of insight in various fields (Liu et al., 2022).

A large number of grey system theory and application research papers can be retrieved in both Chinese and English academic paper databases.

In the database of ISI Web of Science, search according to the English phrases of Grey number, Grey data and Sequence operator, etc. which contained in the article titles, the results are shown in Table 1.1.

Ten grey system related phrases such as grey system, grey number, sequence operator, grey correlation analysis, grey clustering, grey model, GM (1,1), grey prediction, grey decision-making and grey control are input into the China national knowledge infrastructure(CNKI) database for parallel retrieval. The results show that there are 227,374 literatures containing the above phrases are included in the CNKI database from 1982 to 2020. Among them, there are 119,172 journal papers, 4950 conference papers, 231 books, 79 achievements, 40 newspaper articles, 101,905 dissertations,

Table 1.1 Number of papers with grey system related phrases in the titles in ISI Web of Science database

Title Word	Grey number	Grey data	Sequence operator	Grey relation	Grey incidence
Number of papers	425	681	662	3374	406
Title Word	Grey analysis	Grey cluster	Grey clustering	Grey evaluation	Grey model
Number of papers	4976	774	557	1972	5374
Title Word	GM(1,1)	Grey prediction	Grey forecast	Grey decision	Grey control
Number of papers	892	1956	1420	877	1429

Table 1.2 Number of literatures with grey system related phrases in CNKI database from 1982 to 2020

Year	1982–2000	2001	2002	2003	2004	2005	2006
No. of papers	15,276	1856	2384	2891	4151	5445	7199
Year	2007	2008	2009	2010	2011	2012	2013
No. of papers	8821	1.01	1.07	1.19	1.26	1.39	1.42
Year	2014	2015	2016	2017	2018	2019	2020
No. of papers	1.51	1.47	1.58	1.61	1.61	1.58	1.59

including 13,961 doctoral dissertations and 87,944 master's dissertations. See Table 1.2 for the number of documents and achievements containing the above phrases in CNKI database from 1982 to 2020.

As can be seen from Table 1.2, in the 18 years from 1982 to 2000, more than 15,000 grey system papers were included in CNKI database, which is equivalent to the number of papers included in CNKI database of each year since 2014. After entering the new century, the grey system papers included in CNKI database show a rapid growth trend. In 2001, 1856 papers were included in CNKI database. By 2004, the number of papers included in CNKI database had reached 4151, double that of 2001. In 2007, it doubled on the basis of 2004, reaching 8821. Since 2008, more than 10,000 papers have been included in CNKI database every year, and more than 15,000 papers have been included in CNKI database since 2014.

It can be found from the literatures included in CNKI database that a large number of grey system papers have been included in CNKI database in all double first-class universities and double first-class discipline construction universities in China. The top 20 universities of number of journal papers and dissertations of grey system included in CNKI database can be seen in Table 1.3. The data in Table 1.3 fully shows that the grey system theory has played an important role in the training of high-level talents in China.

Table 1.3 Top 20 universities of number of grey system papers included in CNKI database

Name of universities	NCEPU	CAU	SJU	WUT	CSU
No. of papers	4018	2995	2970	2704	2684
Name of universities	BJU	NUAA	JLU	CQU	TJU
No. of papers	2644	2531	2526	2505	2427
Name of universities	HUST	HNU	HHU	ZJU	DUT
No. of papers	2016	1998	1987	1910	1857
Name of universities	HUT	CUMT	DMU	XUAT	HIT
No. of papers	1782	1762	1759	1740	1697

Notes NCEPU North China Electric Power University, CNU: Chang' An University, *SJU* Southwest Jiaotong University; WUT: Wuhan University of Technology, *CSU* Central South University; BJU: Beijing Jiaotong University, *NUAA* Nanjing University of Aeronautics and Astronautics, *JLU* Jilin University, *CQU* Chongqing University, *TJU* Tianjin University, *HUST* Huazhong University of Science and Technology, *HNU* Hunan University, *HHU* Hohai University, *ZJU* Zhejiang University, *DUT* Dalian University of Technology, *HUT* Hefei University of Technology, *CUMT* China University of Mining and Technology, *DMU* Dialian Maritime University, *XUAT* Xi'an University of Architecture and Technology, *HIT* Harbin Institute of Technology)

Among the documents included in CNKI database, there are 37,887 documents marked with national important science and technology plan projects such as NSFC, national key basic research and development plan (973), national high technology research and development plan (863) or national science and technology support plan. See Table 1.4 for details. The data in Table 1.4 shows that the grey system theory has played an important role in promoting China's scientific and technological progress and innovation development. This was fully affirmed by academician Zhao Chunsheng of the Chinese Academy of Sciences (Zhao, 2015).

In the era of big data, the grey system theory based on small data mining has sprung up and become an effective tool for people to extract valuable information from massive data. In the past 40 years, the wide application of grey system methods and models in many fields of social science, natural science and engineering technology has led to innovation and progress in various fields.

Table 1.4 Number of papers of grey system which marked various important national science and technology projects in CNKI database

Programme	NSFC	National key basic research and development plan (973)	National high technology research and development plan (863)	National science and technology support plan	National key R & D plan	National plan for tackling key scientific and technological problems
No. of papers	23,821	1731	1934	2666	1022	582

1.4   Elementary Concepts of Grey System Theory

Many social, economic, agricultural, industrial, ecological and biological systems are named by considering the features of classes of the research objects, while grey systems are labeled using the color of the systems of concern.

In the theory of control, scholars often make use of colors to describe the degree of clearness of available information. For instance, Ashby refers to objects with unknown internal information as black boxes. This terminology has been widely accepted in the scientific community. As another example, as a society moves toward democracy, citizens gradually demand more information regarding policies and the meanings of such policies. That is, citizens want to have an increased degree of information transparency (i.e. white information). Thus, we use "black" to indicate unknown information, "white" to indicate completely known information, and "grey" to convey partially known and partially unknown information. Accordingly, systems with completely known information are regarded as white, while systems with completely unknown information are considered black, and systems with partially known information and partially unknown information are seen as grey.

In this context, incompleteness in information is the fundamental meaning of "grey." However, the meaning of "grey" can be expanded or stretched from different angles and in varied situations (see Table 1.5) (Deng, 1985; Liu et al., 2017; Liu and Lin, 2011).

At this point, the difference between "system" and "box" Must be highlighted. On the one hand, the term "box" is used when one does not pay much attention, or does not attempt, to utilize information regarding the interior characteristics of an object, while focusing mainly on the external characteristics of such an object. In this case, the researcher generally investigates the properties and characteristics of the object through analyzing the input–output relation. On the other hand, the term "system" is employed to indicate the study of the object's structure and functions through the analysis of existing organic connections between the object, relevant factors, its environment, and related laws of change.

The research objects of grey systems theory consist of uncertain systems that are known only partially through small samples and poor information. The theory

Table 1.5 Extensions of the concept of "grey"

Situation/concept	Black	Grey	White
Information	Unknown	Incomplete	Completely known
Appearance	Dark	Blurred	Clear
Processes	New	Changing	Old
Properties	Chaotic	Multivariate	Order
Methods	Negation	Change for the better	Confirmation
Attitude	Letting go	Tolerant	Rigorous
Outcomes	No solution	Multi-solutions	Unique solution

focuses on the generation and excavation of partially known information through grey sequence operators of possibility functions to enable an accurate description and understanding of the material world.

1.5 Fundamental Principles of Grey System Theory

In the process of developing grey systems theory, Julong Deng established six fundamental principles containing intrinsic philosophical intensions, as discussed below (Deng, 1985).

Axiom 1.5.1 (*The Principle of Informational Differences*) . "Difference" implies the existence of information. Each piece of information must carry some kind of "difference".

When we say that object A is different from object B, we mean that there is some special information about object A that is not true for object B. All "differences" between natural objects and events have provided us with elementary information in order for us to understand their nature.

If information "I" has changed our understanding or impression of a complicated matter, then the piece of information "I" is definitely different from what we initially understood the complicated matter to be. Great breakthroughs in science and technology have provided us with necessary information, which we generally call knowledge and tools, to understand and change the world around us. Such advanced information is surely different from pre-scientific information. The more content a piece of information "I" contains, the more the differences from an earlier version of such information will become apparent.

Axiom 1.5.2 (*The Principle of Non-Uniqueness*) The solution to any problem with incomplete and indeterminate information is not unique.

Because of the principle of non-uniqueness, which is a basic law of the application of grey systems theory, one is set free to look at problems with flexibility. With flexibility, one becomes more effective in reaching their goals.

Strategically, the principle of non-uniqueness is realized through the concept of grey target. This concept is a unification of the concept of non-unique target and that of non-restrainable target. For example, on the one hand, if a high school graduate does not plan to enroll in any university except for one specific institution, then his chance of being accepted by a university is greatly limited. On the other hand, if a high school graduate with similar qualifications as the one in the previous example is willing to apply for several universities other than his preferred one, he will be more likely to succeed in being accepted by a university because he has multiple targets, which in turn leads to an improved chance of hitting one of the targets.

The principle of non-uniqueness can be seen as a comprehensive realization that each target can be approached, that any available information can be supplemented, that each plan made earlier can be further modified and improved, that each relationship can be harmonized, that each thinking logic can be multi-directional, that each understanding can be deepened, and that each path can be optimized. When faced with the possibility of multiple solutions, one can locate one or several satisfactory solutions through deterministic analysis and supplementation of information. Therefore, the method of finding solutions on the basis of "non-uniqueness" is one that combines both quantitative and qualitative analysis.

Axiom 1.5.3 (*Principle of Minimal Information*) One characteristic of grey system theory is that it makes the most and best use of the "minimal amount of available information."

The "principle of minimal information" can be seen as a dialectic unification of "a little" and "a lot." One advantage of grey system theory is its ability to handle such uncertain problems with "small data" and/or "poor information." Its foundation of study is the concept of "spaces of limited information." "Minimal amount of information" is the basic territory for grey system theory to show its, power. The amount of acquirable information is the dividing line between "grey" and "not grey". Making sufficient discovery and application of any available "minimal amount of information" is the basic thinking logic of problem-solving used in grey system theory.

Axiom 1.5.4 (*Principle of Recognition Base*). Information is the foundation on which people recognize and understand (nature).

This principle argues that all recognition must be based on information. Without information, there is no way for people to know anything. With complete and deterministic information, we can possibly gain firm understanding of nature. With incomplete and non-deterministic information, it is only possible to obtain incomplete and non-deterministic grey understanding of particular phenomena.

Axiom 1.5.5 (*Principle of New Information Priority*)
The function of new pieces of information is greater than that of old pieces of information.

The "principle of new information priority" is the key idea behind information application in grey system theory. That is, by applying additional weights to new information, one can achieve a better result from grey modeling, grey prediction, grey analysis, grey evaluation, and grey decision making. The belief that "the new replaces the old" reflects our "principle of new information priority." With the availability of new information, the motivation for whitening grey elements is strengthened. The "principle of new information priority" reflects the fact that information in general is time sensitive.

Axiom 1.5.6 *Principle of Absolute Greyness* .
"Incompleteness" of information is absolute. Incompleteness and non-determinism of information have generality.

Completeness of information is relative and temporary. It is the moment when the original non-determinism has just disappeared, and new non-determinism is about to emerge. Human recognition and understanding of the objective world have been improved over time through continued supplementation of information. With endless supply of information, man's recognition and understanding of the world also become endless. That is, greyness of information is absolute and will never disappear.

1.6 Main Contents of Grey System Theory

Through nearly thirty years of development, grey systems theory has been built up as a newly emerging scientific discipline with its very own theoretical structure consisting of systems analysis, evaluation, modeling, prediction, decision-making, control, and techniques of optimization. Its main contents contain (Liu et al., 2017; Liu et al., 2016a, b; Liu, 2004).

(a) The theoretical system developed on the basis of grey algebraic system, grey equations, grey matrices, etc.;
(b) The methodological system established on the basis of sequence operators and generations of grey sequences;
(c) The analysis and evaluation system constructed on the basis of grey incidence spaces and grey cluster evaluations;
(d) The prediction model system centered around GM(1.1);
(e) The decision-making model system represented by multi-attribute intelligent grey target decision models;
(f) The system of combined grey models innovatively developed for producing new and practically useful results; and
(g) The optimization model system, consisting mainly of grey programming, grey input–output analysis, grey game theory, and grey control.

Grey algebraic system, grey matrices, grey equations, etc., constitute the foundation of grey systems theory. In terms of the theoretical beauty and completeness of the theory, there are still a lot of problems left open in this area. In this book, generations of grey sequences are merged into the concept of sequence operators, which mainly include buffer operators (weakening buffer operators, strengthening operators), mean generation operators, ratio generation operators, stepwise ratio generators, accumulating generators, inverse accumulating generators, etc. Grey incidence analysis includes such materials as grey incidence axioms, degree of grey incidence, generalized degree of grey incidence (absolute degree, relative degree, synthetic degree), the degrees of grey incidence based on either similar visual angles or approximate visual angles, grey incidence order, superiority analysis, and others. Grey cluster evaluation includes such contents as grey variable weight clustering, grey fixed weight clustering, cluster evaluations based on (center-point or end-point) triangular whitenization weight functions, and other related materials. Through grey generations or the effect of sequence operators to weaken the randomness, grey prediction models are designed to excavate the hidden laws; and through the inter-change between difference equations and differential equations, a practical jump of using discrete data sequences to establish continuous dynamic differential equations is materialized. Here, GM(1,1) is the central model that has been most widely employed; and discrete grey models are a class of new models we initially developed. In terms of grey predictions, they produce quantitative forecasts on the basis of the GM model. Based on their functions and characteristics, grey predictions can be grouped into sequence predictions, interval predictions, disaster predictions, seasonal disaster predictions, stock-market-like predictions, system predictions, etc. The grey combined models include grey econometric models (G-E), grey Cobb-Douglass Cobb—Douglas models (G-C-D), grey Markov models (G-M), grey-rough mixed models, etc. Grey decision-making includes multi-attribute intelligent grey target models, grey incidence decision-making, grey cluster decision-making, grey situation decisions, grey stratified decisions, etc.

　　The main contents of grey control include the control problems of essential grey systems, the controls composed of grey systems methods, such as grey incidence control, GM(1,1) prediction control, etc. Considering all the feedbacks from the readers of our earlier monograph, Grey Information: Theory and Practical Applications (Liu and Lin, 2006; Liu et al., 2017), we have paid special attention to organize some of the most recent new results obtained by colleagues from around the world in this volume. Also, for the convenience of practical applications, this book is accompanied with a computer software on grey systems modeling, which is designed by Zeng Bo of our research group.

References

Deng, J. L. (1965). Multivariable linear system shunt calibration device of a comprehensive approach. *Acta Automatica Sinica, 3*(1), 13–26.

Deng, J. L. (1982). Control problems of grey systems. *Systems & control letters, 1*(5), 288–294.

Deng, J. L. (1985). *Grey control system.* Wuhan: Press of Huazhong University of Science and Technology.

Liu, S. F. (2004). Appearance and development of grey systems theory. *Journal of Nanjing University of Aeronautics and Astronautics, 36*(2), 267–272.

Liu, S. F., & Lin, Y. edited. (2010). *Advance in grey systems research.* Berlin, Heidelberg: Springer-Verlag.

Liu, S. F., & Lin, Y. (2011). *Grey systems theory and applications.* Berlin, Heidelberg: Springer-Verlag.

Liu, S. F., Tao, L. Y., Xie, N. M., et al. (2016a).On the new model system and framework of grey system theory. *The Journal of Grey System, 28*(1), 1–15.

Liu, S. F., Yang, Y. J., Xie, N. M., Forrest, J. (2016b). New progress of grey system theory in the new millennium. *Grey systems theory and application, 6*(1), 2–31.

Liu, S. F., Yang, Y. J., Forrest, J. (2017). *Grey data analysis.* Singapore: Springer-Verlag.

Liu, S. F., Tao, Y., Xie, N. M., Tao, L. Y., Hu, M. L. (2022). Advance in grey system theory and applications in science and engineering. *Grey systems: Theory and application, 12*(4), 1–18.

Pawlak, Z. (1991). *Rough sets: Theoretical aspects of reasoning about data.* Dordrecht: Kluwer Academic Publishers.

Zadeh, L. A. (1965). Fuzzy sets. *Information and control, 8*, 338–353.

Zhao, C. S. (2015). Book Reviews: Preface to the 7th edition of grey system theory and its applications. *The Journal of Grey System, 27*(1), 127–129.

Chapter 2
Characteristics of Grey System Theory

2.1 A Kind of Poor Data Analysis Method with Strong Penetration

Grey system theory takes the uncertain system with poor information as the research object. It is an interdisciplinary method with strong penetration.

At Nanjing University of Aeronautics and Astronautics, the teaching team of management quantitative method course group led by Professor Liu Sifeng has been committed to the construction of Chinese original grey system theory courses for a long time. With the strong support of peer experts, the grey system theory courses has been selected as the National Excellence Course beginning in 2008, and the National first class courses of online and offline since 2020. The teaching resources including textbooks, videos and modeling software are widely distributed. At the same time, the original elements such as grey sequence operator, grey relational analysis, grey clustering evaluation, grey prediction, grey decision-making and grey linear programming, etc. in the grey system theory and the latest achievements made by the course team and partnerships both at home and abroad are rewritten into teaching cases and injected into the courses of "Operations Research" "Applied Statistics" "Prediction Methods and Technologies" "Theory and Methods on Decision-making" "economic cybernetics" "system modeling and simulation", "input–output analysis" and "econometrics". It enriches the connotation of these courses, and greatly improved the overall construction level of the curriculum group. In 2010, the course team was selected into the national excellence teaching team. In 2018, "The construction of management quantitative method course group and teaching reform led by local original theory" won the prize of national teaching achievement.

S. Liu, *Grey Systems Analysis*, Series on Grey System,
https://doi.org/10.1007/978-981-19-6160-1_2

2.2 Characteristics of Uncertain Systems and the Simplicity Principle in Sciences

The fundamental characteristic of uncertain systems is the incompleteness and inadequacy of their information. Due to the dynamics of system evolution, the biological limitations of the human sensory system, as well as the constraints of relevant economic conditions and technological availabilities, uncertain systems exist commonly (Deng, 1990; Liu, 2021).

2.2.1 Incomplete Information

Incompleteness in information is one of the fundamental characteristics of uncertain systems. The most common situations involving incomplete system information include cases where:

(1) Information about system elements (parameters) is incomplete;
(2) Information on the structure of the system is incomplete;
(3) Information about the boundary of the system is incomplete; and
(4) Information on the system's behaviors is incomplete.

Incomplete information is a common phenomenon in our social, economic, and scientific research activities. For instance, in agricultural production, even if we have exact information regarding plantation, seeds, fertilizers, and irrigation, uncertainties in areas such as labor quality, natural environment characteristics, weather conditions, and the commodity markets make it extremely difficult to precisely predict the production output and consequent economic value of agricultural fields. For biological prevention systems, even if we know the relationship between insects and their natural enemies, it is still really difficult to achieve the expected prevention effects due to uncertainty regarding the relationships between insects and their baits, insects' natural enemies and their baits, and a specific kind of natural enemy with another kind of natural enemy. As for the adjustment and reform of pricing systems, it is often difficult for policy makers to take actions because of the lack of information regarding price elasticity and consumer demand and how price changes on a certain commodity would affect the prices of other commodities. In security markets, even the brightest market analysts cannot be assured of winning constantly due to their inability to correctly predict economic policy and interest rate changes, management changes at various companies, the direction of political changes, investors' behavioral changes in international markets, and the effects of price changes in one block of commodities on another. As for the general economic system, because there are no clear relationships between the "inside" and the "outside" of the system, and between the system itself and its environment, and because the boundaries between the inside and the outside of the system are difficult to define, it is also difficult to analyze the effects of economic input on economic output.

Incompleteness in available information is absolute, while completeness in information is relative. Humans employ their limited cognitive ability to observe the infinite universe in order to try and obtain complete information. However, it is impossible for us to do so. In fact, the concept of large samples in statistics represents the degree of tolerance man has to incompleteness. In theory, when a sample contains at least 30 objects, it is considered "large." However, in some situations, even when a sample contains thousands or several tens of thousands of objects, the true statistical laws of a given system still cannot be successfully uncovered.

2.2.2 Inaccuracies in Data

Another fundamental characteristic of uncertain systems is naturally occurring inaccuracy in available data. In grey systems theory, the meanings of uncertain and inaccurate are roughly the same. Both terms stand for errors or deviations from actual data values. Based on the essence of how uncertainties are caused, inaccuracies can be categorized into three types: the conceptual, level, and prediction type inaccuracies.

(1) The Conceptual Type

Inaccuracies of the conceptual type emanate from the expression of a certain event, object, concept, or wish. For instance, all such frequently used concepts as "large," "small," "many," "few," "high," "low," "fat," "thin," "good," "bad," "young," and "beautiful" are inaccurate due to lack of clear definition. It is very difficult to use exact quantities to express these concepts. As a second example, suppose that a job seeker with an MBA degree wishes to get an annual salary offer of no less than $450,000, or that a manufacturing firm plans to control its rate of defective products to be less than 0.01%. These are all cases of conceptual type inaccuracies.

(2) The Level Type

This kind of data inaccuracy is caused by a change at the level of research or observation. This means that the available data might be accurate when seen at the level of the system of concern, that is, the macroscopic level, or at the level of the whole, that is, the cognitive conceptual level. However, when data are seen at a lower level, that is, a microscopic level, or at a partial localized level of the system, they generally become inaccurate. For example, the height of a person can be measured accurately to the unit of centimeters or millimeters. However, if the measurement has to be accurate to the level of one ten-thousandth micrometers, the former accurate reading will become extremely inaccurate.

(3) The Prediction (or Estimation) Type

Because it is difficult to have complete understanding of the laws of evolution, any prediction of the future tends to be inaccurate. For instance, it is estimated that two years from now, the GDP of a certain country will surpass $10 billion dollars; it is estimated that a certain bank will attract savings from individual residents of between $7 billion and $9 billion for the year 2022; it is predicted that in the coming years the temperature in Leicester, UK, during the month of June will not go beyond 30° C, and so on. All these examples provide uncertain numbers of the prediction type. In statistics, it is often the case that samples are collected to estimate the whole. Therefore, much statistical data are inaccurate. As a matter of fact, no matter what method is used, it is very difficult for anyone to obtain any absolutely accurate (estimated) value. When we draw out plans for the future and make decisions about what course of action to take, we in general have to rely on inaccurate predictions and estimates.

2.2.3 The Scientific Principle of Simplicity

In the history of science, the achievement of simplicity has been a common goal among most scientists. As early as the sixth century BC, natural philosophers had a common wish to understand the material laws of nature: to build knowledge of the material world on the basis of a few common, simple elements. The ancient Pythagoras of Greece introduced the theory of four elements (earth, water, fire, and gas) at around 500 BC. The Greeks believed that all material matters in the universe were composed of these four simple elements. Around the same time, ancient Chinese philosophers also developed a theory of five elements including water, fire, wood, gold, and earth. These are the most primitive and elementary thoughts about simplicity.

The scientific principle of simplicity originates from the simplicity of thinking employed in the process of understanding nature. As the natural sciences matured over time, simplicity became the foundation and guiding principle of scientific research. For example, Newtonian laws of motion unify the macroscopic phenomena of objective movements in their form of extreme simplicity. In his *Mathematical Principles of Natural Philosophy*, Newton pointed out that nature does not do useless work; because nature is fond of simplicity, it does not like to employ extra reasons to flaunt itself. During the Era of relativity, Albert Einstein introduced two criteria for testing a theory: external confirmation and internal completeness, that is, logical simplicity. Einstein believed that a true scientific theory must comply with the principle of simplicity in order to reflect the harmony and orderliness of nature. In the 1870s, Ampere, Weber, Maxwell, and others established theories to explain the phenomenon of electromagnetism based on their different assumptions. Because Maxwell's theory is the one that best complies with the principle of simplicity, it became well accepted. Another example is the well-known Kepler's third law of planetary motion: $T^2 = D^3$. This formula is very concise in form.

According to the dominant principle of synergetics (Haken, 1978), one can transform an original high-dimensional equation into a low-dimensional evolution equation of order-parameters by eliminating the fast-relaxing variables in the high-dimensional nonlinear equation that describes the evolution process of a system. Because the order-parameters dominate the dynamic characteristics of the system near its boundary points, through dominant the evolution equation of order-parameters one can obtain the system's time structure, space structure or time–space structure, so that one can materialize efficient control over the system's behavior.

The simplicity of scientific models is actualized by employing simple expressions and by ignoring unimportant factors of the system of concern. In economics, the methods of using Gini coefficient to describe differences among consumers' incomes (Gini, 1921) and of employing Cobb-Dauglas production function to measure the contribution of advancing technology in economic growth are all introduced on the basis of simplifying realistic systems (Cobb & Douglas, 1928). Modigliani and Brumbergh (1954) use the following model to describe the average propensity to consume:

$$\frac{C_t}{y_t} = a + b\frac{y_0}{y_t}, a > 0, b > 0$$

The curve Alban W. Phillips (1958) employs to describe the relationship between the rate of inflation $\frac{\Delta p}{p}$ and the unemployment rate x is:

$$\frac{\Delta p}{p} = a + b\frac{1}{x}$$

Additionally, the well-known capital asset pricing model (CAPM, William F. Sharpe, 1964) can be seen below:

$$E[r_i] = r_f + \beta_i\big(E[r_m] - r_f\big)$$

Essentially, all of these equations can be reduced to their simplest linear regression model with a few straightforward transformations (Liu, 2021).

2.2.4 Precise Models Suffer from Inaccuracies

When available information is incomplete and the collected data inaccurate, any pursuit of precise models in general becomes meaningless. This fact was well described by Lao Tzu more than two thousand years ago. The principle of incompatibility proposed by L. A. Zadeh, the founder of fuzzy mathematics, also addresses this matter: when the complexity of a system increases, our ability to precisely and meaningfully describe the characteristics of the system decreases accordingly until such a threshold that, as soon as it is surpassed, the preciseness and meaningfulness become two mutually excluding characteristics (Zadeh, 1994). This mutually

Table 2.1 Comparison between the prediction errors of a statistical model and a grey model

Order No	Type	Average error	
		Statistical model	Grey model
1	Horizontal displacement	0.862	0.809
2	Horizontal displacement	0.446	0.232
3	Vertical displacement	1.024	1.029
4	Vertical displacement	0.465	0.449
5	Water level of pressure measurement hole	6.297	3.842
6	Water level of pressure measurement hole	0.204	0.023

antagonistic principle reveals that the pursuit of preciseness can reduce the operationality and meaningfulness of a cognitive outcome. Therefore, precise models are not necessarily an effective means to address complex matters.

In 1994, Jiangping Yue and Xisheng Hua established both theoretically delicate statistical regression model and relatively coarse grey model based on the deformation data and leakage data of a certain large scale hydraulic dam. Their work shows that the grey model provided a better fit than the statistical regression model. When comparing the errors between the predictions of the two models with actual observations, it is found that the prediction accuracy of the grey model is generally better than that of the regression model; see Table 2.1 for details (Yue & Hua, 1994).

In 2001, Dr. Haiqing Guo as well as Zhongru Wu and colleagues respectively established a statistical regression model and a grey time series combined model using the observational data of displacement in the vertical direction of a certain large clay-rock filled dam of inclined walls. They compared the data fitting and predictions of the two models against actual observations and found that the data fitting eof the grey combined model was significantly superior to that of the statistical regression model (Guo et al., 2001).

On the other hand, Xiaobing Li, Haiyan Sun and colleagues employed fuzzy prediction functions (a type of uncertainty prediction) to dynamically trace and precisely control the fuel oil feeding temperature for anode baking. The control effect was clearly better than that obtained by utilizing the traditional PID control method (Li & Sun, 2009).

Finally, Caixing Sun and his research group made use of grey relational analysis, grey clustering, and various new types of grey prediction models to diagnose and predict insulation-related accidents related to electric transformers. Their substantial results indicate that these relatively coarse methods and models are operational and provide efficient results (Li et al., 2002; Sun et al., 2002, 2003).

2.3 Comparison of Several Uncertainty Methods

Probability and statistics, fuzzy mathematics, grey system theory and rough set theory are four of the most widely used research methods in the investigation of uncertain systems. Their research objects contain specific kinds of uncertainty, which represent their commonality. It is precisely the differences among the uncertainties in the research objects that make these four theories of uncertainty distinct from each other.

Probability and statistics study the phenomena of stochastic uncertainty with emphasis placed on revealing historical statistical laws. They investigate the chance of each possible outcome of the stochastic uncertain phenomenon to occur. Their starting point is the availability of large samples, which are required to satisfy a typical form of distribution.

Fuzzy mathematics emphasizes the investigation of problems with cognitive uncertainty, where research objects possess the characteristic of clear intension and unclear extension. For instance, "young man" is a fuzzy concept, because each person knows the intension of "young man." However, if we determine the exact age range within which everybody is young and outside which each person is not young, then we will have great difficulty. That is because the concept of young man does not have a clear extension. In fuzzy mathematics, this kind of cognitive uncertainty problem with clear intension and unclear extension is addressed by making use of experience and the so-called membership function.

Additionally, rough set theory tries to study uncertain systems by using the accuracy mathematical method. The main thought of rough set theory is to describe and address the inaccuracy or uncertain knowledge using a known knowledge library. Professor Z. Pawlak included all the units which cannot be acknowledged to have boundaries. He defined boundary as the difference set between upper approximate set and lower approximate set. The boundary is then described through the upper approximate set approaching the lower approximate set.

The focus of grey system theory, on the other hand, is on the uncertainty problems of small data sets and poor information, which are different to the problems addressed by probability, fuzzy mathematics or rough set theory. It explores and uncovers the realistic laws of evolution, motion of events and materials through information coverage by possibility function, and through the works of sequence operators. One of its characteristics is construct models with small amounts of data. What is clearly different about grey systems theory compared to fuzzy mathematics is that grey system theory emphasizes the investigation of objects that process clear extension and unclear intension. For example, by the year of 2050, China will control its total population within the range of 1.5–1.6 billion people. This range from 1.5 billion to 1.6 billion is a grey concept. Its extension is definite and clear. However, if one inquires further regarding exactly which specific number within the said range it will be, then he will not be able to obtain any meaningful and definite answer. It's a grey number (Deng, 1985; Liu, 2021; Liu et al., 2017).

We summarize the differences among these four main uncertainty research methods in Table 2.2.

Table 2.2 Comparison among the four methods of uncertainty research

Uncertainty research	Grey system	Prob. Statistics	Fuzzy math	Rough set
Research objects	Poor information	Stochastics	Cognitive	Boundary
Basic set	Grey number set	Cantor set	Fuzzy set	Approximate set
Describe method	Possibility func.	Density func.	Membership func.	Upper, lower Appr.
Procedure	Sequence operator	Frequency	Cut set	Dividing
Data requirement	Any distribution	Known distribution	Known membership	Equivalent Rel.
Emphasis	Intension	Intension	Extension	Intension
Objective	Law of reality	Historical law	Cognitive expression	Approx. approaching
Characteristics	Small data	Large sample	Depend on experience	Information form

2.4 Deep Applications of Grey System Theory in the Fields of Social Science, Natural Science and Engineering Technology

2.4.1 Successful Application of Grey System Theory in the Field of Social Sciences

The rapid development of grey system theory in the early stage of its establishment largely benefited from its successful application in the field of economic management, that is, the strong impetus of the urgent need to carry out agricultural zoning and formulate economic development strategic planning all over the country in the 1980s. The reform of the economic system and the adjustment of the statistical system directly affected the integrity and continuity of economic data. The disconnected data posed a big problem for the planners at that time. How to complete the tasks of system analysis and modeling based on small samples and poor information data, so as to obtain the prediction results with high reliability and support the scientific decision-making of governments at all levels? The grey system theory characterized by small sample, poor information data modeling and analysis is just right. At that time, many government departments from the central to local governments tried to use grey system methods and models to analyze economic data and prepare development plans. Professor Deng Julong presided over and completed the research and preparation of the development plan of Yixian County, Hebei Province and Laohekou City, Hubei Province. The author has also presided over and participated in the completion of a number of key bidding projects of the National Development and Reform

Commission of China, the Ministry of Science and Technology of China and the China Association for Science and Technology, as well as the development planning research of Henan Province, Jiangsu Province, Nanjing and Zhongyuan District of Zhengzhou, Hubin District of Sanmenxia, Changge City and Wuzhi county, etc. The data analysis mainly adopts the grey system method and model (Liu & Yang, 1994).

Academician Yang Shanlin and academician Chen Xiaohong of Chinese Academy of engineering, academician Zavadskas of Lithuanian Academy of Sciences and their team have successfully solved many major problems in management practice by using grey system model and method, and achieved a series of research results (Chen, 2018; Jahan & Zavadskas, 2019; Xu, & Yang, 2013).

Emil Scarlat and Camelia Delcea with Bucharest University of Economics of Romania used the methods and models of grey system theory to study the control of economic system, achieved a series of achievements (Delcea et al., 2013; Scarlat & Delcea, 2011), and published a monograph in Romanian.

2.4.2 Deep Application of Grey System Theory in the Field of Natural Science

Enter physics, chemistry, biology, geology, hydrology, crops, and medicine etc. as subject words into CNKI database to search the literature with physics, chemistry and other subjects and accurately containing the phrase "grey system". The results are shown in Table 2.3 (Liu et al., 2022).

Grey system theory has been applied to the fields of physics, chemistry, biology, geology, hydrology, crops, medicine and so on, a large number of valuable research results have been obtained.

For example, in the field of physics

Chen Lei et al. used the grey relational analysis model to study two sky light measurement methods based on ASD ground object spectrometer—standard gray plate inversion measurement method and direct measurement method, and defined the applicable scenarios of different methods (Chen et al., 2011). Wang Yue and Chen Zonghai studied μ particles imaging of cosmic rays by using the method of grey correlation cluster analysis, the efficiency of material differentiation is improved (Wang et al., 2011).

Han Li et al., studied the geophysical characteristics of dynamic compaction fill foundation by using the grey correlation analysis model, and evaluated the quality

Table 2.3 Number of articles containing the phrase "grey system" accurately in various disciplines of Natural Science

Discipline	Physics	Chemistry	Biology	Geology	Hydrology	Crops	Medicine
No. of papers	1793	2621	3313	7280	2688	2562	504

and effect of dynamic compaction by analyzing the correlation between surface wave velocity, resistivity and geophysical characteristic parameters such as soil dry density and water content (Han et al., 2020).

Evans applied the grey system model to study the strength of British steel, and proposed a new method for parameter estimation of Generalized Grey Verhulst model (Evans, 2014).

Shi et al. conducted reliability analysis on passive residual heat removal of AP1000 nuclear power reactor based on grey model (Shi et al., 2017), and Wang Qin et al. used the grey correlation analysis method to study the optimal parameters of arc signal welding process(Wang et al., 2010), both have achieved important results.

In the field of Chemistry

Liu Yaoxin et al. studied the formation reaction of calcium sulphoaluminate in high temperature sulfur fixation phase by using grey correlation analysis and prediction model (Liu, 2007). Pornnapa Kasemsiri et al., used Taguchi method and grey relational analysis model to optimize biodegradable foam composites made from cassava starch, oil palm fiber, chitosan and palm oil (Kasemsiri, 2017).

Gupta et al. applied the grey correlation analysis method to optimize the mechanical properties of hybrid filler pultruded glass fiber composites (Gupta et al., 2019).

Jena et al. applied Taguchi grey correlation analysis to optimize parameters for maximizing photocatalytic behaviour of Zn1-xFexO nanoparticles for methyl orange degradation using Taguchi and Grey relational analysis Approach (Jena et al., 2019).

In the field of Biology

Zhang Fuli et al. studied the effect of BT insect resistant cotton straw returning on soil nutrient characteristics by using grey correlation analysis model. It is considered that straw returning is an ideal way for harmless treatment of Bt transgenic plant straw (Zhang et al., 2020). Yang et al. used the grey correlation analysis model to study the pigment content and standard deviation vegetation index in rice vegetative stage (Yang et al., 2012).

Luo Qin and others used the grey correlation analysis model to study the relationship between trace element content and lead content in the seed body of new irradiated Pleurotus ostreatus, which provided a scientific basis for breeding Pleurotus ostreatus varieties with lower lead content (Luo et al., 2015).

Guo Ruilin has conducted in-depth research on crop grey breeding and cultivated some new crop varieties (Guo, 1995).

Based on hyperspectral data, Jin et al. used grey correlation analysis and partial least square method to estimate the leaf water content of winter wheat (Jin et al., 2013). Wei et al. used the grey correlation analysis method to evaluate the quality of Tibetan highland barley (Wei et al., 2019), has achieved important results.

In the field of Geology and Earth Sciences

Academician Zhao Pengda constructed the theory and method system of quantitative prediction of mineral resources, and put forward "geological anomaly", "mathematical characteristics of geological body", "triple" quantitative metallogenic prediction, research on non-traditional mineral resources, new concepts, new contents and research methods. Two prospective metallogenic belts of copper nickel sulfide were found in Beishan area of Xinjiang and one gold belt was found in East Junggar (Zhao, 2009).

Research on safety analysis, evaluation, excavation and control measures design optimization and real-time monitoring of Geotechnical Engineering (including landslide) by Gao Wei and academician Feng XiaTing (Gao et al., 2004), study on limit displacement discrimination of stability and reliability analysis of surrounding rock of tunnel and underground engineering by Academician Li Xiaohong et al. (Li, 2005), have achieved results of great value.

Peng Fang and Wu Guoping established a new quantitative evaluation method of caprock based on grey programming cluster analysis. They used this method to evaluate 12 kinds of caprock objects in 4 sets of mudstone in 3 main exploration areas of southeast basin of Hainan. The conclusion is consistent with the exploration results (Peng et al., 2005). Liang Bing et al. optimized and ranked the exploration and development potential of complex geological parameter characteristic areas with evaluation index value of interval grey number by establishing a multi index grey correlation degree optimization model (Liang, 2014). Chen Ronghuan and others used the grey system theory to study logging, drilling coring, oil testing and relevant geological data. Through matching, fitting and extracting parameters, they studied and divided formation lithology, physical properties and oil bearing properties by statistical analysis of eigenvalues and their accuracy and resolution, which provided a geological basis for oilfield exploration and development (Chen et al., 2005). Wang yunyun et al. used the grey correlation analysis method to scientifically predict the Yaojialing zinc gold polymetallic deposit (Wang et al., 2013).

Fang Xiaotong and others used the multi-dimensional grey evaluation model to predict the risk of coal and gas outburst, which provided a basis for mine safety production (Fang et al., 2012). Zeng et al., predicted China's shale gas production based on weakening buffer operator and unbiased grey model (Zeng et al., 2018). Kose and Tasci predict geodetic deformation based on multivariable grey prediction model and regression model (Kose & Tasci, 2019).

In the field of hydrology and water resources

Lin Yuezhong and others established the grey prediction model of slope rock mass deformation based on the field slope test data of the Three Gorges, and drew the fitting and prediction curve of slope deformation, which provided a reliable guarantee and theoretical basis for the prediction of slope rock mass deformation (Lin et al., 2005). Academician Xia Jun's research on grey system hydrology (Jia, 1996), academician Wu Zhongru's research on hydraulic structure and dam safety monitoring (Wu et al.,

2012), and research on utilization of water resources by Wang and academician Hipel (Wang, and Hipel, 2011), have achieved a series of important results.

Hao et al. used the grey system model to analyze and predict the hydrological process of karst basin, and obtained high accuracy. They also used the segmented grey model to study the impact of human activities on the hydrological process of karst basin (Hao et al., 2013).

Peng Yong et al. studied the optimization algorithm of cascade reservoir operation based on the combination of grey prediction model and DDDP (Peng, 2018). With limited hydrogeological data, Mahmod et al. used the modified grey model to analyze the groundwater flow in Nubia sandstone area of Halga Oasis, Egypt (Mahmod et al., 2014).

In the field of Medicine

Grey system method and model technology are widely used in modern medical fields such as disease prediction and control, health management evaluation, intelligent diagnosis system construction, drug efficacy evaluation and medical image processing, and have made gratifying achievements, forming a branch field of grey medical research in grey system theory (Zhang, 2015).

Professor Tan xuerui, Dean and doctoral supervisor of Medical College of Shantou University, and his research team have systematically studied the grey correlation methodology of clinical trials with the support of a number of National Natural Science Foundation of China and Guangdong Natural Science Foundation. The new clinical trial methods proposed, such as ergodic grey correlation space theory, polarity analysis theory and method of grey medical correlation factors, axiom system of multi-level grey medical correlation, grey correlation method comparison model, have been applied to many clinical medical disciplines, such as cardiovascular medicine, digestive medicine, neurology, infectious diseases and so on (Tan, 2011).

Wei Hang et al., Established the pattern recognition model of chromatographic fingerprint of traditional Chinese medicine by using the grey system theory. The results of high performance liquid chromatography analysis of 56 batches of different varieties of tangerine showed that the recognition rate exceeded 92.85% for different cultivated varieties of tangerine with very similar chemical composition and content (Wei et al., 2013).

Semra Icer et al. quantitatively graded the ultrasonic images of fatty liver based on grey correlation analysis, and obtained the scientific diagnosis results (Icer et al., 2012). Lai Hsin Yi et al. applied the unsupervised single chain clustering method based on grey correlation analysis to the automatic sorting of spike waves in extracellular electrophysiological records (Lai et al., 2011). Bhupendra Gupta and Mayank Tiwari have achieved good results in breast image brightness preserving contrast enhancement and quality segmentation based on histogram modified grey correlation analysis (Gupta & Tiwari, 2017).

2.4.3 A Large Number of Applications of Grey System Theory in the Field of Engineering Technology

Enter the subject words such as transportation, power and machinery, etc. respectively in the CNKI database, and to search the literatures with transportation, electric power and machinery, etc. as the subject and accurately including the phrase "grey system". The results are shown in Table 2.4 (Liu et al., 2022).

Grey system theory has been applied to the fields of engineering technology such as transportation, electric power and machinery, etc., has achieved thousands or even tens of thousands of research results. Among them, there are more than 5000 papers containing "grey system" in the fields of power, computer and material science. There are more than 10,000 in the field of transportation, more than 20,000 in the field of information science and nearly 30,000 in the field of environmental science.

For example, in the field of transportation

Liu Qiuyan and Zhong Zhangdui comprehensively used grey clustering and rough set model to optimize the planning scheme of railway digital mobile communication system with limited frequency, and improved the accuracy of electrical level and interference matrix estimation (Liu et al., 2010); Gao Fan and Zhang Youpeng designed the grey number of fitness according to the train operation target, and constructed the high-speed train speed controller model based on grey genetic algorithm (Gao et al., 2012); Lu Xiaohong and Wang Changlin studied the modeling and simulation of automatic train speed controller based on predictive grey control (Lu et al., 2013); Based on the data of Britain and the United States, Chirwa et al. used GM (1,1) model to estimate the accident risk (Chirwa et al., 2006); Based on the diagnosis results of three diagnosis methods: fuzzy fault diagnosis method, genetic algorithm and grey system theory, Mi Gensuo et al. constructed the optimal combination model to diagnose the fault of 25 Hz phase sensitive track circuit (Mi et al., 2014).

After comparing the simulation results obtained by artificial neural network, classification and regression tree, k-nearest neighbor method, linear discriminant analysis method, naive Bayesian classifier, quasi optimal algorithm and support vector machine method with the grey correlation classifier algorithm, Twala found that the grey correlation classifier algorithm is most suitable for the modeling and analysis of road traffic accident data in Gauteng Province, South Africa (Twala, 2014).

In the field of Power Engineering

Research of academician Sun Caixin's team on the field of high voltage insulation and Fault Diagnosis Technology (Sun, 2005; Sun et al., 2002, 2003). Academician Li Licheng's research group on Power Grid Engineering, DC transmission and AC/DC parallel power grid operation technology (Huang et al., 2011).

Analysis of oil soluble gas content in power transformers by Liao et al. (2012). According to the measured data of lubricating oil temperature and iron content of wind turbine gearbox, Yang et al. introduced multi-source information, improved the traditional grey system model, predicted the wear trend of wind turbine gearbox,

Table 2.4 Number of articles containing the phrase "grey system" accurately in various disciplines of Engineering Technology

Discipline	Transportation	Power	Machinery	Motive power	Aviation	Architecture	Computer
No. of papers	10,626	5924	2964	2358	1538	4930	6109
Discipline	Electronics	Information	Petroleum	Chemical industry	Material	Irrigation works	Environment
No. of papers	2310	24,545	2932	1121	7465	1678	29,069

and provided a scientific basis for gearbox maintenance and replacement decision-making (Yang et al., 2019).

Ossowski and Korzybski use grey system model to carry out analog circuit fault diagnosis (Ossowski & Korzybski, 2013); Jiang Wei diagnosed the fault of wind turbine drive chain based on grey rough set theory (Jiang, 2012);

Study on Modeling and prediction of non-stationary voltage fluctuations by Dejamkhooy et al. (Dejamkhooy et al., 2017).

In the field of Mechanical Engineering

Academician Jia Zhenyuan's research on shape control machining theory, technology and equipment of high-end equipment and high-performance parts (Jia, 2009). Research on mechanical design and theory, computer aided design and graphics, digital design and manufacturing, etc. by academician Tan Jianrong's group (Fang et al., 2009). Research on submarine noise reduction technology by academician He Lin's group (Liao et al., 2017). Czeslaw Cempel used the grey prediction model to monitor the mechanical vibration state (Cempel, 2008). Wang Xuliang and Nie Hong used the grey system model to predict the fatigue life of mechanical parts, which greatly reduces the prediction error (Wang et al., 2008); Zhang Xueyuan et al. used GM (1,1) model to study the change law of robot emotional state, and realized the emotional robot interaction system (Zhang et al., 2006); Li Tong et al. used the grey prediction model to calculate the fatigue crack growth rate (Li et al., 2010).

Academician Zhang Jie et al. used the grey correlation analysis model to analyze the fault of two tooth difference swing movable teeth transmission, which provided a scientific basis for improving the reliability of two tooth difference swing movable teeth transmission system (Zhang et al., 2012). Xia Xintao and Wang Zhongyu used the grey correlation analysis model to study the relationship between rolling bearing processing quality and vibration, and found that the structural dimension error parameter is the factor that has a great impact on bearing vibration (Xia et al., 2005). Xie Yanmin et al. obtained the best parameters of each factor affecting the robustness of square box by analyzing the\of the grey correlation degree between each factor and the target sequence (Xie et al., 2007).

Prakash et al. study on multi-objective optimization of turning stone powder reinforced aluminum matrix composites based on Taguchi method and grey correlation analysis model (Prakash et al., 2020). Loganathan et al. used the grey correlation analysis model to optimize the input parameters of progressive forming of AA6061 alloy (Loganathan et al., 2020). Pagar and Gawande (Pagar & Gawande, 2020) used the grey correlation analysis method to carry out parametric design analysis on the radial deflection stress of metal expansion bellows. Sharma used Taguchi and grey correlation analysis to study the accuracy and surface roughness of GFRP gears (Sharma et al., 2020). Khan et al. used the grey correlation analysis method to carry out multi-objective optimization of dry, wet and low temperature undercut titanium base alloys (Khan et al., 2020).

In the field of Aerospace

Wang Yanyang and Cao Yihua established a nonlinear online prediction model of China's civil aviation operation risk by using the method of grey neural network (Wang et al., 2010), Yang Tianshe (2008) and Li Peihua (2011) used the grey system model to predict spacecraft faults and achieved high accuracy.

Xie Jianxi et al. solved the optimization decision-making problem of aircraft top-level design scheme by using the grey correlation analysis model (Xie et al., 2004); Zhang Cheng and Ding Songbin et al. studied the aircraft customization scheme based on the grey correlation analysis model (Zhang et al., 2014); Xiao Jun and Zhang Weiwei comprehensively used the grey correlation analysis and fault tree method to study the target crash fault, which provided a theoretical basis for diagnosing the cause of the target crash fault, controlling the occurrence of the fault and improving the system reliability (Xiao et al., 2009).

Yu Fengjie and Ke Yinglin applied the grey clustering decision-making method to the automatic docking and assembly system of aircraft large parts, which improved the system stability, reduced the risk of equipment failure, and controlled the maintenance cost (Yu et al., 2009). Zhang Feng and Wang Pengwei used the grey clustering evaluation model to evaluate the safety of Shipborne aircraft system, which played a positive role in discovering system safety hazards in advance and preventing and reducing accidents (Zhang et al., 2010).

In the field of Intelligent Control

Research on intelligent control theory and robot system, image recognition theory and machine vision application, intelligent control technology of advanced manufacturing equipment, and integrated automatic control system of major projects in power and electrical industry by Academician Wang Yaonan's team (E, J. Q.,et al., 2005). Academician Liu Yexiang of the State Key Laboratory of powder metallurgy of Central South University have used the grey system method and model to study the control of aluminum electrolysis process, and achieved many results (Liu, 2004).

Tian Jianyan et al. established the grey prediction model of billet temperature in heating furnace and put forward the billet temperature control method (Tian et al., 2007); Wang Wei et al. proposed an improved fuzzy expert control method based on combined grey prediction model for the temperature control of coke oven flue with the characteristics of strong nonlinearity, large time delay and multi disturbance (Wang et al., 2010). Combining the traditional feedback control method and grey predictive control, Zhang Guangli et al. designed a self-adjusting grey predictive controller. The simulation results show that the new controller has better dynamic performance and robustness (Zhang et al., 2004).

In view of the randomness, nonlinearity and time variability of the deep-sea walking mechanism in the seabed complex operating environment, and it is difficult to establish an accurate mathematical model, Qiao Guiling et al. proposed a grey prediction fuzzy PID control method to realize the effective control of the deep-sea walking mechanism (Qiao et al., 2009). The research on pneumatic position servo control system based on grey correlation compensation control proposed by

Zhu Jianmin et al. effectively improves the tracking accuracy of traditional control methods for pneumatic position servo control system (Zhu et al., 2012).

In the field of Weapon Equipment Development and Application

Cui Jianpeng et al. studied the selection of surface to air missile weapon system by using multi-objective grey decision model (Cui et al., 2012); Li Xinqi et al. constructed a grey programming model for the optimal configuration of missile nuclear weapons, which provides a theoretical basis for the ordering, storage, position configuration and operational application of missile nuclear weapons (Li et al., 2007).

Han Xiaoming et al. used the grey clustering model to comprehensively evaluate the development scheme of air defense and anti-missile warhead (Han et al., 2014); Yao Junbo and Hu Weiwen applied the grey evaluation model to evaluate the operational effectiveness of over the horizon ground wave radar according to its characteristics and operational tasks (Yao et al., 2008).

Lin Jiajian used the grey relational analysis method to solve the main factors affecting the velocity of explosively formed projectile (EFP), and obtained the results that have important reference value for the design of EFP liner and charge structure (Lin, 2009). Zhao Guogang et al. established the threat assessment model of incoming missile in ship anti-missile operation by using the grey correlation analysis method, which provides a decision-making basis for the ship command and control system to judge the target threat in time (Zhao et al., 2007). And research on radar target tracking by Liu Yi'an et al., (Liu et al., 2006).

References

Cempel, C. (2008). Decomposition of the symptom observation matrix and grey forecasting in vibration condition monitoring of machines. *International Journal of Applied Mathematics and Computer Science*, 18(4), 569–579.

Chen, L., Qin, Y., & Deng, R.R.. (2011). Comparative analysis of the two common methods for skylight measurement based on the ASD spectroradiometer. *Tropical Geography*. 31(2), 182–186.

Chen, X. H. (2018). Analysis on the trend of technology integration and application innovation in the era of digital economy. *Social Scientists*, 8, 8–23.

Chen, D. J., Zhang, Y. L., & Chen, M. Y. (2005). Grey macroscopic control prediction model of systems clouds and applications. *Control and Decision Making, 20*(5), 553–557.

Cobb, C. W., & Douglas, P. H. (1928). A theory of production. *American Economic Review, 18*(Supplement), 139–165.

Cui, J.P., Xin, Y.P., & Liu, X.J. (2012). Research on surface to air missile type selection based on multi-criteria grey decision-making. *Tactical Missile Technology, 1*, 7–10.

Dejamkhooy, A., Dastfan, A., & Ahmadyfard, A. (2017). Modeling and forecasting nonstationary voltage fluctuation based on grey system theory. *IEEE Transactions on Power Delivery, 32*(3), 1212–1219.

Delcea, C., Bradea, I., & Scarlat, E. (2013). A computational grey based model for companies risk forecasting. *The Journal of Grey System, 25*(3), 70–83.

Deng, J. L. (1985). *Grey control system.* Wuhan: Press of Huazhong University of Science and Technology.

Deng, J. L. (1990). *Grey system theory tutorial*. Wuhan: Press of Huazhong University of Science and Technology.

Evans, M. (2014). An alternative approach to estimating the parameters of a generalised grey Verhulst model: An application to steel intensity of use in the UK. *Expert Systems with Applications, 41*(4), 1236–1244.

Fang, H., Tan, J.R., & Yin, G.F. (2009).Research on user demand analysis method of house of quality based on grey theory [J]. *Computer Integrated Manufacturing System, 15*(3), 576–584 + 591.

Fang, X. T., Chen, Y., & Li, S. Q. (2012). Application of multidimensional grey evaluation mMethods in coal and gas outburst prediction. *Industrial Safety and Environmental Protection, 38*(12), 81–83.

Gao, F., Zhang, Y.P., & Gao, P. (2012). Research on speed controller model for high-speed train based on grey genetic algorithm. *Computer Measurement and Control, 20*(5), 272–1275.

Gini, C. (1921). Measurement of Inequality of Incomes. *The Economic Journal (Blackwell Publishing), 31*(121), 124–126.

Guo, H.Q., Wu, Z.R., & Yang, J. (2001). A new combined model for rock-fill dam deformation monitoring. *The Journal of Hohai University, 29*(6), 51–55.

Guo, R. L. (1995). *Crop grey breeding*. China Agricultural Science and Technology Press.

Gupta B., & Tiwari M. (2017). A tool supported approach for brightness preserving contrast enhancement and mass segmentation of mammogram images using histogram modified grey relational analysis. *Multidimensional Systems and Signal Processing, 28*(4), 1549–1567.

Gupta, A., Vaishya, R., & Khan, K., et al. (2019). Multi-response optimization of the mechanical properties of pultruded glass fiber composite using optimized hybrid filler composition by the gray relation grade analysis. *Materials Research Express, 6*(12), 125322.

Han, L., Shuai, H.L., & Ge, Z. H., et al, (2020), Study on geophysical characteristics of dynamic compaction fill foundation. *Low Temperature Building Technology, 42*(01), 124–127.

Han, X. M., Nan, H. Y., & Chen, J. J., et al. (2014). Grey cluster model used to evaluate the development scheme of warhead in antimissile missile of air defense. *Journal of Air Force Engineering University, 1*(1), 29–33.

Hao, Y.H., Cao, B.B., & Chen, X., et al. (2013). A piecewise grey system model for study the effects of anthropogenic activities on Karst hydrological processes. *Water Resources Management, 27*(5), 1207–1220.

Huang, X. B., Ouyang, L. S., Wang, Y. N., & Li, L. C. (2011). Analysis on key influencing factors of transmission line icing. *High Voltage Technology, 37*(7), 1677–1682.

Icer, S., Coskun, A., & Ikizceli, T. (2012). Quantitative Grading Using Grey Relational Analysis on Ultrasonographic Images of a Fatty Liver. *Journal of Medical Systems, 36*(4), 2521–2528.

Jahan, A., & Zavadskas, E. K. (2019). ELECTRE-IDAT for design decision-making problems with interval data and target-based criteria. *Soft Computing, 23*, 129–143.

Jena, M., Manjunatha, C., & Shivaraj, B. W. (2019). Optimization of parameters for maximizing photocatalytic behaviour of $Zn1-xFexO$ nanoparticles for methyl orange degradation using Taguchi and Grey relational analysis Approach. *Material Today Chemistry, 12*, 187–199.

Jiang, W. (2012). An intelligent diagnosis method based on grey rough set theory for wind turbine driving chain. *Power System and Clean Energy, 28*(12), 79–83.

Jin, X., Xu, X., & Song, X. (2013). Estimation of leaf water content in winter wheat using grey relational analysis-partial least squares modeling with hyperspectral data. *Agronomy Journal, 105*(5), 1385–1392.

Kasemsiri, P., Dulsang, N., & Pongsa, U., et al. (2017). Optimization of biodegradable foam composites from cassava starch, oil palm fiber, chitosan and palm oil using taguchi method and grey relational analysis. *Journal of Polymers and The Environment, 25*(2), 378–390.

Khan, M. A., Jaffery, S., & Khan, M., et al. (2020). Multi-objective optimization of turning titanium-based alloy Ti-6Al-4V under dry, wet, and cryogenic conditions using gray relational analysis (GRA). *International Journal of Advanced Manufacturing Technology, 106*(9–10), 3897–3911.

Kose, E., & Tasci, L. (2019). Geodetic deformation forecasting based on multi-variable grey prediction model and regression model. *Grey Systems: Theory and Application, 9*(4), 464–471.

Lai, H.Y., Chen, Y.Y., & Lin, S.H., et al. (2011). Automatic spike sorting for extracellular electrophysiological recording using unsupervised single linkage clustering based on grey relational analysis. *Journal of Neural Engineering, 8*(3), No. 036003.

Li, G. D., Yamaguchi, D., & Nagai, M. (2007). A GM(1,1)–Markov chain combined model with an application to predict the number of Chinese international airlines. *Technological Forecasting and Social Change, 74*(8), 1465–1481.

Li, J., Sun, C. X., & Cheng, W. G. (2002). Study on fault diagnosis of insulation of oil-immersed transformer based on grey cluster theory. *Transactions of China Electro-Technical Society, 17*(4), 80–83.

Li, T., Ren, M. F., & Chen, H. R. (2010). Calculation method of fatigue crack growth rate based on grey system theory. *Mechanical Strength, 32*(3), 472–475.

Liao, J., He, L., Lv, Z. Q., et al. (2017). Research on comprehensive evaluation method of marine steering gear system architecture scheme. *Machine Tools and Hydraulics, 45*(7), 59–63.

Liao, R. J., Yang, J. P., & Grzybowski, S., et al. (2012). Forecasting dissolved gases content in power transformer oil based on weakening buffer operator and least square support vector machine-Markov. *IET Generation, Transmission and Distribution, 6*(2), 142–151.

Lin, Y. Z., Wang, T. C., & Wang, L., et al. (2005). Stability analysis of high excavated slope in Three Gorges Project. *Journal of Tianjing University, 38*(10), 936–940.

Liu, S. F. (2021). *Grey system theory and its applications* (9th ed.). Beijing: Science Press.

Liu, Q.Y., Zhong, Z.D., & Ai Bo. (2010). Research on frequency planning of GSM-R system based on grey cluster theory with rough set. *Journal of the China Railway Society, 32*(5), 53–58

Liu, S. F., & Lin, Y., et al. (2006). On measures of information content of grey numbers. *Kybernetes, 35*(5), 899–904.

Liu, S. F., Yang, Y. J., Forrest, J. (2017). *Grey data analysis*. Singapore: Springer-Verlag

Liu, S. F., Tao, Y., Xie, N. M., Tao, L. Y., Hu, M. L. (2022). Advance in grey system theory and applications in the fields of natural sciences and engineering-technology. Grey Systems-Theory and Application, 12(4), 1–18.

Loganathan, D.; Kumar, S S.; Ramadoss, R.(2020). Grey relational analysis-based optimization of input parameters of incremental forming process applied to the AA6061 alloy. *Transactions of Famena, 44*(1), 93–104

Luo, Q., Jiang, Z.H., & Xiao, S.X.(2015). Grey relational analysis of trace element content and lead content in new irradiated seeds of Pleurotus ostreatus. *Journal of Mountain Agricultural Biology, 34*(04), 18–21

Mahmod, W. E., & Watanabe, K. (2014). Modified Grey Model and its application to groundwater flow analysis with limited hydrogeological data: A case study of the Nubian Sandstone, Kharga Oasis. Egypt. *Environmental Monitoring and Assessment, 186*(2), 1063–1081.

Mi, G.S., Yang, R.X., & Liang, L. (2014). Research on the method of diagnosis of complex fault in track circuit based on combined model. *Journal of the China Railway Society, 36*(10), 65–69

Modigliani, F., & Brumbergh, R. (1954). Utility analysis and the consumption function: An interpretation of cross-section data. In: K. Kurihara (Ed.), Post Keynesian Economics, George Allen and Unwin

Ossowski, M., & Korzybski, M. (2013). Data mining based algorithm for analog circuits fault diagnosis. *Przeglad Elektrotechniczny, 89*(2a), 285–287.

Pagar, N. D., & Gawande, S. H. (2020). Parametric design analysis of meridional deflection stresses in metal expansion bellows using gray relational grade. *Journal of the Brazilian Society of Mechanical Sciences and Engineering, 42*(5), 256

Phillips, A. W. H. (1958). The Relation Between Unemployment and the Rate of Change of Money Wage Rates in the United Kingdom, 1861–1957. *Economica, 25*(2), 283–299.

Prakash, K. S; Gopal, P. M.; Karthik, S.(2020). Multi-objective optimization using Taguchi based grey relational analysis in turning of Rock dust reinforced Aluminum MMC. *Measurement, 157*, 107664

Scarlat, E., & Delcea, C. (2011). Complete analysis of bankruptcy syndrome using grey systems theory. *Grey Systems: Theory and Application, 1*(1), 19–32.

Sharma, A., Aggarwal, M. L., & Singh, L. (2020). Investigation of GFRP gear accuracy and surface roughness using Taguchi and grey relational analysis[J]. *Journal of Advanced Manufacturing Systems, 19*(1), 147–165.

Sharpe, W. F. (1964). Capital asset prices: A theory of market equilibrium under conditions of risk. *Journal of Finance, 19*(3), 425–442.

Shi, Q., Tao, Z., & Shahzad, M. A.. (2017). Reliability analysis on passive residual heat removal of AP1000 based on grey model. *ATW-International Journal for Nuclear Power, 62*(6), 408–417

Sun, C.X. (2005). Present situation and prospect of online monitoring and diagnosis technology of the state of power transmission and transformation equipment. *China Power, 38*(2), 1–7

Sun, C. X., Bi, W. M., & Zhou, Q. (2003). New gray prediction parameter model and its application in electrical insulation fault prediction. *Control Theory and Applications, 20*(5), 798–801.

Sun, C. X., Li, J., & Zheng, H. P. (2002). A new method of faulty insulation diagnosis in power transformer based on degree of area incidence analysis. *Power System Technology, 26*(7), 24–29.

Twala, B. (2014). Extracting grey relational systems from incomplete road traffic accidents data: The case of Gauteng Province in South Africa. *Expert Systems, 31*(3), 220–231

Wang, Q., Kuang, L. Z., & Zeng, S. B. (2010). Grey relational analysis of optimal parameters of welding process based on arc signal. *Electric Welding Machine, 40*(3), 75–78.

Wang, Y., Cheng, Z. H., Wang, H. Y., et al. (2011). Application of gray relational cluster method in Muon tomography. *Nuclear Electronics & Detection Technology., 31*(8), 871–873.

Wang, Y. M., Dang, Y. G., & Wang, Z. X. (2008). Optimization of the background values of non-equal-distant GM(1,1) model. *Management Science of China, 16*(4), 159–162.

Wang, Y. Y., Zhou, T. F., Zhang, M. M., et al. (2013). Application of grey relational analysis in Yaojialing Zn-Au polymetallic deposit prediction. *Journal of Hefei University of Technology., 36*(10), 1236–1241.

Wei, H., Lin, X., Zhang, Y. et al. (2013). Research on the application of grey system theory in the pattern recognition for chromatographic fingerprints of traditional Chinese medicine. *Chinese Journal of Chromatography. 31*(2), 127–132.

Wei, N., & Zhang, T. L. (2019). Evaluation of Tibetan highland barley quality by grey relational analysis. *Bangladesh Journal of Botany, 48*(3), 817–826.

Wu, Z.R., Xu, B., & Gu, C.S., et al. (2012).On comprehensive evaluation methods of the service condition of the dam. *China Science: Technological Sciences, 42*(11), 1243–1254.

Xia, X.T., & Wang, Z.Y., Chang, H. (2005). Degree of grey incidence for the quality of processing and vibration of Rolling bearing. *Journal of Aerospace Power, 20*(2), 250–254.

Xie, Y. M., Yu, H. P., & Chen, J. et al. (2007). Design for square-cup deep-drawing based on grey systems theory. *Journal of Mechanical Engineering, 27*(3), 10–15.

Xie, J. X., Song, B. F., & Liu, D. X. (2004). Grey relational analysis method for optimal decision of aircraft top-level design scheme. *Journal of Systems Engineering, 19*(4), 350–354.

Xu, G.F., & Yang, S.L. (2013). Application of Grey Theory in fault diagnosis of sucker rod pumping wells. *Journal of Hefei University of Technology, 36*(10) : 1265–1268.

Yang, Z.W., Hsu, M.H., & Chen, J.C. (2012). Grey relational analysis of pigment levels and the normalized difference vegetation index during the vegetative phase of paddy rice. *The Journal of Grey System, 24*(3): 275–284.

Yang, X. Y., Fang, Z. G., & Yang, Y. J. (2019). A novel multi-information fusion grey model and its application in wear trend prediction of wind turbines. *Applied Mathematical Modelling, 71*, 543–557.

Yu, F.J., Ke, Y.L., & Ying, Z.(2009).Decision of fault repair in aircraft assembly system of automated docking. *Computer Integrated Manufacturing System, 15*(9): 1823–1830.

Yue, J. P., & Hua, X. S. (1994). An grey regression model and accuracy analysis. *Dam and Safety, 2*, 23–28.

Zadeh, L. A. (1994). Soft computing and fuzzy logic. *IEEE Software, 11*(6), 48–56.

Zeng, B., Duan, H. M., Bai, Y., et al. (2018). Forecasting the output of shale gas in China using an unbiased grey model and weakening buffer operator. *Energy, 151*, 238–249.

Zhang, G. L., Fu, Y., & Yang, R.Q.(2004). Novel self-adjustable grey prediction controller. *Control and Decision, 19*(2), 212–215.

Zhang, F., Wang, P.W., Xiao, Z.R., et al (2010).Application of grey theory in safety evaluation of carier aircraft system. *Aircraft Design, 30*(3), 56–61.

Zhang, C., & Ding, S.B., Wang, B. (2014). Study on customization model of aircraft based on grey incidence analysis. *Traffic Information and Safety, 32*(4), 131–136.

Zhang, F. L., Yin, Q., & Wang, D., et al. (2020). Effects of Bt insect resistant cotton straw returning on soil nutrient characteristics. *Journal of Biosafety, 29*(01), 69–77.

Zhang, J., Liang, S. M., & Zhou, R. G., et al. (2012). Fault Tree Analysis of Two Teeth Difference Swing Movable Teeth Transmission Based on Grey Correlation. *Machinery Design & Manufacture, No., 6*, 183–185.

Zhao, G. G., Sun, Y. K., & Xu, Y. J., et al. (2007). Grey decision analysis of threat assessment in surface ship anti missile operation. *Tactical Missile Technology, No., 3*, 32–35.

Zhu, J. M., Huang, Z. W., & Zhai, D. T., et al. (2012). Research on grey forecasting PID control simulation based on strengthening buffer operator. *Journal of Shanghai University of Science and Technology, 34*(4), 327–332.

Chapter 3
Grey Numbers and Their Operations

3.1 Grey Numbers

A grey system is described with grey numbers, grey sequences, grey equations, or grey matrices. Here, grey numbers are the elementary "atoms" or "cells", and their exact values are unknown. In applications, a grey number stands for an indeterminate number that takes its possible value within an interval or a general set of numbers. A grey number is generally represented using the symbol " $\otimes$." There are several types of grey numbers, as discussed below.

(1) Grey numbers with only a lower bound: This kind of grey number $\otimes$ is represented as $\otimes \in [a, \infty)$ or $\otimes(\underline{a})$, where $\underline{a}$ stands for the definite, known lower bound of the grey number $\otimes$. The interval $[\underline{a}, \infty)$ is referred to as the field of $\otimes$.

For example, the weight of a celestial body which is far away from the Earth is a grey number containing only a lower bound, because the weight of the celestial body must be greater than zero. However, the exact value of the weight cannot be obtained through normal means. If we use the symbol $\otimes$ to represent the weight of the celestial body, we then have that $\otimes \in [0, \infty)$.

(2) Grey numbers with only an upper bound: This kind of grey number $\otimes$ is written as $\otimes \in (-\infty, \overline{a}]$ or $\otimes(\overline{a})$, where $\overline{a}$ stands for the definite, known upper bound of $\otimes$.

A grey number containing only an upper bound is a grey number with a negative value, but its asssbsolute value is infinitely great. For example, the opposite number of the weight of the celestial body mentioned above is a grey number with only an upper bound.

(3) Interval grey numbers: This kind of grey number $\otimes$ has both a lower $\underline{a}$ and an upper bound $\overline{a}$, written $\otimes \in [\underline{a}, \overline{a}]$.

For example, for an investment opportunity, there always exists an upper limit representing the maximum amount of money that can be mobilized. For an electrical equipment, there must be a maximum critical value for the equipment to function normally. The critical value could be for a maximum voltage

S. Liu, *Grey Systems Analysis*, Series on Grey System,
https://doi.org/10.1007/978-981-19-6160-1_3

or for a maximum amount of current allowed to be applied to the equipment. At the same time, the values of investment, voltage, and current are all greater than zero. Therefore, the amount of dollars that can be used for a specific investment opportunity, and the voltage and the current requirements for the electrical equipment are all examples of interval grey numbers.

(4) Continuous and discrete grey numbers: This kind of grey number takes only a finite number or a countable number of potential values and is known as discrete. If a grey number can potentially take any value within an interval, then it is known as continuous.

For example, if a person's age is between 30 and 35, his or her age could be one of the values 30, 31, 32, 33, 34, 35. Thus, age is a discrete grey number. As for a person's height and weight, they are continuous grey numbers.

(5) Black and white numbers: Black numbers are represented as $\otimes \in (-\infty, +\infty)$; that is, when $\otimes$ has neither an upper nor a lower bound, then $\otimes$ is known as a black number. When $\otimes \in [\underline{a}, \overline{a}]$ and $\underline{a} = \overline{a}$, $\otimes$ is known as a white number.

For the sake of parsimony, in our discussion we treat black and white numbers as special grey numbers.

(6) Essential and non-essential grey numbers: The former stands for a grey number that temporarily cannot be represented by a white number; the latter entails a grey number that can be represented by a white number obtained either through experience or through a certain method. The definite white number is referred to as the whitenization (value) of the grey number, denoted $\tilde{\otimes}$. Also, we use $\otimes(a)$ to represent grey number(s) with a as its whitenization.

A grey number is an uncertain number with its value in a specific range. The range can be regarded as a cover of the grey number. Therefore, an interval grey number $\otimes \in [\underline{a}, \overline{a}], \underline{a} < \overline{a}$ is very different from an interval number $[\underline{a}, \overline{a}], \underline{a} < \overline{a}$. An interval grey number $\otimes \in [\underline{a}, \overline{a}], \underline{a} < \overline{a}$ is only one value in interval $[\underline{a}, \overline{a}], \underline{a} < \overline{a}$. However, an interval number $[\underline{a}, \overline{a}], \underline{a} < \overline{a}$ is the whole interval $[\underline{a}, \overline{a}], \underline{a} < \overline{a}$.

3.2 The Whitenization of a Grey Number and Degree of Greyness

When a type of grey number vibrates around a certain fixed value, the whitenization of this kind of grey number is relatively easy. One can simply use that fixed value as its whitenization. A grey number that vibrates around a can be written as $\otimes(a) = a + \delta_a$ or $\otimes(a) \in (-, a, +)$, where δ_a stands for the vibration. In this case, the whitenized value is $\tilde{\otimes}(a) = a$.

For the general interval grey number $\otimes \in [a, b]$, we can take its whitenization value $\tilde{\otimes}$ as indicated in (3.1), based on the possible value information:

$$\tilde{\otimes} = \alpha a + (1 - \alpha)b, \alpha \in [0, 1] \tag{3.1}$$

Here, α is called the positioned coefficient of the interval grey number $\otimes \in [a, b]$ (Liu, 1989).

Definition 3.2.1 The whitenization of the form $\tilde{\otimes} = \alpha a + (1 - \alpha)b, \alpha \in [0, 1]$ is called a whitenization with positioned coefficient α.

Definition 3.2.2 Mean whitenization occurs when $\alpha = \frac{1}{2}$. When the distribution of an interval grey number is unknown, mean whitenization is often employed.

Definition 3.2.3 Take the interval grey numbers $\otimes_1 \in [a, b], \otimes_2 \in [a, b]$; let $\tilde{\otimes}_1 = \alpha a + (1 - \alpha)b, \alpha \in [0, 1]$; and $\tilde{\otimes}_2 = \beta a + (1 - \beta)b, \beta \in [0, 1]$.

If $\alpha = \beta$, we say that both $\otimes_1$ and $\otimes_2$ are synchronous. If $\alpha \neq \beta$, we say that the grey numbers $\otimes_1$ and $\otimes_2$ are non-synchronous. When two grey numbers $\otimes_1$ and $\otimes_2$ have the same value range in interval [a, b], it is only when they are synchronous that it is possible to have $\otimes_1 = \otimes_2$.

When the distribution of a grey number is known, mean whitenization is not used. For instance, a certain person's age is within the range of 35–45 years old. Thus, $\otimes \in [35, 45]$ is a grey number. It is also known that the person in question finished their 12 years of pre-college education and entered college at the end of 1990s. Hence, the chance of the person to be around 40 years old in 2022 is quite good. For this grey number, it is not reasonable for us to employ mean whitenization.

When the value information of a grey number is known to a certain extent, we can use a possibility function to describe the possibility of the grey number has taking its potential values.

The possibility function is different from the membership function in fuzzy mathematics. The membership function describes the degree to which an object belongs to a certain set. However, the possibility function describes the possibility that a grey number can take a certain value, or the possibility that a certain value is the true value of a grey number. The possibility function is similar to the density function of probability distribution, but there are essential differences between the two concepts. A grey number described by the possibility function is a number with incomplete value information. Once a number with complete value information can be treated as a random variable with a certain probability distribution, it is no longer a grey number with poor value information:

For any conceptual type of grey number that represents wishes, its possibility function generally increases monotonically. In Fig. 3.1, the possibility function $f(x)$ stands for, say, the grey number of the amount of funds for a research project (in ten thousand dollars) and its degree of preference. A straight line stands for the "normal desire," that is, the degree of preference is directly proportional to the amount of funds, with different slopes representing different intensities of desire. In particular, $f_1(x)$ represents a relatively mild intensity of desire, where a funds in amount of $100,000 is not enough, a funds in the amount of $200,000 will be more satisfying, and a funds of $300,000 will be quite adequate. $f_2(x)$ stands for a desire with more intensity, where a funds in the amount of $350,000 is only about 40% satisfactory. The curve of $f_3(x)$ means that even for a funds in the amount of $400,000, the

Fig. 3.1 Different types of possibility functions

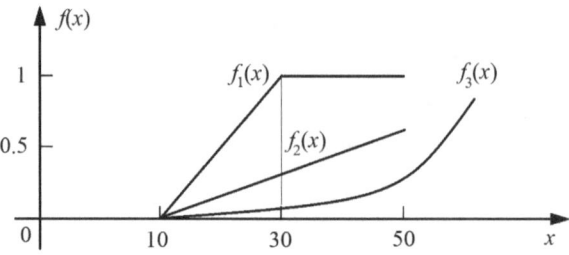

degree of satisfaction is only about 20%. To be satisfied, the amount of funds has to be somewhere around $800,000.

Generally speaking, the possibility function of a grey number is designed according to what is known to the researcher. Therefore, it does not have a fixed form. The start and end of the curve should have its significance. For instance, in a trade negotiation, there is a process of changing from a grey state to a white state. The eventual agreed upon deal will be somewhere between the ask and the bid. Thus, the relevant possibility function should start at the level of the ask (or the bid) and end at the level of bid (or the ask).

The typical possibility function is a continuous function with fixed starting and ending points so that the left-hand side increases and the right-hand side decreases, as seen in Fig. 3.2a, where:

$$f_1(x) = \begin{cases} L(x), & x \in [a_1, b_1) \\ 1, & x \in [b_1, b_2] \\ R(x), & x \in (b_2, a_2] \end{cases}.$$

For the convenience of computer programming and computation, in practical applications the left- and right-hand functions $L(x)$ and $R(x)$ are generally simplified into straight lines, as seen in Fig. 3.2b, where:

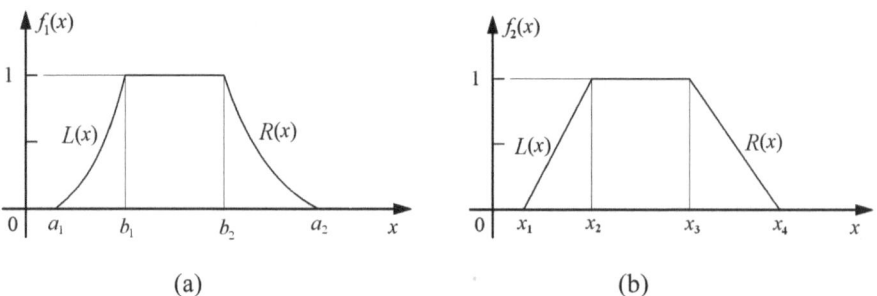

(a)   (b)

Fig. 3.2 Typical possibility function

$$f_2(x) = \begin{cases} L(x) = \frac{x-x_1}{x_2-x_1} & x \in [x_1, x_2) \\ 1, & x \in [x_2, x_3] \\ R(x) = \frac{x_4-x}{x_4-x_3} & x \in (x_3, x_4] \end{cases}.$$

Definition 3.2.4 For the possibility function shown in Fig. 3.2a, the following representation is referred to as the degree of greyness of $\otimes$ (Deng, 1985):

$$g^{\circ} = \frac{2|b_1 - b_2|}{b_1 + b_2} + \max\left\{ \frac{|a_1 - b_1|}{b_1}, \frac{|a_2 - b_2|}{b_2} \right\} \tag{3.2}$$

The expression g° is a sum of two parts. The first part represents the greyness of the grey number as affected by the size of the peak area under the curve of the possibility function, while the second part shows the effect of the size of the area under the curves of $L(x)$ and $R(x)$. Generally, the greater the peak area and the area under $L(x)$ and $R(x)$, the greater the value of g°. When $\max\left\{ \frac{|a_1 - b_1|}{b_1}, \frac{|a_2 - b_2|}{b_2} \right\} = 0$, $g^{\circ} = \frac{2|b_1 - b_2|}{b_1 + b_2}$. In this case, the possibility function is a horizontal line. When $\frac{2|b_1 - b_2|}{b_1 + b_2} = 0$, grey number $\otimes$ is a grey number with its basic value $b = b_1 = b_2$. When $g^{\circ} = 0$, $\otimes$ is a white number.

3.3 Degree of Greyness Defined by Axioms

Professor Julong Deng (1985) provided a definition of degree of greyness of a grey number with a typical possibility function, as shown in Fig. 3.2a. In 1996, Professor Sifeng Liu established an axiomatic definition of degree of greyness by using the length $l(\otimes)$ of the grey number interval and its kernel $\hat{\otimes}$ (Liu, 1996):

$$g^{\circ}(\otimes) = \frac{l(\otimes)}{\hat{\otimes}} \tag{3.3}$$

Such a definition is valid on the basis of the postulates of non-negativity, zero greyness, infinite greyness, and scalar multiplication. However, the concept of greyness as defined in Eqs. (3.2) and (3.3) suffers from the following problems:

(1) When the length $l(\otimes)$ of the grey interval approaches infinity, the degree of greyness as defined in both (3.2) and (3.3) is likely to approach infinity.
(2) Grey numbers centered at zero will not have greyness. In this case, in Eq. (3.2), one has $b_1 = b_2 = 0$; and in Eq. (3.3), one faces $\hat{\otimes} = 0$. That is, neither (3.2) nor (3.3) is meaningful.

A grey number is a way to express the behavioral characteristics of a specific grey system (Deng, 1990). The greyness of grey numbers reflects the degree to which the researcher understands the uncertainty involved in such numbers (Liu et al., 1999; Chen, 2001). Therefore, the magnitude of the greyness of a grey number should be closely related to the background on which the grey number is come from, or to the field of discourse within which the said number becomes grey. If this background, or field of discourse, and the characteristics of a grey system are not detailed, there is no means through which to discuss the degree of greyness of a given grey number. With this understanding in place, let Ω be the field of discourse within which grey number $\otimes$ is come from, and $\mu(\otimes)$ is the measure of the number field from which $\otimes$ takes its value. Then, the degree of greyness $g^\circ(\otimes)$ of grey number $\otimes$ should satisfy the axioms below.

Axiom 3.3.1 $0 \leq g^\circ(\otimes) \leq 1$. That is, the degree of greyness of any grey number has to be within the range of 0–1.

Axiom 3.3.2
Any $\otimes \in [\underline{a}, \overline{a}], \underline{a} \leq \overline{a}$, when $a = \overline{a}$, $g^\circ(\otimes) = 0$. That is, white numbers contain no ambiguity, so their degree of greyness is 0.

Axiom 3.3.3 $g^\circ(\Omega) = 1$. That is, because the background Ω within which grey number $\otimes$ is come from is generally known. Therefore, Ω does not contain any useful information leading to the greatest level of uncertainty.

Axiom 3.3.4 $g^\circ(\otimes)$ is directly proportional to $\mu(\otimes)$ and inversely proportional to $\mu(\Omega)$.

Definition 3.3.1 The following equation is called the degree of greyness of grey number $\otimes$:

$$g^\circ(\otimes) = \mu(\otimes)/\mu(\Omega) \tag{3.4}$$

Ω is the field of discourse of grey number $\otimes$, and μ is the measure of field Ω (Liu et al., 2010a).

Theorem 3.3.1 *The degree of greyness of grey numbers satisfies the following properties*:

(1) If $\otimes_1 \subset \otimes_2$, then $g^\circ(\otimes_1) \leq g^\circ(\otimes_2)$.
(2) $g^\circ(\otimes_1 \cup \otimes_2) \geq g^\circ(\otimes_k), k = 1, 2$, where $\otimes_1 \cup \otimes_2 = \{\xi | \xi \in [a, b] \text{ or } \xi \in [c, d]\}$ is the union of grey numbers $\otimes_1 \in [a, b], a < b$ and $\otimes_2 \in [c, d], c < d$.
(3) $g^\circ(\otimes_1 \cap \otimes_2) \leq g^\circ(\otimes_k), k = 1, 2$, where $\otimes_1 \cap \otimes_2 = \{\xi | \xi \in [a, b] \text{ and } \xi \in [c, d]\}$ is the interaction between grey numbers $\otimes_1 \in [a, b], a < b$ and $\otimes_2 \in [c, d], c < d$.
(4) If $\otimes_1 \subset \otimes_2$, then $g^\circ(\otimes_1 \cup \otimes_2) = g^\circ(\otimes_2), g^\circ(\otimes_1 \cap \otimes_2) = g^\circ(\otimes_1)$.
(5) If $\mu(\Omega) = 1$ and the measures of $\otimes_1$ and $\otimes_2$ are independent of μ, then
(6) $g^\circ(\otimes_1 \cap \otimes_2) = g^\circ(\otimes_1) \cdot g^\circ(\otimes_2)$; and
(7) $g^\circ(\otimes_1 \cup \otimes_2) = g^\circ(\otimes_1) + g^\circ(\otimes_2) - g^0(\otimes_1) \cdot g^\circ(\otimes_2)$.

Proof The conclusions (1)-(4) obviously.Therefore, the proof is omitted because all of them can be exported directly from Definition 3.3.1.

For 5 1°), from $\mu(\Omega) = 1$ and the assumption that measures of $\otimes_1$ and $\otimes_2$ are independent of μ, we have:

$$g^{\circ}(\otimes_1 \cap \otimes_2) = \mu(\otimes_1 \cap \otimes_2) = \mu(\otimes_1) \cdot \mu(\otimes_2) = g^{\circ}(\otimes_1) \cdot g^{\circ}(\otimes_2)$$

Similarly, for 2°, we have:

$$g^{\circ}(\otimes_1 \cup \otimes_2) = \mu(\otimes_1 \cup \otimes_2) = \mu(\otimes_1) + \mu(\otimes_2) - \mu(\otimes_1) \cdot \mu(\otimes_2)$$
$$= g^{\circ}(\otimes_1) + g^{\circ}(\otimes_2) - g^{\circ}(\otimes_1) \cdot g^{\circ}(\otimes_2) \text{ QED.}$$

The way in which grey numbers are combined affects the degree of greyness and the reliability of the resultant grey number. Generally, when grey numbers are "unioned" together, the resultant degree of greyness and reliability of the new information increase; when grey numbers are intersected together, the resultant degree of greyness drops and the reliability of the combined information decreases. When solving practical problems and processing a large amount of grey numbers, it is advisable to combine the numbers at several different levels so that useful information can be extracted at individual levels. Additionally, in the process of combining grey numbers, "union" and "intersection" operations should be done at individual and other levels in order to guarantee that the extracted information satisfies pre-determined requirements in terms of reliability and degree of greyness.

3.4 The Operations of Interval Grey Numbers

In what follows, let us look at the operations of interval grey numbers. Given grey numbers $\otimes_1 \in [a, b]$, $a < b$ and $\otimes_2 \in [c, d]$, $c < d$ let us use * to represent an operation between $\otimes_1$ and $\otimes_2$. If $\otimes_3 = \otimes_1 * \otimes_2$, then $\otimes_3$ should also be an interval grey number satisfying $\otimes_3 \in [e, f]$, $e < f$ and for any $\tilde{\otimes}_1$ and $\tilde{\otimes}_2$, $\tilde{\otimes}_1 * \tilde{\otimes}_1 \in [e, f]$. The operation rules of interval grey numbers are discussed below (Deng, 1985).

Rule 3.4.1 (Additive operation). Assume that $\otimes_1 \in [a, b]$, $a < b$; $\otimes_2 \in [c, d]$, $c < d$, then the following equation is called the sum of $\otimes_1$ and $\otimes_2$:

$$\otimes_1 + \otimes_2 \in [a + c, b + d] \tag{3.5}$$

Example 3.4.1 Assume that $\otimes_1 \in [3, 4]$, $\otimes_2 \in [5, 8]$, then $\otimes_1 + \otimes_2 \in [8, 12]$.

Rule 3.4.2 (Additive inverse). Assume that $\otimes \in [a, b]$, $a < b$, then the additive inverse of $\otimes$ is given by:

$$-\otimes \in [-b, -a] \tag{3.6}$$

Example 3.4.2 Assume that $\otimes \in [3, 4]$, then $-\otimes \in [-4, -3]$.

Rule 3.4.3 (Subtraction operation). Assume that $\otimes_1 \in [a, b]$, $a < b$; $\otimes_2 \in [c, d]$, $c < d$, then the following is called the deviation $\otimes_1$ minus $\otimes_2$:

$$\otimes_1 - \otimes_2 = \otimes_1 + (-\otimes_2) \in [a - d, b - c] \tag{3.7}$$

Example 3.4.3 Assume that $\otimes_1 \in [3, 4]$, $\otimes_2 \in [1, 2]$, then:

$$\otimes_1 - \otimes_2 \in [3 - 2, 4 - 1] = [1, 3], \otimes_2 - \otimes_1 \in [1 - 4, 2 - 3] = [-3, -1].$$

Rule 3.4.4 (Multiplication operation). Assume that $\otimes_1 \in [a, b]$, $a < b$; $\otimes_2 \in [c, d]$, $c < d$, then the following equation is called the product of $\otimes_1$ and $\otimes_2$:

$$\otimes_1 \cdot \otimes_2 \in [\min\{ac, ad, bc, bd\}, \max\{ac, ad, bc, bd\}] \tag{3.8}$$

Example 3.4.4 Assume that $\otimes_1 \in [3, 4]$, $\otimes_2 \in [5, 10]$, then:

$$\otimes_1 \cdot \otimes_2 \in [\min\{15, 30, 20, 40\}, \max\{15, 30, 20, 40\}] = [15, 40].$$

Rule 3.4.5 (Reciprocal). Assume that $\otimes \in [a, b]$, $a < b$, $a \neq 0$, $b \neq 0$, $ab > 0$, then the following equation is called the reciprocal of $\otimes$:

$$\otimes^{-1} \in \left[\frac{1}{b}, \frac{1}{a}\right] \tag{3.9}$$

Example 3.4.5 Assume that $\otimes \in [2, 4]$ then $\otimes^{-1} \in [0.25, 0.5]$.

Rule 3.4.6 (Division). Assume that $\otimes_1 \in [a, b]$, $a < b$; $\otimes_2 \in [c, d]$, $c < d$ and $c \neq 0$, $d \neq 0$, $cd > 0$, then the following is called the quotient of $\otimes_1$ division by $\otimes_2$:

$$\otimes_1 / \otimes_2 = \otimes_1 \times \otimes_2^{-1} \in \left[\min\left\{\frac{a}{c}, \frac{a}{d}, \frac{b}{c}, \frac{b}{d}\right\}, \max\left\{\frac{a}{c}, \frac{a}{d}, \frac{b}{c}, \frac{b}{d}\right\}\right] \tag{3.10}$$

Example 3.4.6 Assume that $\otimes_1 \in [3, 4]$, $\otimes_2 \in [5, 10]$, then:

$$\otimes_1 / \otimes_2 \in \left[\min\left\{\frac{3}{5}, \frac{3}{10}, \frac{4}{5}, \frac{4}{10}\right\}, \max\left\{\frac{3}{5}, \frac{3}{10}, \frac{4}{5}, \frac{4}{10}\right\}\right] = [0.3, 0.8].$$

Rule 3.4.7 (Scalar multiplication). Let $\otimes \in [a, b]$, $a < b$, and k a positive real number, then the following is called the product of scalar k with grey number $\otimes$:

$$k \cdot \otimes \in [ka, kb] \tag{3.11}$$

Example 3.4.7 Assume that $\otimes \in [2, 4]$, and k $= 5$, then $5 \times \otimes \in [10, 20]$.

Rule 3.4.8 (Power,). Let $\otimes \in [a, b]$, $a < b$, kk a positive real number, then the following equation is called the kth power of the grey number $\otimes$:

$$\otimes^k \in \left[a^k, b^k\right] \tag{3.12}$$

Example 3.4.8 Assume that $\otimes \in [2, 4]$, and k $= 5$, then $\otimes^5 \in [32, 1024]$.

3.5 General Grey Numbers and Their Operations

3.5.1 Reduced Form of Interval Grey Numbers

As the basis of grey system theory, grey numbers, grey number operations and grey algebraic systems have received much attention from grey system scholars over the past years. In the 1980s, we put forward the concept of mean whitenization of grey numbers (Liu, 1989), and based on this concept we developed a new algebraic system for grey numbers.

According to the standard definition of degree of greyness of grey numbers (Liu, 1996, 2006; Yang, 2007, Yang and Liu, 2011), it is possible to address grey intervals after the operation of grey numbers, with the help of the concept of degree of greyness (Jiang et al., 2017, 2021).

In this section, a definition for grey "kernel" is put forward. The axioms for operation of grey numbers and a grey algebraic system is built based on grey "kernel" and the degree of greyness of grey numbers. Also, the properties of the operation are discussed with regards to how the operation of grey numbers can be transformed to the operation of real numbers. Thus, to a certain extent the problem for setting up the operation of grey numbers and grey algebraic systems is solved.

Definition 3.5.1 (The "Kernel" of Grey Number)

(1) Suppose an interval grey number $\otimes \in [\underline{a}, \overline{a}], \underline{a} < \overline{a}$. In case of a lack of distributing information of the values of grey number $\otimes$, $\hat{\otimes} = \frac{1}{2}(\underline{a} + \overline{a})$ is called the "kernel" of grey number $\otimes$.
(2) If a grey number $\otimes$ is a discrete number and $a_i \in [\underline{a}, \overline{a}](i = 1, 2, \ldots n)$ are all the possible values for grey number $\otimes$, then $\hat{\otimes} = \frac{1}{n} \sum_{i=1}^{n} a_i$ is called the "kernel" of grey number $\otimes$.
(3) Suppose that grey number $\otimes \in [\underline{a}, \overline{a}], \underline{a} < \overline{a}$ is a random grey numbers with value distribution information. Then $\hat{\otimes} = E(\otimes)$ is called the "kernel" of grey number $\otimes$ (Liu et al., 2010a, 2010b).

$\hat{\otimes}$, the "kernel" of grey number $\otimes$, is the representation of grey numbers $\otimes$, which cannot be exchangeable in the course of transforming the operation of grey numbers to operation of real numbers. In fact, the "kernel" of grey number $\otimes$, as a real number, can be completely operated by the operation of real numbers, such as plus, minus, multiplication, division, power, extract,, and so on. Also, it is reasonable to take the operation results of the "kernels" as the "kernel" of operation results of grey numbers.

Definition 3.5.2 Let $\hat{\otimes}$ and $g°$ be the kernel and the degree of greyness of a grey number $\otimes$, respectively. Then $\hat{\otimes}_{(g)}$ is called the reduced form of grey number $\otimes$. The reduced form $\hat{\otimes}_{(g)}$ contains all the information of grey number $\otimes \in [\underline{a}, \overline{a}], \underline{a} < \overline{a}$.

Proposition 3.5.1 *For interval grey numbers, there is an one-to-one correspondence between the reduced form $\hat{\otimes}_{(g)}$ and grey number $\otimes \in [\underline{a}, \overline{a}], \underline{a} < \overline{a}$.*

In fact, for any chosen grey number $\otimes \in [\underline{a}, \overline{a}], \underline{a} < \overline{a}$, one can compute $\hat{\otimes}_{(g)}$ through both $\hat{\otimes}$ and $g°$. On the other hand, when $\hat{\otimes}_{(g)}$ is given, one can determine the position of $\otimes$ from $\hat{\otimes}$. Therefore, from the definition of degree of greyness $g°$, one can compute the measure of the grey number $\otimes$ and consequently the upper and lower bounds $\overline{a}$ and $\underline{a}$, which provides detailed information for $\otimes \in [\underline{a}, \overline{a}], \underline{a} < \overline{a}$.

Example 3.5.1 Assume that the grey numbers $\otimes_1 = [-2, -1], \otimes_2 = [8, 18], \otimes_3 = [-2, 18]$ all on background $\Omega \in [-2, 20]$. Take the length of grey interval as the measure of grey numbers, and calculate the reduced forms of $\otimes_1, \otimes_2, \otimes_3$.

Solution The measures of $\Omega, \otimes_1, \otimes_2, \otimes_3$ are $\mu(\Omega) = 20 - (-2) = 22, \mu(\otimes_1) = 1, \mu(\otimes_2) = 10, \mu(\otimes_3) = 20$. Then we can get to the kernels and the degree of greyness of $\otimes_1, \otimes_2, \otimes_3$ as follows:
$\hat{\otimes}_1 = -1.5 \ \hat{\otimes}_2 = 13, \ \hat{\otimes}_3 = 8; g_1°(\otimes_1) = 0.045, g_2°(\otimes_2) = 0.45, g_3°(\otimes_3) = 0.91$.
Therefore, we obtained:

$$\otimes_1 = -1.5_{(0.045)}, \quad \otimes_2 = 13_{(0.5)}, \quad \otimes_3 = 8_{(0.91)}.$$

3.5.2 General Grey Number and Its Reduced Form

Definition 3.5.3 (Basic element of grey number). Together, an interval grey number and a white number are called the basic element of a grey number.

Definition 3.5.4 (General grey number). Let $g^{\pm} \in \mathfrak{R}$ be an unknown real number within a union set of closed or open grey intervals, where:

$$g^* \in \bigcup_{i=1}^{n} [\underline{a}_i, \overline{a}_i] \tag{3.13}$$

If $i = 1, 2, \ldots, n$, n is an integer and $0 < n < \infty$, $\underline{a}_i, \overline{a}_i \in \mathbb{R}$ and $\overline{a}_{i-1} \leq \underline{a}_i \leq \overline{a}_i \leq \underline{a}_{i+1}$, for any grey interval $\otimes_i \in [\underline{a}_i, \overline{a}_i] \subset \bigcup_{i=1}^{n} [\underline{a}_i, \overline{a}_i]$, then $g^{\pm}$ is called a general grey number. $g^- = \inf_{\underline{a} \in g^+} \underline{a}_i$ and $g^+ = \sup_{\overline{a}_i \in g^{\pm}} \overline{a}_i$ are called the lower and upper limits of $g^{\pm}$ (Liu et al., 2012).

Definition 3.5.5 (The "Kernel." of General Grey Number)

(1) For a general grey number $g^{\pm} \in \bigcup_{i=1}^{n} [\underline{a}_i, \overline{a}_i]$, the following is called the "kernel" of a general grey number:

$$\hat{g} = \frac{1}{n} \sum_{i=1}^{n} \hat{a}_i \qquad (3.14)$$

(2) If the probability distribution of $g^{\pm} \in [\underline{a}_i, \overline{a}_i] | (i = 1, 2, \ldots, n)$ is known, assume that p_i is the probability for $g^{\pm} \in [\underline{a}_i, \overline{a}_i] | (i = 1, 2, \ldots, n)$, $\hat{a}_i$ the "kernel" of grey interval $\otimes_i \in [\underline{a}_i, \overline{a}_i]$, and the following conditions hold:

$p_i > 0, i = 1, 2, \ldots, n$; and

$$\sum_{i=1}^{n} p_i = 1.$$

Then, the "kernel" $\hat{g}$ of general grey number $g^{\pm} \in \bigcup_{i=1}^{n} [\underline{a}_i, \overline{a}_i]$ can be defined as follows:

$$\hat{g} = \sum_{i=1}^{n} p_i \hat{a}_i \qquad (3.15)$$

Definition 3.5.6 (The degree of greyness of a general grey number). Suppose that the background which makes a general grey number $g^{\pm} \in \bigcup_{i=1}^{n} [\underline{a}_i, \overline{a}_i]$ come into being is Ω, μ is the measure of Ω, and $\otimes_i \in [\underline{a}_i, \overline{a}_i]$, $i = 1, 2, \ldots, n$ are basic elements of general grey number $g^{\pm} \in \bigcup_{i=1}^{n} [\underline{a}_i, \overline{a}_i]$. Then the following is called the degree of greyness of general grey number $g^{\pm} \in \bigcup_{i=1}^{n} [\underline{a}_i, \overline{a}_i]$, also denoted as g° for short (Liu et al., 2012):

$$g^{\circ}(g^{\pm}) = \frac{1}{\hat{g}} \sum_{i=1}^{n} \hat{a}_i \mu(\otimes_i) \bigg/ \mu(\Omega) \qquad (3.16)$$

Definition 3.5.7 (The reduced form of general grey number). If $\hat{g}$ is the "kernel" of a general grey number $g^{\pm} \in \bigcup_{i=1}^{n} [\underline{a}_i, \overline{a}_i]$ and g° is the degree of greyness of this general grey number, then, $\hat{g}_{(g^{\circ})}$ is called the reduced form of a general grey number.

The reduced form $\hat{g}_{(g°)}$ of a general grey number contains important information regarding the values of general grey number $g^\pm \in \bigcup_{i=1}^{n} [\underline{a}_i, \overline{a}_i]$. If all the $\hat{a}_i$ and $\mu(\otimes_1)(i = 1, 2, \ldots, n)$ are known, then the reduced form of grey number $\hat{g}_{(g°)}$ contains all the information regarding the values of general grey numbers $g^\pm \in \bigcup_{i=1}^{n} [\underline{a}_i, \overline{a}_i]$.

Example 3.5.2 Let us take a mixed general grey number $g^\pm = \otimes_1 \cup \otimes_2 \cup 2 \cup \otimes_4 \cup 6$, where $\otimes_1 \in [1, 3]$, $\otimes_2 \in [2, 4]$, $\otimes_4 \in [5, 9]$. Assume that the background or field which makes general grey number $g^\pm$ come into being is $\Omega = [0, 32]$. If we take the length of the interval as the measure of these grey numbers, try and work out the reduced forms of general grey number $g^\pm$.

Solution $\hat{\otimes}_1 = 2$, $\hat{\otimes}_2 = 3$, $\hat{\otimes}_4 = 7$, thus, the kernel of general grey number $g^\pm$ is as follows:

$$\hat{g} = \frac{1}{5}(\hat{\otimes}_1 + \hat{\otimes}_2 + 2 + \hat{\otimes}_4 + 6) = \frac{1}{5}(2 + 3 + 2 + 7 + 6) = 4.$$

From that, $\mu(\otimes_1) = 2$, $\mu(\otimes_2) = 2$, $\mu(\otimes_4) = 4$, $\mu(2) = \mu(6) = 0$, we have:

$$g°(g^\pm) = \frac{1}{\hat{g}}\sum_{i=1}^{5} \hat{\otimes}_i \mu(\otimes_i) \bigg/ \mu(\Omega)$$

$$= \frac{1}{4}(2 \times 2 + 3 \times 2 + 2 \times 0 + 7 \times 4 + 6 \times 0)/32 \approx 0.297.$$

Therefore, the reduced forms of general grey number $g^\pm$ is $4_{(0.297)}$. When the probability distribution of $g^\pm$ is known, assume that:

$$p_1 = 0.1, \; p_2 = 0.2, \; p_3 = 0.3, \; p_4 = 0.3, \; p_5 = 0.1.$$

Then: $\hat{g} = \sum_{i=1}^{n} p_i \cdot \hat{\otimes}_i = (0.1 \cdot 2 + 0.2 \cdot 3 + 0.3 \cdot 2 + 0.3 \cdot 7 + 0.1 \cdot 6) = 4.1$

Therefore, the reduced form of general grey number $g^\pm$ is $4.1_{(0.297)}$.

3.5.3 Synthesis of Degree of Greyness and Operations of General Grey Numbers

Axiom 3.5.1 (The synthesis axiom of degree of greyness) When plus and minus are operated on n general grey numbers of $g_1^\pm, g_2^\pm, \ldots, g_n^\pm$, then the degree of greyness $g°$ of the operation results in $g^\pm$, which can be arrived at as follows:

$$g^{\circ} = \frac{1}{\sum_{i=1}^{n} \hat{g}_i} \sum_{i=1}^{n} g_i^{\circ} \hat{g}_i = \sum_{i=1}^{n} w_i g_i^{\circ} \qquad (3.17)$$

where $w_i = \frac{\hat{g}_i}{\sum_{i=1}^{n} \hat{g}_i}, i = 1, 2, \ldots, n$, are the weights of g_i°.

One can arrive the conclusion as following Proposition 3.5.2 through Axiom 3.5.1.

Proposition 3.5.2 When sums and subtractions are operated on n general grey numbers of $g_1^{\pm}, g_2^{\pm}, \ldots, g_n^{\pm}$, g° is the degree of greyness of the operation result $g^{\pm}$; if $g_m^{\circ} = \min_{1 \le i \le n}\{g_i^{\circ}\}$, $g_M^{\circ} = \max_{1 \le i \le n}\{g_i^{\circ}\}$, then:

$$g_m^{\circ} \le g^{\circ} \le g_M^{\circ} \qquad (3.18)$$

Axiom 3.5.2 (The Unreduction Aaxiom of Degree of Greyness) When divisions and multiplications are operated on n general grey numbers, the degree of greyness g° of the operation result $g^{\pm}$ is not less than g_M°, the maximum value of the degree of greyness $g_1^{\circ}, g_2^{\circ}, \ldots, g_n^{\circ}$ of n general grey numbers $g_1^{\pm}, g_2^{\pm}, \ldots, g_n^{\pm}$.

Usually, g_M°, the maximum number of the degree of greyness of n general grey numbers is taken as the degree of greyness of the operation results. One can arrive at this conclusion through Proposition 3.5.3 below.

Proposition 3.5.3 When divisions and multiplications are operated on n general grey numbers with the same degree of greyness, then the degree of greyness of the operation result holds the line.

Proposition 3.5.4 When divisions and multiplications are operated on a white number and a general grey number, the degree of greyness of the result is equal to the degree of greyness of the general grey number.

Suppose that $g_1^{\pm}, g_2^{\pm}$ are two general grey numbers; $\hat{g}_1, \hat{g}_2$ are their kernels, respectively, and g_1°, g_2° are their degrees of greyness, respectively. Then, the following rules come into existence according to Axioms 3.5.1 and 3.5.2:

$$\textbf{Rule 1} \quad \hat{g}_{1(g_1^{\circ})} + \hat{g}_{2(g_2^{\circ})} = \left(\hat{g}_1 + \hat{g}_2\right)_{(w_1 g_1^{\circ} + w_2 g_2^{\circ})} \qquad (3.19)$$

$$\textbf{Rule 2} - \hat{g}_{1(g_1^{\circ})} = \left(-\hat{g}_1\right)_{(g_1^{\circ})} \qquad (3.20)$$

$$\textbf{Rule 3} \quad \hat{g}_{1(g_1^{\circ})} - \hat{g}_{2(g_2^{\circ})} = \left(\hat{g}_1 - \hat{g}_2\right)_{(w_1 g_1^{\circ} + w_2 g_2^{\circ})} \qquad (3.21)$$

$$\textbf{Rule 4} \quad \hat{g}_{1(g_1^{\circ})} \times \hat{g}_{2(g_2^{\circ})} = \left(\hat{g}_1 \times \hat{g}_2\right)_{(g_1^{\circ} \vee g_2^{\circ})} \qquad (3.22)$$

$$\textbf{Rule 5 If } \hat{g}_1 \ne 0, \text{ then } 1/\hat{g}_{1(g_1^{\circ})} = \left(1/\hat{g}_1\right)_{(g_1^{\circ})} \qquad (3.23)$$

Rule 6 $\hat{g}_2 \neq 0$, then $\hat{g}_{1(g_1^\circ)} \div \hat{g}_{2(g_2^\circ)} = (\hat{g}_1 \div \hat{g}_2)_{(g_1^\circ \vee g_2^\circ)}$ (3.24)

Rule 7 k, is a real number, then $k \quad \hat{g}_{(g_1^i)} = (k \cdot \hat{g})_{(g_1^i)}$ (3.25)

The operations of general grey numbers can be extended to cases where many general grey numbers must be operated. In such cases, we can take the operation results of the "kernels" as the "kernel" of operation results of general grey numbers. We can then get the degree of greyness of the results according to Axioms 3.5.1 and 3.5.2, and, thus, we can arrive at the reduced forms of the results.

Example 3.5.3 Take two mixed general grey numbers $g_1^\pm = \otimes_1 \cup \otimes_2 \cup 2 \cup \otimes_4 \cup 6$ and $g_2^\pm = \otimes_6 \cup 20 \cup \otimes_8 \cup \otimes_9$, where $\otimes_1 \in [1, 3]$, $\otimes_2 \in [2, 4]$, $\otimes_4 \in [5, 9]$, $\otimes_6 \in [12, 16]$, $\otimes_8 \in [11, 15]$, $\otimes_9 \in [15, 19]$. Assume that the background or field which makes general grey number $g_1^\pm$ come into being is $\Omega = [0, 32]$, and the background or field which makes general grey number $g_2^\pm$ come into being is $\Omega = [10, 60]$. Try and calculate the values of $g_3^\pm = g_1^\pm + g_2^\pm$, $g_4^\pm = g_1^\pm - g_2^\pm$, $g_5^\pm = g_1^\pm \times g_2^\pm$, and $g_6^\pm = g_1^\pm \div g_2^\pm$.

Solution First, calculate the reduced forms of $g_1^\pm$ and $g_2^\pm$. From Example 3.5.2, we have $g_1^\pm = 4_{(0.297)}$. From that, $\hat{\otimes}_6 = 14$, $\hat{\otimes}_8 = 13$, $\hat{\otimes}_9 = 17$, and $\mu(\otimes_6) = 4$, $\mu(\otimes_7) = 0$, $\mu(\otimes_8) = 4$, $\mu(\otimes_9) = 4$, we have:

$$\hat{g}_2 = \frac{1}{4}(\hat{\otimes}_6 + 20 + \hat{\otimes}_8 + \hat{\otimes}_9) = \frac{1}{4}(14 + 20 + 13 + 17) = 16$$

and

$$g_2^\circ(g^\pm) = \frac{1}{\hat{g}_2} \sum_{i=1}^{4} \hat{\otimes}_i \mu(\otimes_i) \bigg/ \mu(\Omega_2)$$

$$= \frac{1}{16}(14 \times 4 + 20 \times 0 + 13 \times 4 + 17 \times 4)/50 = 0.22.$$

Thus, the reduced form of general grey number $g_2^\pm$ is $16_{(0.22)}$. With the reduced forms, as well as $w_1 = \frac{4}{20} = 0.2$, $w_2 = \frac{16}{20} = 0.8$, it is possible for us to get the following results:

$$g_3^\pm = g_1^\pm + g_2^\pm = (\hat{g}_1 + \hat{g}_2)_{(w_1 g_1^\circ + w_2 g_2^\circ)} = (4 + 16)_{(0.2 \times 0.297 + 0.8 \times 0.22)} = 20_{0.235}$$

$$g_4^\pm = g_1^\pm - g_2^\pm = (\hat{g}_1 - \hat{g}_2)_{(g_1^i \vee g_2^i)} = (4 - 16)_{(0.2 \times 0.297 + 0.8 \times 0.22)} = (-12)_{0.235}$$

$$g_5^\pm = g_1^\pm \times g_2^\pm = (\hat{g}_1 \times \hat{g}_2)_{(g_1^\circ \vee g_2^\circ)} = (4 \times 16)_{(0.297 \vee 0.22)} = 64_{0.297}$$

$$g_6^\pm = g_1^\pm \div g_2^\pm = \left(\hat{g}_1 \div \hat{g}_2\right)_{\left(g_1^\circ \vee g_2^\circ\right)} = (4 \div 16)_{(0.297 \vee 0.22)} = \left(\frac{1}{4}\right)_{0.297}$$

Definition 3.5.8 Assume that $F\left(g^\pm\right)$ is a set of general grey numbers, and that $g_i^\pm, g_j^\pm \in F\left(g^\pm\right)$. If $g_i^\pm + g_j^\pm$, $g_i^\pm - g_j^\pm$, $g_i^\pm \cdot g_j^\pm$, and $g_i^\pm \div g_j^\pm$ all belong to $F\left(g^\pm\right)$ (when division is considered, the conditions in rule 6 need to be satisfied), then $F\left(g^\pm\right)$ is called a field of general grey numbers.

Theorem 3.5.1 *The totality of all general grey numbers constitutes a field of general grey numbers.*

Definition 3.5.9 *Assume that $R\left(g^\pm\right)$ is a set of general grey numbers. If for $g_i^\pm, g_j^\pm$ and $g_k^\pm \in R\left(g^\pm\right)$, the following hold true:*

(1) $g_i^\pm + g_j^\pm = g_j^\pm + g_i^\pm$

(2) $\left(g_i^\pm + g_j^\pm\right) + g_k^\pm = g_i^\pm + \left(g_j^\pm + g_k^\pm\right)$;

(3) There exists a zero element $0 \in R\left(g^\pm\right)$, such that $g_i^\pm + 0 = g_i^\pm$;

(4) For any $g_i^\pm \in R\left(g^\pm\right)$, there exists a $-g_i^\pm \in R\left(g^\pm\right)$, such that $g_i^\pm + \left(-g_i^\pm\right) = 0$;

(5) $\left(g_i^\pm \cdot g_j^\pm\right) \cdot g_k^\pm = g_i^\pm \cdot \left(g_j^\pm \cdot g_k^\pm\right)$;

(6) There exists a unit element $1 \in R\left(g^\pm\right)$, such that $1 \cdot g_i^\pm = g_i^\pm \cdot 1 = g_i^\pm$;

(7) $\left(g_i^\pm + g_j^\pm\right) \cdot g_k^\pm = g_i^\pm \cdot g_k^\pm + g_j^\pm \cdot g_k^\pm$ and

(8) $g_i^\pm \cdot \left(g_j^\pm + g_k^\pm\right) = g_i^\pm \cdot g_j^\pm + g_i^\pm \cdot g_k^\pm$.

Thus, $R\left(g^\pm\right)$ is called a linear space of general grey numbers.

Theorem 3.5.2 *The totality of all synchronous general grey numbers constitutes a linear space.*

A grey number is the most elementary component of grey system theory and forms the basis for studying the quantitative relations of a grey system. The operation of grey numbers is the starting point for grey maths, and it has much significance in the development of grey system theory. On the basis of intensifying the effect and significance of the "kernel" of general grey numbers, and with the degree of greyness of general grey numbers as a link, the operation of grey numbers has been translated into the operation of real numbers. Therefore, to a certain extent the problem of operation of grey numbers has been solved, and a grey algebraic system based on this operation has been constructed. The operation of grey numbers defined in this chapter can be extended to grey algebraic equations, grey differential equations and grey matrix operations. This is a development of great significance to the study of grey input–output models and grey programming, which has been progressing slowly due to the difficulty of grey number operations.

The calculation of degree of greyness of general grey numbers relates to the field Ω of general grey numbers. Thus, the field Ω must be considered in order to translate the reduced form of general grey number to its common form. Researchers tend to pay attention only to the operation of general grey numbers and ignore the field of the results, which creates difficulties in reverting general grey numbers. However, the reduced form of a general grey number provides relevant information about the "kernel" and degree of greyness, So that we can know what we know. This is similar to the digital characteristics of a random variable such as mean and variance, which hold the distribution information of the random variable. The "kernel" and degree of greyness arising from the reduced form are very important as they allow us to learn the value information of a general grey number.

References

Chen, S. L. (2001). Grey element and grey element space in grey system. *Journal of Kunming University of Science and Technology, 26*(2), 92–95.

Deng, J. L. (1985). *Grey control systems.* Press of Huazhong University of Science and Technology.

Deng, J. L. (1990). *Grey system theory tutorial.* Press of Huazhong University of Science and Technology.

Jiang, S. Q., Liu, S. F., Fang, Z. G., & Liu, Z. X. (2017). Study on distance measuring and sorting method of general grey number. *Grey systems: Theory and application, 7*(3), 320–328.

Jiang, S. Q., Liu, S. F., & Liu, Z. X. (2021). General grey number decision-making model and its application based on intuitionistic grey number set. *Grey systems: Theory and application, 11*(4), 556–570.

Liu, S. F. (1989). On Perron-Frobenius theorem of grey nonnegative matrix. *Journal of Grey System, 1*(2), 157–166.

Liu, S. F. (1996). Axiom on grey degree. *The Journal of Grey System, 8*(4), 397–400.

Liu, S. F. (2006). On index system and mathematical model for evaluation of scientific and technical strength. *Kybernetes: The International Journal of Systems & Cybernetics, 35*(7–8), 1256–1264.

Liu, S. F., Fang, Z. G., & Xie, N. M. (2010a). Algorithm rules of interval grey numbers based on the "Kernel" and the degree of greyness of grey numbers. *Systems Engineering and Electronics, 32*(2), 313–316.

Liu, S. F., Yuan, W. F., & Sheng, K. Q. (2010b). Multi-attribute intelligent grey target decision model. *Control and Decision, 25*(8),1159–1163

Liu, S. F., Guo, T. B., & Dang, Y. G. (1999). *Grey systems theory and its applications* (2nd ed.). Beijing: Science Press.

Liu, S. F., Fang, Z. G., Yang, Y. J., & Forrest, J. (2012). General grey numbers and its operations. *Grey Systems: Theory and Application, 2*(3), 341–349.

Yang, Y. (2007). Extended grey numbers and their operations. In *Proceedings of 2007 IEEE International Conference on Systems, Man and Cybernetics*. Montreal, Canada, pp. 2181–2186.

Yang, Y., & Liu, S. (2011). Reliability of operations of grey numbers using kernels. *Grey systems: Theory and application. 1*(1), 57–71.

Chapter 4
Sequence Operators and Grey Data Mining

4.1 Introduction

One of the main tasks of grey systems theory is to uncover the mathematical relationships between different system variables and the laws of change of certain system variables themselves based on the available data of characteristic behaviors of social, economic and ecological systems, for example. Grey systems theory looks at each stochastic variable as a grey quantity that varies within a fixed region and within a certain time frame, and each stochastic process as a grey process.

When investigating the behavioral characteristics of a system, what is available is often a sequence of definite white numbers. There is no substantial difference between whether we treat the sequence as a trajectory or actualization of a stochastic process, or as whitenized values of a grey process. However, to uncover the laws of evolution of systems' behavioral characteristics, different methods are developed using different thinking logics. For instance, the theory of stochastic investigates statistical laws on the basis of probabilities borrowed from prior knowledge. This methodology generally requires large amounts of data. However, even with large amounts of data there is no guarantee that any of the desired laws can be successfully uncovered. That is because the number of basic forms of distribution considered in this methodology is very limited. It is often extremely difficult to deal with non-typical distribution processes. Nonetheless, grey systems theory uncovers laws of change by excavating and organizing the available raw data, representing an approach of finding data out of data through grey sequence operators. Grey systems theory believes that a system possesses overall functions and properties, even if the expression of such an objective system might be complicated, and its data chaotic. Therefore, there must be internal laws governing the existence of the system and its operation. The key is to choose an appropriate method to excavate the internal laws and make use of such laws. For any given grey sequence, its implicit pattern can always be revealed through weakening the explicit randomness.

S. Liu, *Grey Systems Analysis*, Series on Grey System,
https://doi.org/10.1007/978-981-19-6160-1_4

For example, the following sequence does not clearly show any regularity or pattern:

$$X^{(0)} = (1, 2, 1.5, 3) = \left(x^{(0)}(1), x^{(0)}(2), x^{(0)}(3), x^{(0)}(4)\right).$$

Now, we depict the data set with the graph in Fig. 4.1. From this graph, it can be seen that the curve of $X^{(0)}$ undulates with relatively large amplitude. If we apply the accumulating operator once to the original data set $X^{(0)}$, and denote the resultant sequence as $X^{(1)}$, then we have:

$$X^{(1)} = (1, 3, 4.5, 7.5) = \left(x^{(1)}(1), x^{(1)}(2), x^{(1)}(3), x^{(1)}(4)\right)$$

where for k $= 1, 2, 3, 4, x^{(1)}(k) = \sum_{i=1}^{k} x^{(0)}(i)$.

Now, the processed sequence $X^{(1)}$ clearly shows a growing tendency (see Fig. 4.2 for more details).

Fig. 4.1 The curve of $X^{(0)}$

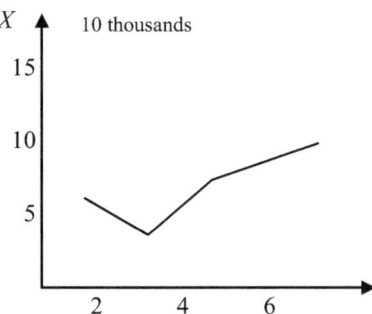

Fig. 4.2 The curve of $X^{(1)}$

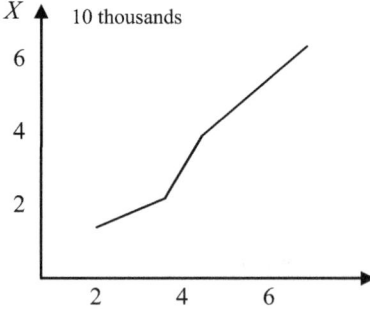

4.2 Systems Under Shocking Disturbances and Buffer Operators

4.2.1 The Trap for Shocking Disturbed System Forecasting

Behavioral prediction of problems under the influence of shocking disturbances has always been a difficult problem. For such predictions, any theory on how to choose models would lose its validity. This is because the problem to be address here is not about which model is the best; instead, when a system is severely impacted by shocks, the available behavioral data of the past no long represent the current state of the system. In this case, the available data of the system's behavior can no longer truthfully reflect the law of change of the system (Liu 2021; Liu et al., 2017).

Definition 4.2.1 Assume that

$$X^{(0)} = \left(x^{(0)}(1), x^{(0)}(2), \ldots, x^{(0)}(n) \right)$$

stands for a sequence of a system's true behaviors. If the observed behaviors of the system are

$$x = (x(1), x(2), \ldots, x(n))$$
$$= \left(x^{(0)}(1) + \varepsilon_1, x^{(0)}(2) + \varepsilon_2, \ldots, x^{(0)}(n) + \varepsilon_n \right) = X^{(0)} + \varepsilon$$

where $\varepsilon = (\varepsilon_1, \varepsilon_2, \ldots, \varepsilon_n)$ is a term for the shocking disturbance, then X is called a shock-disturbed sequence (Liu, 1991).

To correctly uncover and recognize the true behavior sequence $X^{(0)}$ of the system from the shock-disturbed sequence X, one first has to go over the hurdle ε. If we directly established our model and made our predictions using the severely affected data X without first cleaning up the disturbance, then our predictions would most likely fail. This is because the model would not have described the true state $X^{(0)}$ of change of the underlying system.

The wide spread existence of severely shocked systems often causes quantitative predictions to disagree with the outcomes of intuitive qualitative analyses. Hence, there is a need to seek an organic equilibrium between quantitative predictions and qualitative analyses, by eliminating shock wave disturbances in order to recover the true state of the systems' behavioral data. This way the accuracy of the consequent predictions can be greatly improved, which is one of the most important tasks performed by grey systems scientists. To this end, the discussion in this section is centered around the overall goal of reaching $X^{(0)}$ from X.

4.2.2   Axioms of Buffer Operators

Definition 4.2.2 Assume that $X = (x(1), x(2), \ldots, x(n))$ is a system's behavior data sequence.

(1) If $\forall k = 2, 3, \ldots, n, x(k) - x(k-1) > 0$, then X is referred to as a monotonic increasing sequence;

(2) If the inequality sign in (1) is inversed, then X is referred to as a monotonic decreasing sequence;

(3) If there are $k, k^1 \in \{2, 3, \ldots, n\}$ such that $x(k) - x(k-1) > 0, x(k') - x(k'-1) < 0$, then X is referred to as a random vibrating or fluctuating sequence. If $M = \max\{x(k)k = 1, 2, \ldots, n\}$ and $m = \min\{x(k)k = 1, 2, \ldots, n\}$, then $M - m$ is referred to as the amplitude of sequence X.

Definition 4.2.3 Assume that X is a data sequence of a system's behavior, D an operator to work on X, and after being applied by the operator D, X becomes the following sequence:

$$XD = (x(1)d, x(2)d, \ldots, x(n)d)$$

where D is referred to as a sequence operator and XD the first order sequence of operator D (Liu, 1991). If D_1, D_2, and D_3 are all sequence operators, then $D_1 D_2$ is referred to as a second order sequence operator, and

$$XD_1 D_2 = (x(1)d_1 d_2, x(2)d_1 d_2, \ldots, x(n)d_1 d_2)$$

a second order sequence of $D_1 D_2$. Similarly, $D_1 D_2 D_3$ is referred to as a third order sequence operator and

$$XD_1 D_2 D_3 = (x(1)d_1 d_2 d_3, x(2)d_1 d_2 d_3, \ldots, x(n)d_1 d_2 d_3)$$

a third order sequence of $D_1 D_2 D_3$.

Axiom 4.2.1 (*Fixed Point*). Assume that X is a data sequence of a system's behavior and D a sequence operator. Then D satisfies $x(n)d = x(n)$.

This fixed point axiom means that under the effect of a sequence operator, data point $x(n)$ remains unchanged, and this is the last entry of the system's behavior data sequence. Based on the conclusions of relevant qualitative analysis, we can also leave several of the last entries of the data unchanged by the operator D, say,

$$x(j)d \neq x(j) \text{ and } x(i)d = x(i)$$

$$\text{for } j = 1, 2, \ldots, k - 1; i = k, k + 1, \ldots, n.$$

Axiom 4.2.2 (*In accordance with information*). The sequence operator must be defined in accordance with information in the data sequence X. That is, each entry value $x(k), k = 1, 2, \ldots, n$, in the data sequence X of the system's behavior should sufficiently participate in the entire process of application of the operator.

This axiom requires that any sequence operator be defined by using known information of the given sequence. It cannot be produced without referencing available raw data (Liu, 1991).

Axiom 4.2.3 (*Expressed normality*). Each $x(k)d, k = 1, 2, \ldots, n$, is expressed by a uniform, elementary analytic representation of $x(1), x(2), \ldots, x(n)$ (Liu, 1991).

This last axiom requires that the procedure of applying sequence operators be clear, normalized, and uniform, so that it can be conveniently carried out on computers.

Definition 4.2.4 Any sequence operator satisfying these three axioms is referred to as a buffer operator; the first order, second order, third order, ..., sequences obtained by applying a buffer operator are referred to as the first order, second order, third order, ..., buffered sequences.

Definition 4.2.5 For a raw data sequence X and a buffer operator D, when X is respectively an increasing, decreasing, or fluctuating sequence:

(1) If the buffered sequence XD increases, decreases, or fluctuates slower or with smaller amplitude, respectively, than the original sequence X, then D is referred to as a weakening operator.
(2) If the buffered operator XD increases, decreases, or fluctuates faster or with larger amplitude, respectively, than the original sequence X, then D is referred to as a strengthening operator (Liu, 1991).

4.2.3   Properties of Buffer Operators

Theorem 4.2.1 *Assume that X is a monotonic increasing sequence , then:*

(1) If D is a weakening operator $x(k)d \geq x(k), k = 1, 2, \ldots, n$;
(2) If D is a strengthening operator $x(k)d \leq x(k), k = 1, 2, \ldots, n$ (Liu, 1991).

Proof Assume that

$$r(k) = \frac{x(n) - x(k)}{n - k + 1}, k = 1, 2, 3, \ldots$$

is the average increasing rate from $x(k)$ to $x(n)$ in the sequence X of raw data, and

$$r(k)d = \frac{x(n)d - x(k)d}{n - k + 1}, k = 1, 2, 3, \ldots$$

is the average increasing rate from $x(k)d$ to $x(n)d$ in the buffered sequence XD. Given the condition that

$$x(n)d = x(n)$$

It follows that

$$r(k) - r(k)d = \frac{[x(n) - x(k)] - [x(n)d - x(k)d]}{n - k + 1} = \frac{x(k)d - x(k)}{n - k + 1}$$

If D is a weakening operator, then, $r(k) \geq r(k)d$, that is $r(k) - r(k)d \geq 0$. Therefore $x(k)d - x(k) \geq 0$, that is, $x(k)d \geq x(k)$ and vice versa.

If D is a strengthening operator, then $r(k) \leq r(k)d$, that is $r(k) - r(k)d \leq 0$. Therefore $x(k)d - x(k) \leq 0$, that is, $x(k) \geq x(k)d$ and vice versa.

Theorem 4.2.2 *Assume that X is a monotonic decreasing sequence , then:*

(1) If D is a weakening operator $\Leftrightarrow x(k)d \leq x(k), k = 1, 2, \ldots, n$;
(2) If D is a strengthening operator $\Leftrightarrow x(k)d \geq x(k), k = 1, 2, \ldots, n$ (Liu, 1991).

Theorem 4.2.3 *Assume that X is a fluctuating sequence and XD a buffered sequence, then:*

(1) If D is a weakening operator, then $\max_{1 \leq k \leq n}\{x(k)\} \geq \max_{1 \leq k \leq n}\{x(k)d\}$ and $\min_{1 \leq k \leq n}\{x(k)\} \leq \min_{1 \leq k \leq n}\{x(k)d\}$;
(2) If D is a strengthening operator, then $\max_{1 \leq k \leq n}\{x(k)\} \leq \max_{1 \leq k \leq n}\{x(k)d\}$ and $\min_{1 \leq k \leq n}\{x(k)\} \geq \min_{1 \leq k \leq n}\{x(k)d\}$.

For detailed proofs and relevant discussions of these theorems, please consult Liu and Lin (2006, pp. 64–67). What theorem implies is that each monotonic increasing sequence expands under the effect of a weakening operator and shrinks under a strengthening operator. What theorem indicates is that each monotonic decreasing sequence shrinks under the effect of a weakening operator and expands under a strengthening operator.

4.3 Construction of Practically Useful Buffer Operators

4.3.1 Weakening Buffer Operators

Theorem 4.3.1 *Given a raw data sequence* $X = (x(1), x(2), \ldots, x(n))$, *let* $XD = (x(1)d, x(2)d, \ldots, x(n)d)$, *where*

$$x(k)d = \frac{1}{n - k + 1}[x(k) + x(k + 1) + \cdots + x(n)], k = 1, 2, \ldots, n \qquad (4.1)$$

Then D is always a weakening operator regardless of whether X is a monotonic increasing, decreasing, or vibrating sequence. This operator is referred to as an average weakening buffer operator (AWBO) (Liu, 1991).

The weakening operator D in Theorem 4.3.1 possesses some very good properties and has been applied widely in modeling and prediction of systems with interference of uncontrollable shock waves.

Corollary 4.3.1 *For the weakening operator* D *as defined in* Theorem 4.3.1, *let:*

$$XD^2 = XDD = \left(x(1)d^2, x(2)d^2, \ldots, x(n)d^2\right)$$
$$x(k)d^2 = \frac{1}{n - k + 1}[x(k)d + x(k + 1)d + \cdots + x(n)d]; k = 1, 2, \ldots, n \quad (4.2)$$

Then D^2 is always a second-order weakening operator for monotonic increasing, monotonic decreasing, and fluctuating sequences.

Example 4.3.1 Let $X = (36.5, 54.3, 80.1, 109.8, 143.2)$ and D and D^2 as defined in Theorem 4.3.1 and Corollary 4.3.1 respectively, calculate the buffered sequence XD and XD^2.

Solution Here $n = 5$, from Formula 4.1, we have:

$$
\begin{aligned}
x(1)d &= \frac{1}{n - k + 1}[x(k) + x(k + 1) + \cdots + x(n)] \\
&= \frac{1}{5 - 1 + 1}[x(1) + x(2) + \cdots + x(5)] \\
&= \frac{1}{5 - 1 + 1}[36.5 + 54.3 + 80.1 + 109.8 + 143.2] = 84.78 \\
x(2)d &= \frac{1}{n - k + 1}[x(k) + x(k + 1) + \cdots + x(n)] \\
&= \frac{1}{5 - 2 + 1}[x(2) + \cdots + x(5)]
\end{aligned}
$$

$$= \frac{1}{4}[54.3 + 80.1 + 109.8 + 143.2] = 96.85$$

$$x(3)d = \frac{1}{5-3+1}[x(3) + x(4) + x(5)]$$

$$= \frac{1}{3}[80.1 + 109.8 + 143.2] = 111.03$$

$$x(4)d = \frac{1}{5-4+1}[x(4) + x(5)] = \frac{1}{2}[109.8 + 143.2] = 126.5$$

$$x(5)d = 143.2$$

Therefore:

$$XD = (84.78, 96.85, 111.03, 126.5, 143.2)$$

Similarly, we can obtained the second-order buffered sequence XD^2 as follows:

$$XD^2 = (112.47, 119.4, 126.91, 134.85, 143.2).$$

Theorem 4.3.2 *Assume that $X = (x(1), x(2), \ldots, x(n))$ is a sequence of raw data, $\omega = (\omega_1, \omega_2, \ldots, \omega_n)$ is a weight vector, and $\omega_1 > 0, i = 1, 2, \ldots, n$. Let:*

$$XD = (x(1)d, x(2)d, \ldots, x(n)d)$$

where

$$x(k)d = \frac{\omega_k x(k) + \omega_{k+1} x(k+1) + \cdots + \omega_n x(n)}{\omega_k + \omega_{k+1} + \cdots + \omega_n} = \frac{1}{\sum_{i=k}^{n} \omega_i} \sum_{i=k}^{n} \omega_i x(i) \quad , (k = 1, 2, \ldots, n) \qquad (4.3)$$

Then D is always a weakening operator regardless of whether X is a monotonic increasing, decreasing, or vibrating sequence (Dang et al., 2004). This operator D is called as a weighted average (or mean) weakening buffer operator ($WAWBO$).

Corollary 4.3.2 *For the weighted average weakening operator D as defined in Theorem 4.3.2, let:*

$$\omega = (1, 1, \ldots, 1).$$

Then:

$$\frac{1}{\sum_{i=k}^{n} \omega_i} \sum_{i=k}^{n} \omega_i x(i) = \frac{1}{n-k+1} \sum_{i=k}^{n} x(i)$$

That is, the average weakening buffer operator (AWBO) is a special case of the weighted average weakening buffer operator ($WAWBO$).

Theorem 4.3.3 *Assume that* $X = (x(1), x(2), \ldots, x(n))$ *is a sequence of raw data,* $\omega = (\omega_1, \omega_2, \ldots, \omega_n)$ *is a weight vector, and* $\omega_1 > 0, i = 1, 2, \ldots, n$. *Let:*

$$XD = (x(1)d, x(2)d, \ldots, x(n)d)$$

where

$$x(k)d = \left[x(k)^{\omega_k} \cdot x(k+1)^{\omega_{k+1}} \cdots x(n)^{\omega_n} \right]^{\frac{1}{\omega_k + \omega_{k+1} + \cdots + \omega_n}} = \left[\prod_{i=k}^{n} x(i)^{\omega_i} \right]^{\frac{1}{n}\omega_i}, \quad (k = 1, 2, \ldots, n) \quad (4.4)$$

Then D is always a weakening operator, regardless of whether X is a monotonic increasing, decreasing, or vibrating sequence (Dang et al., 2004).

This operator D is called as a weighted geometric average weakening buffer operator ($WGAWBO$)

Example 4.3.2 From 1983 to 1986, the overall business revenue of private enterprises in Changge county, located in the Henan Province of The People's Republic of China, was recorded as:

$$X = (10155, 12588, 23480, 35388)$$

This showed a tendency of rapid growth. The average rate of revenue growth for these years was 51.6%, and the average rate of revenue growth from 1984 to 1986 was 67.7%. The people involved in the economic planning of the county, including politicians, scholars, policy makers, and residents, commonly believed that the overall revenue of private enterprises in this county would not be able to keep up with this record speed of growth in the coming years. If relevant data had been used to build models and make predictions, nobody would have accepted the resultant conclusions. After numerous rigorous analyses and discussions, all parties involved recognized that the reason for such a high growth rate between 1983 and 1986 was mainly a low baseline. Such a low baseline had been a consequence of the fact that, in the past, policies relevant to private enterprises had been neither existent, nor encouraged. To weaken the growth rate of the sequence of the raw data, it was necessary to artificially add all favorable environmental factors to past years' data, and such environmental factors were created based on the introduction of relevant policies for the development of private enterprise in recent years. With this goal in mind, we introduced the second-order weakening operator, as defined in Theorem 4.3.1, and obtained the following second-order buffered sequence:

$$XD^2 = (27260, 29547, 32411, 35388).$$

As a result, the consequent modeling based on XD^2 produced credible predictions for the county's business revenue growth between 1987 and 2000.

4.3.2 Strengthening Buffer Operators

Theorem 4.3.4 *Assume that* $X = (x(1), x(2) \ldots, x(n))$ *is a sequence of raw data, and* $XD = (x(1)d, x(2)d, \ldots, x(n)d)$, *where D is defined as follows:*

$$x(k)d = \frac{x(1) + x(2) + \cdots + x(k-1) + kx(k)}{2k-1}; k = 1, 2, \ldots, n-1 \quad (4.5)$$

If $x(n)d = x(n)$, then D is a strengthening buffer operator regardless of whether the raw data sequence X is a monotonic, increasing or decreasing sequence (Liu, 1991).

Theorem 4.3.5 *Assume that* $X = (x(1), x(2), \ldots, x(n))$ *is a sequence of raw data, and* D_i *is a sequence operator defined by:*

$$x(k)d_i = \frac{x(k-1) + x(k)}{2}; k = 2, 3, \ldots, n; i = 1, 2 \quad (4.6)$$

If $x(1)d_1 = \alpha x(1), \alpha \in [0, 1], x(1)d_2 = (1 + \alpha)x(1), \alpha \in [0, 1]$, and $x(n)d_i = x(n), i = 1, 2$, then D_1 is a strengthening buffer operator for monotonic increasing sequences, and D_2 a weakening buffer operator for monotonic decreasing sequences (Liu, 1991).

Both D_1 and D_2 are called even strengthening buffer operators (ESBO).

Theorem 4.3.6 *For a given increasing or decreasing sequence X of raw data, the operator D is defined as follows:*

$$x(k)d = \frac{[x(k) + x(k+1) + \cdots x(n)]/(n-k+1)}{x(n)} \cdot x(k); k = 1, 2, \ldots n \quad (4.7)$$

D is a strengthening buffer operator, and is called average strengthening buffer operator (ASBO).

Theorem 4.3.7 *Assume that* $X = (x(1), x(2), \ldots, x(n))$ *is a sequence of raw data,* $\omega = (\omega_1, \omega_2, \ldots, \omega_n)$ *is a weight vector, and* $\omega_1 > 0, i = 1, 2, \ldots, n$. *Let* $XD = (x(1)d, x(2)d, \ldots, x(n)d)$, *where D is defined as follows:*

$$x(k)d = \frac{(\omega_k + \omega_{k+1} + \cdots + \omega_n)(x(k))^2}{\omega_k x(k) + \omega_{k+1} x(k+1) + \cdots + \omega_n x(n)} = \frac{\sum_{i=k}^{n} \omega_i (x(k))^2}{\sum_{i=k}^{n} \omega_i x(i)}, (k = 1, 2, \ldots, n) \quad (4.8)$$

D is a strengthening buffer operator regardless of whether the raw data sequence X is a monotonic increasing, decreasing, or vibrating sequence (Dang et al., 2005). D is called a weighted average strengthening buffer operator (WASBO).

4.3.3 The General Form of Buffer Operator

Theorem 4.3.8 *Assume that* $X = (x(1), x(2), \ldots, x(n))$ *is a sequence of raw data,* $\omega = (\omega_1, \omega_2, \ldots, \omega_n)$ *is a weight vector, and* $\omega_1 > 0, i = 1, 2, \ldots$. *Let* $XD = (x(1)d, x(2)d, \ldots, x(n)d)$, *where* D *is defined as follows:*

$$x(k)d = x(k) \cdot \left[x(k) \bigg/ \frac{\omega_k x(k) + \omega_{k+1} x(k+1) + \cdots + \omega_n x(n)}{\omega_k + \omega_{k+1} + \cdots + \omega_n} \right]^{\alpha}$$

$$= x(k) \cdot \left[x(k) \bigg/ \frac{1}{\sum_{i=k}^{n} \omega_i} \sum_{i=k}^{n} \omega_i x(i) \right]^{\alpha} \tag{4.9}$$

Then:

(1) When $\alpha < 0$, D is a weakening operator regardless of whether X is a monotonic increasing or decreasing sequence.
(2) When $\alpha > 0$, D is a strengthening buffer operator regardless of whether the raw data sequence X is a monotonic increasing or decreasing sequence.
(3) When $\alpha = 0$, D is an identical operator (Wei et al., 2011).

D is called the general form of buffer operator (GFBO).

Corollary 4.3.3 *Take* $\alpha = -1$ *in* Theorem 4.3.6, *then* Formula (4.8) *changes to* (4.2). *That is, the weighted average weakening buffer operator* (*WAWBO*)*is a special case of the general form of buffer operator (GFBO).*

Corollary 4.3.4 *Take* $\alpha = 1$ *in* Theorem 4.3.6, *then* Formula (4.8) *changes to* (4.7). *That is, the weighted average strengthening buffer operator (WASBO) is a special case of the general form of buffer operator (GFBO).*

The buffer operator concept has been employed not only in grey systems modeling, but also in other kinds of model building. Generally, before building a mathematical model based on qualitative analysis and its conclusions, one applies a buffer operator on the original data sequence. This is done to soften or eliminate the effects of shock-disturbances on the behavior sequence of a given system. By doing so, expected results are often obtained.

Example 4.3.3 From 1996 to 1999, the annual gross revenues produced by the agricultural, forestry, animal husbandry, and fishery sectors in the area of Nanjing were (in 0.1 billion yuan):

$$X = (91.9895, 94.2439, 96.9644, 98.9199)$$

The growth rate shown in X is very slow, as it represents an average of about 2.4% annually. Such a slow growth rate was not aligned with the fast advances of the overall annual economic development of the area. If such a slow growth continued in these economic sectors, it would have caused imbalances in the development of the overall economic structure of the region and sustained regional economic growth would have been adversely affected. In 2000, Nanjing City gradually adjusted the economic structure of the countryside to counteract slow economic growth. In order to accurately control that economic development tendency in a timely fashion, there was a need to produce scientifically reasonable economic forecasts. To achieve this goal we had to address available data where slow growth was recorded. This would allow the resultant predictions to possess practical value in the realm of economic forecast and pro-growth government intervention. By applying the strengthening operator in Theorem 2.12 twice on the available data sequence, we obtained the following second order buffered data sequence:

$$XD^2 = (79.5513, 85.5446, 93.1686, 98.9199)$$

A GM(1,1) model (for details, see Liu and Lin (2006), or Sect. 4.1 in this book) based on this buffered sequence provided:

$$\frac{dX^{(1)}}{dt} - 0.0720X^{(1)} = 77.1389$$

The time response function was as follows:

$$\widehat{X^{(1)}}(k+1) = 1150.7003e^{0.0720k} - 1071.1503.$$

Based on this equation, the computational simulation results, effectiveness of the data fit, and prediction efficacy are given in Tables 4.1 and 4.2.

Table 4.1 The effectiveness of the simulation results

Year	Strengthened data $x^{(0)}(k)$	Simulated data $\hat{x}^{(0)}(k)$	Error $\varepsilon(k) = \hat{x}^{(0)}(k) - x^{(0)}(k)$	Relative error$\Delta_k = \frac{|\varepsilon(k)|}{x^{(0)}(k)}$
1997	85.5446	85.9245	0.3799	0.4441
1998	93.1686	92.3407	−0.8279	0.8886
1999	98.9199	99.2359	0.316	0.3195

Table 4.2 The efficacy of the predictions

Year	Actual data $x^{(0)}(k)$	Predictions $\hat{x}^{(0)}(k)$	Error $\varepsilon(k) = \hat{x}^{(0)}(k) - x^{(0)}(k)$	Relative error$\Delta_k = \frac{|\varepsilon(k)|}{x^{(0)}(k)}$ (%)
2000	106.3412	106.6460	0.3048	0.2866
2001	113.29	114.6094	1.3194	1.1646
2005		152.8703		

Tables 4.1 and 4.2 show that by employing the buffered data using a strengthening operator to establish our model, the simulated results and corresponding predictions are quite good. In particular, for 2000 and 2001, predicted values reached an accuracy rate of over 98% compared to the actual data for those years.

4.4 Average Operator

Due to various obstacles that are difficult to overcome, available data sequences may or may not contain missing entries. Nevertheless, even if data sequences are complete without any missing entries, systems' behaviors can change suddenly at any point in time, and corresponding entries in data sequences can become out of the ordinary. This can create great difficulties for the researcher. For example, if abnormal entries are removed, blank entries are created. Hence, how to effectively fill blanks in data sequences naturally becomes one of the first questions one has to address when processing available data. Data generation using averages is another frequently used method to create new data, fill a vacant entry in the available data sequence, and construct new sequences.

Assume that $x = (x(1), x(2), \ldots, x(k), x(k+1), \ldots, x(n))$ is a sequence of raw data. Then, entry $x(k)$ is referred to the preceding value and $x(k+1)$ the succeeding value. If $x(n)$ stands for a piece of , then for any $k \leq n - 1$, $x(k)$ will be seen as a piece of old information. If the sequence X has a blank entry at location k, denoted $\emptyset(k)$, then the entries $x(k-1)$ and $x(k+1)$ will be referred to as $\emptyset(k)$'s boundary values, with $x(k-1)$ being the preceding boundary and $x(k+1)$ the succeeding boundary. If a value $x(k)$ at the location of $\emptyset(k)$) is generated on the basis of $x(k-1)$ and $x(k+1)$, then the established value $x(k)$ is referred to as an internal point of the interval $[x(k-1), x(k+1)]$.

Definition 4.4.1 Assume that $x(k-1)$ and $x(k+1)$ are two entries in a data sequence X. If $x(k-1)$ stands for a piece of old information and $x(k+1)$ a piece of new information, the sequence operator D is defined as:

$$x(k)d = x^*(k) = \alpha x(k+1) + (1-\alpha)x(k-1), \text{ for } \alpha \in [0, 1] \quad (4.9)$$

D is called a non-adjacent neighbor generating operator. The new value $x^*(k)$ is referred to as generated by the new and old information under the generation coefficient (weight) α. When $\alpha > 0.5$, the generation of $x^*(k)$ is seen with more weight placed on the new information than the old information. When $\alpha < 0.5$, the generation of $x^*(k)$ is seen with more weight placed on the old information than the new information. If $\alpha = 0.5$, then the value $x^*(k)$ is seen as generated without preference.

In terms of stable time series, the exponential smoothing method employed in smooth prediction, focuses on the generation of predictions with more preference given to old information than new information. This is because the smoothing value

$$s_k^{(1)} = \alpha x_k + (1 - \alpha)s_{k-1}^{(1)}$$

stands for a weighted sum of old and new information, with the weight α taking value from the range of 0.1–0.3.

Definition 4.4.2 Assume that sequence X has a blank entry $\emptyset(k))$ at location k. This blank entry $\emptyset(k)$ is filled by using the sequence operator D, which is defined as follows:

$$x(k)d = x^*(k) = 0.5x(k - 1) + 0.5x(k + 1)$$

If $x(k - 1)$ and $x(k + 1)$ are the adjacent neighbors of the location k, then D will be referred to as mean generation operator by using the non-adjacent neighbor. If $x(k + 1)$ stands for a piece of new information, then the non-adjacent neighbor mean generation operator is an equal weight generation operator of new and old information. This kind of operator is employed when it is difficult to determine the degree of influence of new and old information on the missing value $x(k)$.

Definition 4.4.3 For a given sequence $X = (x(1), x(2), \ldots, x(n))$, the sequence operator D is defined as:

$$x(k)d = x^*(k) = 0.5x(k) + 0.5x(k - 1) \qquad (4.11)$$

In this case, D is referred to as even generation operator by adjacent neighbor.

The sequence worked by even generation operator by adjacent neighbor is referred to as a sequence of even generated by adjacent neighbor. In grey systems modeling, the sequence of even generated by adjacent neighbor is often employed. It provides a method of constructing new sequences based on available time series data.

For the sequence X of length n, as given above, if Z stands for the sequence of even generated by adjacent neighbor, then the length of $Z = (z(2), z(3), \ldots, z(n))$ is $n - 2$, where $z(1)$ cannot be generated based on what is given in X.

4.5 The Quasi-Smooth Sequence and Stepwise Ratio Operator

Definition 4.5.1 Assume that, $X = (x(1), x(2), \ldots, x(n))$, $x(k) \geq 0, k = 1, 2, \ldots, n$, then the following is referred to as the smoothness ratio of the sequence X (Deng, 1985):

$$\rho(k) = \frac{x(k)}{\sum_{i=1}^{k-1} x(i)} ; k = 2, 3, \ldots, n \tag{4.12}$$

The concept of smoothness ratio reflects the smoothness of a sequence from a special angle. In particular, it uses the ratio $\rho(k)$ of the kth data value $x(k)$ over the sum $\sum_{i=1}^{k-1} x(i)$ of the previous values to check whether or not the changes in the data points of X are stable. The more stable the changes of the data points in sequence X are, the smaller the smoothness ratio $\rho(k)$.

Definition 4.5.2 If a sequence $X = (x(1), x(2), \ldots, x(n))$, $x(k) \geq 0, k = 1, 2, \ldots, n$, satisfies the following, then X is referred to as a quasi-smooth sequence:

1. $\frac{\rho(k+1)}{\rho(k)} < 1; k = 2, 3, \ldots, n-1$;
2. $\rho(k) \in [0, \varepsilon], k = 3, 4, \ldots, n$; and;
3. $\varepsilon < 0.5$.

Quasi-smooth conditions are very important criteria, which are employed to check whether a sequence can be used to build a grey model.

If the first entry $x(1)$ or the last entry $x(n)$ of a sequence are blank, that is, $x(1) = \emptyset(1)$ or $x(n) = \emptyset(n)$, we cannot fill these missing entries by using the method of adjacent neighbor mean generation operator. In this case, the operator of stepwise ratio is often employed.

Definition 4.5.3 Assume that a sequence $X = (x(1), x(2), \ldots, x(n))_1$ $x(k) \geq 0, k = 1, 2, \ldots, n$, then the following is referred to as the operator of stepwise ratios of X (Deng, 1985):

$$x(k)d = \sigma(k) = \frac{x(k)}{x(k-1)} ; k = 2, 3, \ldots, n \tag{4.13}$$

The missing entry $x(1) = \emptyset(1)$ can be generated by using the operator of stepwise ratio of its right-hand side neighbors, and $x(n) = \emptyset(n)$ its left-hand side neighbors. The sequence obtained by filling all its missing entries using the operators of stepwise ratio is referred to as stepwise ratio generated.

Proposition 4.5.1 Assume that a sequence $X = (x(1), x(2), \ldots, x(n))$, $x(k) \geq 0, k = 1, 2, \ldots, n$ and $x(1) = \emptyset(1)$ or $x(n) = \emptyset(n)$ If both x(1) and x(n) are generated by operator of stepwise ratio, then:

$$x(1) = x(2)/\sigma(3), x(n) = x(n-1)\sigma(n-1)$$

Proposition 4.5.2 Stepwise ratio $\sigma(k+1)$ and smoothness ratio as defined in Formulas (4.11) and (4.12), respectively, satisfy the relation as follows:

$$\sigma(k+1) = \frac{\rho(k+1)}{\rho(k)}(1+\rho(k)); k = 2, 3, \ldots, n \qquad (4.14)$$

Proposition 4.5.3 *If* $X = (x(1), x(2), \ldots, x(n))$ *is an increasing sequence, and satisfies the following conditions:*

(1) For any $k = 2, 3, \ldots, n, \sigma(k) < 2$; and
(2) $\frac{\rho(k+1)}{\rho(k)} < 1$.

then for any $\varepsilon \in [0, 1]$ and $k = 2, 3, \ldots, n$, when $\rho(k) \in [0, \varepsilon]$, we have $\sigma(k+1) \in [1, 1+\varepsilon]$.

4.6 Accumulating and Inverse Accumulating Operators

Accumulating operator is a method employed to mine the law implied in a grey data sequence. It plays an extremely important role in grey system modelling. Through the accumulating operator method, one can potentially uncover a development tendency existing in the process of accumulated grey quantities. This allows the characteristics and laws of integration hidden in chaotic original data to be sufficiently revealed. For instance, when looking at the financial outflows of a family, if we do our computations on a daily basis, we may not see obvious patterns. However, if our calculations are done on a monthly basis, some patterns of spending, which are somehow related to the monthly income of the family, will likely emerge.

The inverse accumulating operator is often employed to acquire additional insights from a small amount of available information. It plays the role of recovery from the acts of the accumulating operator and is its inverse operation. In particular,

Definition 4.6.1 For an original sequence $x^{(0)} = \left(x^{(0)}(1), x^{(0)}(2), \ldots, x^{(0)}(n)\right)$, D is a sequence operator defined as follows:

$$x^{(0)}D = \left(x^{(0)}(1)d, x^{(0)}(2)d, \ldots, x^{(0)}(n)d\right), \text{ where}$$

$$x^{(0)}(k)d = \sum_{i=1}^{k} x^{(0)}(i); k = 1, 2, \cdots, n \qquad (4.15)$$

Here, D is called a once accumulating generation operator of $X^{(0)}$, denoted as 1-AGO. And $X^{(0)}D$, the sequence worked by accumulating operator D on $X^{(0)}$, is denoted as $X^{(1)}$ for parsimony:

$$x^{(0)}D = X^{(1)} = \left(x^{(0)}(1)d, x^{(0)}(2)d, \ldots, x^{(0)}(n)d\right)$$

If the accumulating operator D is applied r times on $X^{(0)}$, we obtain:

$$X^{(0)} D^r = X^{(r)} = \left(x^{(r)}(1), x^{(r)}(2), \ldots, x^{(r)}(n) \right)$$

where

$$x^{(r)}(k) = \sum_{i=1}^{k} x^{(r-1)}(i); k = 1, 2, \ldots n \qquad (4.16)$$

D^r is denoted as r-AGO (Deng, 1985). Corresponding to the accumulating operator, the inverse accumulating operator D is defined below.

Definition 4.6.2 For an original sequence $x^{(0)} = \left(x^{(0)}(1), x^{(0)}(2), \ldots, x^{(0)}(n) \right)$, D is a sequence operator defined as follows:

$$X^{(0)} D = \left(x^{(0)}(1)d, x^{(0)}(2)d, \ldots, x^{(0)}(n)d \right), \text{ where}$$

$$x^{(0)}(k)d = x^{(0)}(k) - x^{(0)}(k-1); k = 2, \ldots, n \qquad (4.17)$$

D is called an inverse accumulating generation operator of $X^{(0)}$, denoted as 1-IAGO. In $X^{(0)} D$, the sequence works by inverse accumulating operator D on $X^{(0)}$, and is denoted as $\alpha^{(1)} X^{(0)}$.

If the inverse accumulating operator D is applied r times on $X^{(0)}$, we write conventionally:

$$X^{(0)} D^r = \alpha^{(r)} X^{(0)} = \left(\alpha^{(r)} x^{(0)}(1), \alpha^{(r)} x^{(0)}(2), \ldots, \alpha^{(r)} x^{(0)}(n) \right)$$

where $\alpha^{(r)} x^{(0)}(k) = \alpha^{(r-1)} x^{(0)}(k) - \alpha^{(r-1)} x^{(0)}(k-1); k = 1, 2, \ldots, n$ (Deng, 1985).

Proposition 4.6.1 For an original sequence $X^{(0)} = \left(x^{(0)}(1), x^{(0)}(2), \ldots, x^{(0)}(n) \right)$, if both $X^{(r)}$ and $\alpha^{(r)}$ are defined according to Definitions 4.6.1 and 4.6.2, then:

$$\alpha^{(r)} X^{(r)} = X^{(0)}$$

Example 3.6.1 If $X = (5.3, 7.6, 10.4, 13.8, 18.1)$, calculate the 1-AGO $X^{(1)}$, 2-AGO $X^{(2)}$ and 1-IAGO $\alpha^{(1)} X^{(0)}$.

Solution The results are shown in Table 4.3.

Table 4.3 The 1-AGO, 2-AGO and 1-IAGO of $X^{(0)}$

$X^{(0)}$	5.3	7.6	10.4	13.8	18.1
$X^{(1)}$	5.3	12.9	23.3	37.1	55.2
$X^{(2)}$	5.3	18.2	41.5	78.6	133.8
$\alpha^{(1)} X^{(0)}$	5.3	2.3	2.8	3.4	4.3

4.7   Exponentiality of Accumulating Sequence

After applying the accumulating operator a few times, the general non-negative quasi-smooth sequence will show the pattern of exponential growth with decreased randomness. The smoother the original sequence is, the more obvious an exponential growth pattern in the first order accumulation sequence will appear.

Example 4.7.1 The sales quantity of cars from 2010 to 2015 in a city located in southeast of China is as follows:

$$X^{(0)} = \left\{ x^{(0)}(k) \right\}_1^6 = (50810, 46110, 51177, 93775, 110574, 110524)$$

The 1-AGO sequence of $X^{(0)}$ is:

$$X^{(1)} = \left\{ x^{(1)}(k) \right\}_1^6 = (50810, 96920, 148097, 241872, 352446, 462970)$$

The Figures of $X^{(0)}$ and $X^{(1)}$ are shown in Figs. 4.3 and 4.4, respectively.

For the curve shown in Fig. 4.3, it is difficult to find a simple curve as the approximation of $X^{(0)}$. However, the curve shown in Fig. 4.4 is very close to an exponential growth curve. $X^{(1)}$ can be fitted with an exponential curve.

Fig. 4.3   The curve of $X^{(0)}$

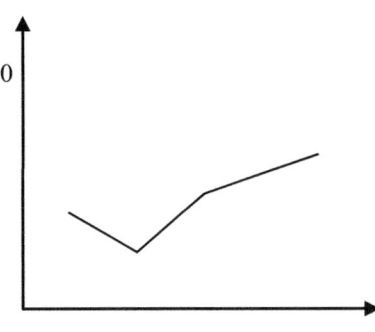

Fig. 4.4   The curve of $X^{(1)}$

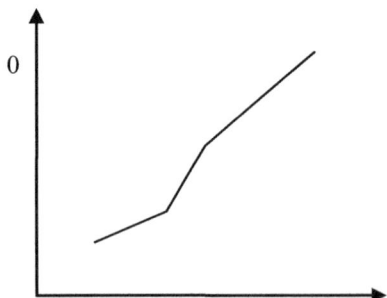

Definition 4.7.1 Assume that $X(t) = ce^{at} + b, c, a \neq 0$ is a continuous exponential function, then:

(1) $X(t)$ is referred to as homogeneous exponential function, if $b = 0$;
(2) $X(t)$ is referred to as non-homogeneous exponential function, if $b \neq 0$.

Definition 4.7.2 If a sequence $X(t) = ce^{at} + b_{\alpha} c, a \neq 0$ satisfies:

(1) $x(k) = ce^{ak}, c, a \neq 0$, for $k = 1, 2, \ldots, n$, then X is referred to as a homogeneous exponential sequence; and
(2) $x(k) = ceak + b, c, a, b \neq 0$, for $k = 1, 2, \ldots, n$, then X is referred to as a non-homogeneous sequence.

Theorem 4.7.1 *A sequence $X = (x(1), x(2), \ldots, x(n))$ is a homogeneous exponential sequence if, and only if, for $k = 1, 2, \ldots, n, \sigma(k)$ is a constant.*

Proof

(1) Assume that $\forall k = 1, 2, \ldots, n, x(k) = ce^{ak}, c, a \neq 0$, then:

$$\sigma(k) = \frac{x(k)}{x(k-1)} = \frac{ce^{ak}}{ce^{a(k-1)}} = e^a = const$$

(2) Assume that $\forall k = 1, 2, \ldots, n, \sigma(k) = const = e^a$, then:

$$x(k) = e^a x(k-1) = e^{2a} x(k-2) = \cdots = x(1)e^{a(k-1)}$$

Definition 4.7.3 *For the given sequence $X = (x(1), x(2), \ldots, x(n))$,*

(1) if $\forall k, \sigma(k) \in (0, 1)$, then X is referred to as satisfying the law of negative grey exponent;
(2) if $\forall k, \sigma(k) \in (1, b)$, for some $b > 1$, then X is referred to as satisfying the law of positive grey exponent;
(3) if $\forall k, \sigma(k) \in [a, b], b - a = \delta$, then X is referred to as satisfying the law of grey exponent with the absolute degree of greyness δ; and
(4) if $\delta < 0.5$, then X is referred to as satisfying the law of quasi-exponent.

Theorem 4.7.2 *Assume that $X^{(0)}$ is a non-negative quasi-smooth sequence . Then, the sequence $X^{(1)}$, generated by applying accumulating operator once on $X^{(0)}$, satisfies the law of quasi-exponent.*

Proof According to the definition of quasi-smooth sequence and.

$$\sigma^{(1)}(k) = \frac{x^{(1)}(k)}{x^{(1)}(k-1)} = \frac{x^{(0)}(k) + x^{(1)}(k-1)}{x^{(1)}(k-1)} = 1 + \rho(k)$$

We have

$$\forall k, \rho(k) < 0.5$$

Therefore

$$\sigma^{(1)}(k) \in [1, 1.5), \delta < 0.5$$

Thus, $X^{(1)}$ is a sequence that satisfies the law of quasi-exponent.

Theorem 4.7.2 is the theoretical foundation of grey systems modeling. In fact, because economic, ecological and agricultural systems (among others) can be seen as energy systems, and given that the accumulation and release of energy generally satisfy an exponential law, this explains why exponential modeling of grey systems theory has found an extremely wide range of applications.

Theorem 4.7.3 *Assume that $X^{(0)}$ is a non-negative sequence. If $X^{(r)}$ satisfies a law of exponent, and the stepwise ratio of $X^{(r)}$ is given by $\sigma^{(r)}(k) = \sigma$, then according to Deng (1985):*

(1) $\sigma^{(r+1)}(k) = \frac{1-\sigma^k}{1-\sigma^{k-1}}$;
(2) When $\sigma \in (0, 1)$, $\lim\limits_{k\to\infty} \sigma^{(r+1)}(k) = 1$; and for each k, $\sigma^{(r+1)}(k) \in (1, 1+\sigma]$;
(3) When $\sigma > 1$, $\lim\limits_{k\to\infty} \sigma^{(r+1)}(k) = \sigma$; and for each k, $\sigma^{(r+1)}(k) \in (\sigma, 1+\sigma]$.

Proof

(1) Assume that $X^{(r)}$ satisfies a law of exponent, and $\forall k$, $\sigma^{(r)}(k) = \frac{x^{(r)}(k)}{x^{(r)}(k-1)} = \sigma$,

then $\forall k$,

$$x^{(r)}(k) = \sigma x^{(r)}(k-1) = \sigma^2 x^{(r)}(k-2) = \cdots = \sigma^{(k-1)}x^{(r)}(1)$$

$$X^{(r)} = (x^{(r)}(1), \sigma x^{(r)}(1), \sigma^2 x^{(r)}(1), \ldots, \sigma^{(n-1)}x^{(r)}(1))$$

$$X^{(r+1)} = (x^{(r)}(1), (1+\sigma)x^{(r)}(1),$$
$$(1+\sigma+\sigma^2)x^{(r)}(1), \ldots, (1+\sigma+\cdots+\sigma^{(n-1)})x^{(r)}(1))$$

Therefore

$$\sigma^{(r+1)}(k) = \frac{x^{(r+1)}(k)}{x^{(r+1)}(k-1)} = \frac{(1+\sigma+\cdots+\sigma^{k-1})x^{(r)}(1)}{(1+\sigma+\cdots+\sigma^{k-2})x^{(r)}(1)} = \frac{\frac{1-\sigma^k}{1-\sigma}}{\frac{1-\sigma^{k-1}}{1-\sigma}} = \frac{1-\sigma^k}{1-\sigma^{k-1}}$$

(2) When $\sigma \in (0, 1)$, $\sigma^{(r+1)}(k)$ will decrease as k increases.

$$k = 2$$

$$\sigma^{(r+1)}(2) = \frac{x^{(r+1)}(2)}{x^{(r+1)}(1)} = 1 + \sigma$$

$$k \to \infty$$

$$\sigma^{(r+1)}(k) = \frac{1 - \sigma^k}{1 - \sigma^{k-1}} \to 1$$

Therefore $\forall\, k$,

$$\sigma^{(r+1)}(k) \in [1, 1 + \sigma]$$

(3) When $\sigma > 1$, $\sigma^{(r+1)}(k)$ will decrease as k increases.

$$k = 2$$

$$\sigma^{(r+1)}(2) = 1 + \sigma$$

$$k \to \infty$$

$$\sigma^{(r+1)}(k) = \frac{1 - \sigma^k}{1 - \sigma^{k-1}} \to \sigma$$

Therefore $\forall\, k$,

$$\sigma^{(r+1)}(k) \in (\sigma, 1 + \sigma]$$

The Theorem 4.7.3 says that if the rth accumulating sequence of $X^{(0)}$ satisfies an obvious law of exponent, additional application of the accumulating operator will destroy the pattern of exponent. In practical applications, if the rth accumulating sequence of $X^{(0)}$ satisfies the law of quasi-exponent, we generally stop applying the accumulating operator. To this end, Theorem 4.7.2 implies that only one application of the accumulating operator is needed for a non-negative quasi-smooth sequence before establishing an exponential model.

References

Dang, Y. G., Liu, S. F., Liu, B., & Tang, X. W. (2004). Research on weakening buffer operator. *Chinese Journal of Management Science, 12*(2), 108–111.

Dang, Y. G., Liu, B., & Guang, Y. Q. (2005). On strengthening buffer operators. *Control and Decision, 20*(12), 1332–1336.

Deng, J. L. (1985). *Grey control systems.* Press of Huazhong University of Science and Technology.

Liu, S. F. (1991). The three axioms of buffer operator and their application. *Journal of Grey System, 3*(1), 39–48.

Liu, S. F. (2021). *Grey system theory and its application* (9th ed.). Beijing: Science Press.

Liu, S. F., & Lin, Y. (2006). *Grey information: Theory and practical applications.* Springer-Verlag.

Liu, S. F., Yang, Y. J., & Forrest, J. (2017). *Grey data analysis.* Singapore: Springer-Verlag.

Wei, Y., Kong, X., & Hu, D. (2011). A kind of universal constructor method for buffer operators. *Grey systems: Theory and application, 1*(2), 178–185.

Chapter 5
Grey Relational Analysis Models

5.1 Introduction

Any given system, such as a social, economic, agricultural, ecological, and educational system, will encompass different kinds of factors. It is the result of the mutual interactions of these factors that determines the development tendency and behavior of the system. It is often the case that, among all the factors, investigators will need to know which ones are primary and which ones are secondary. Primary factors have dominant effects on the development of systems. Such factors drive the development of systems positively and must be strengthened. Conversely, secondary factors exert less influence on the development of systems. They tend to pose obstacles for the development of systems and, therefore, must be weakened. For instance, there are generally many influencing factors on the overall performance of an economic system. In order to realize the production of additional output with less input, systems analysis must be conducted prudently and a key part of this analysis is the identification of primary and secondary factors.

Regression analysis, variance analysis, and main component analysis are the most commonly employed methods for conducting systems analysis. However, these methods suffer from the following weaknesses:

(1) Large samples are needed in order to produce reliable conclusions.
(2) Available data need to satisfy some typical types of probability distribution; linear relationships between factors and system behaviors are assumed, while no interactions can be found between factors. Generally, these requirements are difficult to satisfy.
(3) The amount of computation is large and generally done by using computers.
(4) At times quantitative conclusions do not resonate with qualitative analysis outcomes so that the laws governing system development are distorted or misunderstood.

In fact, when available data are small it is extremely difficult to apply such traditional methods of statistics to analyze such data. This is because small data do

S. Liu, *Grey Systems Analysis*, Series on Grey System,
https://doi.org/10.1007/978-981-19-6160-1_5

not satisfy the modelling conditions of traditional methods; they contain relatively large amounts of grey information and do not follow any conventional probability distribution.

The Grey Relational Analysis (GRA) model is a new method to analyze systems where statistical methods do not seem appropriate. It can be applied to large or small samples and does not have conventional distribution requirements. Additionally, the amount of computation involved is small and can be carried out conveniently, without issues of disagreement between quantitative and qualitative conclusions.

The basic idea of grey relational analysis is to use the degree of similarity of the geometric curves of available data sequences to determine whether or not their connections are close. The more similar the curves, the closer the relational between sequences, and vice versa.

A number of scholars have conducted meaningful research focused on the construction and properties of GRA models, and such researchers have achieved valuable results. For example, Zhang et al. (1996) has analyzed the predominant point trend of Deng's (1985) GRA model. They has introduced grey relation entropy to improve the traditional model, and has proposed a new method to calculate degree of grey relational. Xiao and Colleagues (2006) have constructed a weighted degree of grey relational through the weighted compound of relational coefficient of each point. Zhao et al. (1998a) have introduced Euclid nearness into grey relational analysis, and have established the Euclid relational degree model based on the measurement of nearness of factor points through calculating nearness. Furthermore, Zhao et al. (1998b) have defined a GRA model according to upper and lower boundaries of distances between grey factor points. The authors have also demonstrated that their GRA model as well as Deng's (1985) GRA model through weighted relational analysis and the Euclid relational degree model are three special types of GRA model. Shi (1995) has proposed extreme difference relation according to the difference between distance of maximum value and distance of sequences, complementing Deng's (1985) relational coefficient. Zhang et al. (2007) have integrated the method of discrimination coefficient correction, the entropy weight method and the projection method to advance Deng's GRA model. Zhao et al. (2007) has introduced variant coefficient to relational analysis, and improved Deng's GRA model through weighted values of variant coefficient and relational coefficient. Further, Zhou et al. (2005) defined relational coefficient with the application of generalized distance in fuzzy math to measure the difference between reference sequence and compared sequence. Peng (2008) has extended Deng's GRA model to second-order trend relational analysis model through second-order difference. Finally, Wang (1989) has proposed the B-type relational degree model, Tang (1995) has developed the T-type relational degree model, and Dang and Liu (2004) has proposed the gradient relational degree model as well as its improved version. Among these models, the GRA model proposed by Professor Deng (1985) is the most influential one.

Thus, research based on early GRA models relies on relational coefficients of particular points to the absolute degree, relative degree, and synthetic degree of the original GRA model, which in turn is based on integral or overall perspectives. Such research also includes GRA models that measure similarity based on nearness to the models, which consider similarity and nearness, respectively. Additionally, research objects extend from the analysis of relationship among curves to those of relationships among curved surfaces, analysis of relationships in three-dimensional space and even the analysis of relationships among super surfaces in n-dimensional space. However, the study of high-dimensional models is still in its infancy. Indeed, many practical and scientific problems are yet to receive research attention and there is a need to focus on analysis methods based on panel data, matrix data, matrix sequence data and high-dimensional data. The absolute degree of GRA model, which extends definite integral models to multiple integral ones, can be used for relational analysis of high-dimensional data. However, the testing and specific quantitative standards of GRA models require additional research (Liu et al., 2013, 2016; Liu, 2021)

5.2 Grey Relational Factors and Set of Grey Relational Operators

When analyzing a system, one must choose the mapping variable to reflect the characteristics of such a system, and determine the factors that influence the behavior of the system. If a quantitative analysis is considered, one needs to process the chosen mapping variable and the effective factors using sequence operators so that the available data are converted to their relevant non-dimensional values of roughly equal magnitudes.

Definition 5.2.1 Assume that X_i is a system factor and its observation value at the ordinal position k is $x_i(k)$, $k = 1, 2, \ldots, n$, then $X_i = (x_i(1), x_i(2), \ldots, x_i(n))$ is referred to as the behavioral sequence of factor X_i.

If k stands for the time order, then $x_i(k)$ is referred to as the observational value of factor X_i at time moment k, and $X_i = (x_i(1), x_i(2), \ldots, x_i(n))$ is the behavioral time sequence (or series) of X_i.

If k stands for an index ordinal number and $x_i(k)$ the observational value of the kth index of factor X_i, then $X_i = (x_i(1), x_i(2), \ldots, x_i(n))$ is referred to as the behavioral index sequence of factor X_i.

If k stands for the ordinal number of the observed object and $x_i(k)$ is the observed value of the kth object of factor X_i, then $X_i = (x_i(1), x_i(2), \ldots, x_i(n))$ is referred to as the horizontal sequence of factor X_i's behavior.

For example, if X_i represents an economic factor, k time, and $x_i(k)$ the observed value of factor X_i at time moment k, then $X_i = (x_i(1), x_i(2), \ldots, x_i(n))$ is a time series of economic behaviors. If k is the ordinal number of an index, then $X_i = (x_i(1), x_i(2), \ldots, x_i(n))$ is the index sequence of an economic behavior. If k represents the ordinal number of different economic regions or departments, then $X_i = (x_i(1), x_i(2), \ldots, x_i(n))$ is a horizontal sequence of an economic behavior. No matter what kinds of sequence data are available, they can be employed in relational analysis.

Definition 5.2.2 Let $X_i = (x_i(1), x_i(2), \ldots, x_i(n))$ be the behavioral sequence of factor X_i, and D_1 a sequence operator such that $X_i D_1 = (x_i(1)d_1, x_i(2)d_1, \ldots, x_i(n)d_1)$, where:

$$x_i(k)d_1 = x_i(k)/x_i(1), x_i(1) \neq 0, \quad k = 1, 2, \ldots, n \qquad (5.1)$$

Then D_1 is referred to as an initialising operator and $X_i D_1$ is its image, called initial image of X_i (Deng, 1985).

Example 5.2.1 Let $X = (3.2, 3.7, 4.5, 4.9, 5.6)$, and calculate the initial image of X.

Solution: From Formula 5.1, we have:

$$x(1)d_1 = x(1)/x(1) = 1, x(2)d_1 = x(2)/x(1) = 3.7 - 3.2 = 1.15625.$$

Similarly,

$$x(3)d_1 = 1.40625, \ x(4)d_1 = 1.53125, \ x(5)d_1 = 1.75.$$

Therefore:

$$XD_1 = (x(1)d_1, x(2)d_1, x(3)d_1, x(4)d_1, x(5)d_1) = (1, 1.15625, 1.40625, 1.53125, 1.75).$$

Definition 5.2.3 Let $X_i = (x_i(1), x_i(2), \ldots, x_i(n))$ be the behavioral sequence of factor X_i. Sequence operator D_2 satisfies $X_i D_2 = (x_i(1)d_2, x_i(2)d_2, \ldots, x_i(n)d_2)$, and:

$$x_i(k)d_2 = \frac{x_i(k)}{\overline{X_i}}, \ \overline{X_i} = \frac{1}{n}\sum_{k=1}^{n} x_i(k), \quad k = 1, 2, \ldots, n \qquad (5.2)$$

Here, D_2 is referred to as an averaging operator and $X_i D_2$ is its image, called the average image of X_i (Deng, 1985).

Example 5.2.2 Let X be the same as Example 5.2.1 and calculate the average image of X.

Solution: From Formula 5.2, we have:

$$\overline{X} = \frac{1}{5}\sum_{k=1}^{5} x(k) = 4.38, x(1)d_2 = x(1)/\overline{X} = 0.73, x(2)d_2 = x(2)/\overline{X} = 0.84.$$

Similarly:

$$x(3)d_2 = 1.03, x(4)d_2 = 1.12, x(5)d_2 = 1.28.$$

Therefore:

$$XD_2 = (x(1)d_2, x(2)d_2, x(3)d_2, x(4)d_2, x(5)d_2) = (0.73, 0.84, 1.03, 1.12, 1.28).$$

Definition 5.2.4 Let $X_i = (x_i(1), x_i(2), \ldots, x_i(n))$ be the behavioral sequence of factor X_i. Sequence operator D_3 satisfies $X_i D_3 = (x_i(1)d_3, x_i(2)d_3, \ldots, x_i(n)d_3)$, and:

$$x_i(k)d_3 = \frac{x_i(k) - \min\limits_{k} x_i(k)}{\max\limits_{k} x_i(k) - \min\limits_{k} x_i(k)}; \quad k = 1, 2, \ldots, n \quad (5.3)$$

D_3 is referred to as an interval operator and $X_i D_3$ is its image, called the interval image of X_i (Deng, 1985).

Example 5.2.3 Let X be the same as Example 5.2.1, and calculate the interval image of X.

Solution: $\min\limits_{k} x(k) = 3.2$, $\max\limits_{k} x(k) = 5.6$. From Formula 5.3, we have:

$$x(1)d_3 = 0, x(2)d_3 = 0.208.$$
$$x(3)d_3 = 0.542, x(4)d_3 = 0.708, x(5)d_3 = 1.$$

Therefore:

$$XD_3 = (x(1)d_3, x(2)d_3, x(3)d_3, x(4)d_3, x(5)d_3) = (0, 0.208, 0.542, 0.708, 1).$$

As usual, D_1, D_2, D_3 should not be mixed or overlapped. Only one of them can be selected according to a particular situation.

Definition 5.2.5 Let $X_i = (x_i(1), x_i(2), \ldots, x_i(n))$ be the behavioral sequence of factor X_i. The behavioral sequence of factor X_i satisfies $x_i(k) \in [0, 1]$, $i = 1, 2, \ldots, n$, sequence operator D_4 satisfies $X_i D_4 = (x_i(1)d_4, x_i(2)d_4, \ldots, x_i(n)d_4)$, and:

$$x_i(k)d_4 = 1 - x_i(k), \quad k = 1, 2, \ldots, n \quad (5.4)$$

Then D_4 is referred to as a reversing operator and $X_i D_4$ is its image, called the reverse image of X_i (Deng, 1985).

Definition 5.2.6 Let $X_i = (x_i(1), x_i(2), \ldots, x_i(n))$ be the behavioral sequence of factor X_i. Sequence operator D_5 satisfies $X_i D_5 = (x_i(1)d_5, x_i(2)d_5, \ldots, x_i(n)d_5)$, and:

$$x_i(k)d_5 = 1/x_i(k), x_i(k) \neq 0, \quad k = 1, 2, \ldots, n \quad (5.5)$$

Here, D_5 is referred to as a reciprocating operator with $X_i D_5$ as its image, called the reciprocal image of X_i (Deng, 1985).

Let X_0 be the sequence of a system's behavioral characteristics, which is increasing, and X_i the behavioral sequence of a relevant factor. If X_i is also an increasing sequence, then both X_i and X_0 have a positive or direct relational relationship. If X_i is a decreasing sequence, then both X_i and X_0 have a negative or inverse relational relationship.

The negative relationship will be transformed to a positive relationship if affected by reversing operator D_4 or reciprocating operator D_5. Here, D_4 and D_5 should not be mixed or overlapped either.

Definition 5.2.7 The set $D = \{D_i | i = 1, 2, 3, 4, 5\}$ is referred to as the set of grey relational operators.

Definition 5.2.8 If X stands for the set of all system factors and D the set of grey relational operators, then (X, D) is referred to as the space of grey relational factors of a system.

5.3 Grey Relational Axioms and Deng's Grey Relational Analysis Model

Given the sequence $X = (x(1), x(2), \ldots, x(n))$, we can image the corresponding zigzagged line of the plane $X = \{x(k) + (t - k)(x(k + 1) - x(k)) | k = 1, 2, \ldots, n - 1; t \in [k, k + 1]\}$. Without causing confusion, the same symbol is used for both the sequence and its zigzagged line. For parsimony, we will not distinguish between the two in our discussions.

Definition 5.3.1 The given sequence $X = (x(1), x(2), \ldots, x(n))$, $\alpha = \frac{x(s) - x(k)}{s - k}$, $s > k, k = 1, 2, \ldots, n - 1$, is referred to as the slope of X on interval $[k, s]$, and $\alpha = \frac{1}{n-1}(x(n) - x(1))$ the average slope of X.

Theorem 5.3.1 *Assume that X_i and X_j are non-negative increasing sequences such that $X_j = X_i + c$, where c is a nonzero constant. Let D_1 be an initialing operator, $Y_i = X_i D_1$ and $Y_j = X_j D_1$. If α_i and α_j are respectively the average slopes of X_i and X_j, and β_i and β_j the average slopes of Y_i and Y_j, then, the following must be true: $\alpha_i = \alpha_j$; when $c < 0$, $\beta_i < \beta_j$; and when $c > 0$, $\beta_i > \beta_j$.*

What is meant here is that when the absolute amount of increase of two increasing sequences are the same, the sequence with the smallest initial value will increase faster than the other. To maintain the same relative rate of increase, the absolute amount of increase of the sequence with the greatest initial value must be greater than that of the sequence with the smallest initial value.

Definition 5.3.2 Let $X_0 = (x_0(1), x_0(2), \ldots, x_0(n))$ be a data sequence of a system's behavioral characteristic and the following are relevant factor sequences:

$$X_1 = (x_1(1), x_1(2), \ldots, x_1(n))$$

$$\cdots\cdots\cdots\cdots\cdots\cdots\cdots\cdots\cdots\cdots\cdots\cdots$$

$$X_i = (x_i(1), x_i(2), \ldots, x_i(n))$$

$$\cdots\cdots\cdots\cdots\cdots\cdots\cdots\cdots\cdots\cdots\cdots\cdots$$

$$X_m = (x_m(1), x_m(2), \ldots, x_m(n))$$

Given real numbers $\gamma(x_0(k), x_i(k))$, $i = 1, 2, \ldots, m$, and $k = 1, 2, \ldots, n$, if the following satisfies conditions of normality (1) and closeness (2) below:

$$\gamma(X_0, X_i) = \frac{1}{n} \sum_{k=1}^{n} \gamma(x_0(k), x_i(k)).$$

(1) Normality: $0 < \gamma(X_0, X_i) \leq 1, \gamma(X_0, X_i) = 1 \Leftrightarrow X_0 = X_i$; and
(2) Closeness: the smaller $|x_0(k) - x_i(k)|$, the greater $\gamma(x_0(k), x_i(k))$.

In this case, $\gamma(X_0, X_i)$ is referred to as the Deng's grey relational degree between X_i and X_0, $\gamma(x_0(k), x_i(k))$ as the Deng's grey relational coefficient of X_i and X_0 at point k (Deng, 1985).

Theorem 5.3.2 *Given a system's behavioral sequences* $X_0 = (x_0(1), x_0(2), \ldots, x_0(n))$ *and* $X_i = (x_i(1), x_i(2), \ldots, x_i(n))$, $i = 1, 2, \ldots, m$, *for* $\xi \in (0, 1)$, *it is possible to define:*

$$\gamma(x_0(k), x_i(k)) = \frac{\min\limits_{i} \min\limits_{k} |x_0(k) - x_i(k)| + \xi \max\limits_{i} \max\limits_{k} |x_0(k) - x_i(k)|}{|x_0(k) - x_i(k)| + \xi \max\limits_{i} \max\limits_{k} |x_0(k) - x_i(k)|} \quad (5.6)$$

and:

$$\gamma(X_0, X_i) = \frac{1}{n} \sum_{k=1}^{n} \gamma(x_0(k), x_i(k)) \quad (5.7)$$

In this case, $\gamma(X_0, X_i)$ is the Deng's grey relational degree between X_0 and X_i, where ξ is known as the distinguishing coefficient (Deng, 1985).

The Deng's grey relational degree of $\gamma(X_0, X_i)$ is commonly written as γ_{0i}, and the Deng's grey relational coefficient of $\gamma(x_0(k), x_i(k))$ as $\gamma_{0i}(k)$.

Based on Theorem 5.3.1, the computation steps of the Deng's grey relational degree between X_0 and X_i can be accomplished as explained below.

Step 1: Calculate the initial image (or average image) of X_0 and X_i, $i = 1, 2, \ldots, m$, where:

$$X_i' = X_i/x_i(1) = \left(x_i'(1), x_i'(2), \ldots, x_i'(n)\right) \quad i = 0, 1, 2, \ldots, m.$$

Step 2: Compute the difference sequences of X_0' and $X_i', i = 1, 2, \ldots, m$, and write as:

$$\Delta_i(k) = x_0'(k) - x_j'(k)|, \quad \Delta_i = \left(\Delta_j(1), \Delta_i(2), \ldots, \Delta_i(n)\right) \quad i = 1, 2, \ldots, m.$$

Step 3: Find the maximum and minimum differences, and denote as:

$$M = \max_i \max_k \Lambda_i(k), \quad m = \min_i \min_k \Delta_i(k).$$

Step 4: Calculate the Deng's grey relational coefficients:

$$\gamma_{0i}(k) = \frac{m + \xi M}{\Delta_i(k) + \xi M}, \quad \xi \in (0, 1) \quad k = 1, 2, \ldots, n; \quad i = 1, 2, \ldots, m.$$

Step 5: Compute the Deng's grey relational degree:

$$\gamma_{0i} = \frac{1}{n} \sum_{k=1}^{n} \gamma_{0i}(k); \quad i = 1, 2, \ldots, m.$$

Example 5.3.1 Let

$$
\begin{aligned}
X_0 &= (x_0(1), x_0(2), x_0(3), x_0(4), x_0(5)) \\
&= (12011.65, 7568.15, 3969.87, 2630.42, 2933.20) \\
X_1 &= (x_1(1), x_1(2), x_1(3), x_1(4), x_1(5)) \\
&= (127467, 73378, 47472, 28728, 24063) \\
X_2 &= (x_2(1), x_2(2), x_2(3), x_2(4), x_2(5)) \\
&= (281.02, 197.78, 97.88, 55.50, 62.02) \\
X_3 &= (x_3(1), x_3(2), x_3(3), x_3(4), x_3(5)) \\
&= (2.50, 2.65, 2.50, 2.31, 2.05) \\
X_4 &= (x_4(1), x_4(2), x_4(3), x_4(4), x_4(5)) \\
&= (391, 423, 262, 497, 104)
\end{aligned}
$$

where X_0 is the sequence of the regional GDP of the Suzhou, Wuxi, Changzhou, Zhenjiang and Yangzhou in Jiangsu Province in 2012, unit: 100 million yuan. And

X_1 is the sequence of the number of people engaged in R&D activities of the above five cities, unit: number of people.

X_2 is the sequence of the R&D expenditure of the above five cities, unit: 100 million yuan.

X_3 is the sequence of the R&D expenditure/regional GDP of the above five cities, unit: %.

X_4 is the sequence of the number of invention patents authorized of the above five cities, unit: number of items.

Data sources: China Statistical Yearbook 2013.

Calculate the Deng's grey relational degree between X_i, i = 1, 2, 3, 4 and X_0 (Liu, 2021).

Solution: Take X_0 as the system's behavioral characteristics sequence.

Step 1: Calculate the mean image of X_i, i = 0, 1, 2, 3, 4

From $X_i' = X_i/\overline{X}_i = (x_i'(1), x_i'(2), x_i'(3), x_i'(4), x_i'(5))$; $i = 0, 1, 2, 3, 4$, we have:

$$X_0' = X_0/\overline{X}_0 = (2.0629, 1.2998, 0.6818, 0.4518, 0.5038)$$
$$X_1' = X_1/\overline{X}_1 = (2.1166, 1.2185, 0.7883, 0.4770, 0.3996)$$
$$X_2' = X_2/\overline{X}_2 = (2.0241, 1.4245, 0.7050, 0.3997, 0.4467)$$
$$X_3' = X_3/\overline{X}_3 = (1.0408, 1.1032, 1.0408, 0.9617, 0.8535)$$
$$X_4' = X_4/\overline{X}_4 = (1.1658, 1.2612, 0.7812, 1.4818, 0.3101)$$

Step 2: Compute the difference sequences.

From
$\Delta_i(k) = |x_0'(k) - x_i'(k)|$; i = 1, 2, 3, 4, it follows that:

$$\Delta_1 = (0.0531, 0.0813, 0.1065, 0.0252, 0.1042)$$
$$\Delta_2 = (0.0388, 0.1247, 0.0232, 0.0521, 0.0571)$$
$$\Delta_3 = (1.0221, 0.1966, 0.3590, 0.5099, 0.3497)$$
$$\Delta_4 = (0.8971, 0.0386, 0.0994, 1.0300, 0.1937)$$

Step 3: Find the maximum and minimum differences.

$$M = \max_i \max_k \Delta_i(k) = 1.0300$$
$$m = \min_i \min_k \Delta_i(k) = 0.0232$$

Step 4: Calculate the Deng's relational coefficients.

Let $\xi = 0.5$, it follows that:

$$\gamma_{0i}(k) = \frac{m + \xi M}{\Delta_i(k) + \xi M} = \frac{0.5382}{\Delta_i(k) + 0.5150}; \quad i = 1, 2, 3, 4; \quad k = 1, 2, \ldots, 5$$

Therefore:

$$\gamma_{01}(1) = 0.9474, \gamma_{01}(2) = 0.9026, \gamma_{01}(3) = 0.8660,$$

$$\gamma_{01}(4) = 0.9963,\ \gamma_{01}(5) = 0.8692$$

$$\gamma_{02}(1) = 0.9718,\ \gamma_{02}(2) = 0.8413,\ \gamma_{02}(3) = 1.0000,$$

$$\gamma_{02}(4) = 0.9490,\ \gamma_{02}(5) = 0.9407$$

$$\gamma_{03}(1) = 0.3501,\ \gamma_{03}(2) = 0.7563,\ \gamma_{03}(3) = 0.6158,$$

$$\gamma_{03}(4) = 0.5251,\ \gamma_{03}(5) = 0.6224$$

$$\gamma_{04}(1) = 0.3811,\ \gamma_{04}(2) = 0.9722,\ \gamma_{04}(3) = 0.8760,$$

$$\gamma_{04}(4) = 0.3483,\ \gamma_{04}(5) = 0.7594$$

Step 5: Compute the Deng's grey relational degrees.

$$\gamma_{01} = \frac{1}{5}\sum_{k=1}^{5}\gamma_{01}(k) = 0.9163$$

$$\gamma_{02} = \frac{1}{5}\sum_{k=1}^{5}\gamma_{02}(k) = 0.9406$$

$$\gamma_{03} = \frac{1}{5}\sum_{k=1}^{5}\gamma_{03}(k) = 0.5739$$

$$\gamma_{04} = \frac{1}{5}\sum_{k=1}^{5}\gamma_{04}(k) = 0.6674$$

According to the calculation results based on the data of five cities in Jiangsu Province, both of the R&D expenditure of X_2 and the number of people engaged in R&D activities of X_1 have great impact on the regional GDP of X_0. Note that both X_2 and X_1 are input factors of R&D, it can be seen that scientific and technological funds and personnel investment play an important role in regional economic development.

5.4   Grey Absolute Relational Degree

Proposition 5.4.1 *Let $X_i = (x_i(1), x_i(2), \ldots, x_i(n))$ be the data sequence of a system's behavior, $X_i - x_i(1)$ denote the zigzagged line $(x_i(1) - x_i(1), x_i(2) - x_i(1), \ldots, x_i(n) - x_i(1))$, and let*

$$s_i = \int_{1}^{n} (X_i - x_i(1))dt \tag{5.8}$$

Then, when X_i increases, $s_i \geq 0$; when X_i decreases, $s_i \leq 0$; and when X_i vibrates, the sign of s_i varies.

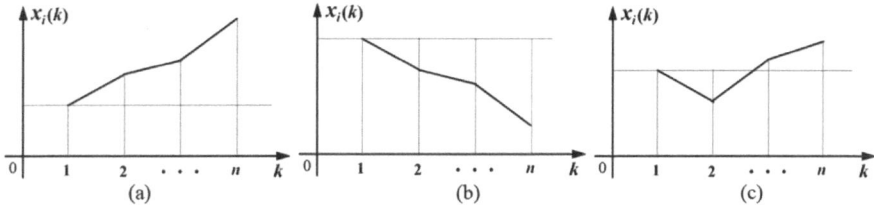

Fig. 5.1 The zigzagged line of Proposition 5.1

The results of Proposition 5.4.1 *are represented in* Fig. 5.1, *where (a) shows the case where the sequence increases; (b) the situation where* X_i *decreases; and (c) the scenario where* X_i *vibrates.*

Definition 5.4.1 Let $X_i = (x_i(1), x_i(2), \ldots, x_i(n))$ be the data sequence of a system's behavior and D the sequence operator which satisfies $X_i D = (x_i(1)d, x_i(2)d, \ldots, x_i(n)d)$ and $x_i(k)d = x_i(k) - x_i(1)$, $k = 1, 2, \ldots, n$. Then D is referred to as a zero-starting point operator and $X_i D$ is the image of X_i. $X_i D$ is often written as $X_i D = X_i^0 = (x_i^0(1), x_i^0(2), \ldots, x_i^0(n))$.

Proposition 5.4.2 *Assume that the* images *of the zero-starting point of two behavioral sequences* X_i *and* X_j *are respectively* $X_i^0 = (x_i^0(1), x_i^0(2), \ldots, x_i^0(n))$ *and* $X_j^0 = (x_j^0(1), x_j^0(2), \ldots, x_j^0(n))$. *Let:*

$$s_i - s_j = \int_1^n (X_i^0 - X_j^0)dt \tag{5.9}$$

and

$$S_i - S_j = \int_1^n \left(X_i - X_j\right)dt \tag{5.10}$$

Then, when X_i^0 is entirely located above X_j^0, $s_i - s_j \geq 0$; when X_i^0 is entirely underneath X_j^0, $s_i - s_j \leq 0$; and when X_i^0 and X_j^0 alternate their positions, the sign of $s_i - s_j$ is not fixed.

As shown in Fig. 5.2, when X_i^0 is entirely located above X_j^0 (Fig. 5.2a), the shaded area is positive so that $s_i - s_j \geq 0$. When X_i^0 and X_j^0 alternate their positions (Fig. 5.2b), the sign of $s_i - s_j$ is not fixed. Similarly, We can discuss the sign of X_i as $s_i - s_j$.

Definition 5.4.2 The sum of time intervals between consecutive observation values of a sequence X_i is called the length of X_i. It should be noted that two sequences with the same length may not have the same number of data. For example:

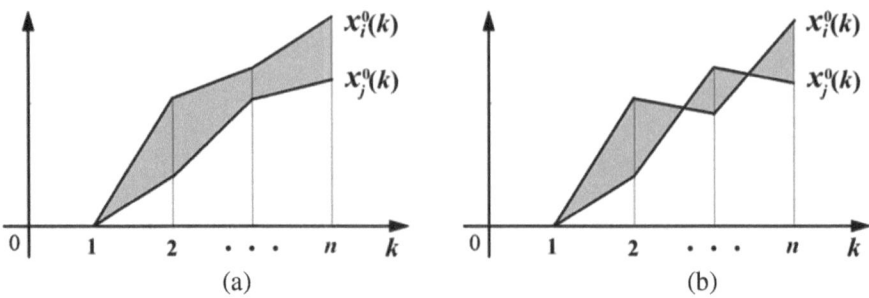

Fig. 5.2 A description of the relationship between X_i^0 and X_j^0

$$X_1 = (x_1(1), x_1(3), x_1(6))$$
$$X_2 = (x_2(1), x_2(3), x_2(5), x_2(6))$$
$$X_3 = (x_3(1), x_3(2), x_3(3), x_3(4), x_3(5), x_3(6))$$

The lengths of X_1, X_2, X_3 are all 6, but X_1 has 3 data, X_2 has 4 data, and X_3 has 6 data.

Definition 5.4.3 Let X_i and X_j be two sequences of the same length, and s_i and s_j are defined as above. Then, the following is referred to as the grey absolute relational degree between X_i and X_j, or absolute relational degree for short (Liu, 1991):

$$\varepsilon_{ij} = \frac{1 + |s_i| + |s_j|}{1 + |s_i| + |s_j| + |s_i - s_j|} \qquad (5.11)$$

As for sequences of different lengths, the concept of absolute relational degree can be defined by either shortening the longest sequence or by prolonging the shortest sequence using appropriate methods. This procedure will ensure that the sequences have the same length. However, by doing so, the ultimate value of the absolute relational degree will be affected.

Proposition 5.4.3 *Assume that X_i and X_j are two sequences with the same length. Let $X_i' = X_i - a, X_j' = X_j - b$, where a, b are real numbers. Denote ε_{0i} as the grey absolute relational degree between X_i' and X_j', then $\varepsilon_{0i} = \varepsilon_{0i}$. In fact, when X_i and X_j have been transformed, the values of S_i, S_j, and $S_i - S_j$ are not changed. Therefore, the value of absolute relational degree does not change.*

Definition 5.4.4 If the time intervals of any two consecutive observation values of a sequence X_i with the same length, then X_i is called an equal- time-interval sequence.

Lemma 5.4.1 *Assume that X_i is an equal-time-interval sequence. If the length of time-interval $l \neq 1$, then following can transform X_i into an l-time-interval sequence:*

$$t : T \rightarrow T$$

$$t \mapsto t/l$$

Lemma 5.4.2 *Assume that X_i and X_j are 1-time-interval sequences of the same length, and the following are zero-starting point images of X_i and X_j:*

$$X_i^0 = \left(x_i^0(1), x_i^0(2), \ldots, x_i^0(n)\right)$$
$$X_j^0 = \left(x_j^0(1), x_j^0(2), \ldots, x_j^0(n)\right)$$

Then, according to Liu (1991):

$$|s_i| = \left|\sum_{k=2}^{n-1} x_i^0(k) + \frac{1}{2}x_i^0(n)\right|$$

$$|s_j| = \left|\sum_{k=2}^{n-1} x_j^0(k) + \frac{1}{2}x_j^0(n)\right|$$

$$|s_i - s_j| = \left|\sum_{k=2}^{n-1} (x_i^0(k) - x_j^0(k)) + \frac{1}{2}(x_i^0(n) - x_j^0(n))\right|.$$

Theorem 5.4.1 *Assume that X_i and X_j are two sequences with the same length, same time distances from one moment to another, and equal time moment intervals. Then, the grey absolute relational degree can also be computed as follows* (Liu, 1991):

$$\varepsilon_{ij} = [1 + \left|\sum_{k=2}^{n-1} x_i^0(k) + \frac{1}{2}x_j^0(n)\right| + \left|\sum_{k=2}^{n-1} x_j^0(k) + \frac{1}{2}x_j^0(n)\right|]$$

$$\times [1 + \left|\sum_{k=2}^{n-1} x_i^0(k) + \frac{1}{2}x_i^0(n)\right| + \left|\sum_{k=2}^{n-1} x_j^0(k) + \frac{1}{2}x_j^0(n)\right|$$

$$+ \left|\sum_{k=2}^{n-1} (x_i^0(k) - x_j^0(k)) + \frac{1}{2}(x_i^0(n) - x_j^0(n))\right|]^{-1}$$

Example 5.4.1 Calculate the absolute relational degree ε_{01} of sequences X_0 and X_1. Let sequences X_0 and X_1 be as follows:

$$X_0 = (x_0(1), x_0(2), x_0(3), x_0(4), x_0(5), x_0(7)) = (10, 9, 15, 14, 14, 16)$$
$$X_1 = (x_1(1), x_1(3), x_1(7)) = (46, 70, 98$$

Solution
Step 1: Transform X_1 into a sequence with the same corresponding time-intervals as X_0.

$$x_1(2) = \frac{1}{2}(x_1(1) + x_1(3)) = \frac{1}{2}(46 + 70) = 58$$

$$x_1(5) = \frac{1}{2}(x_1(3) + x_1(7)) = \frac{1}{2}(70 + 98) = 84$$

$$x_1(4) = \frac{1}{2}(x_1(3) + x_1(5)) = \frac{1}{2}(70 + 84) = 77$$

Thus, we have a new sequence X_1 in place of the original X_1:

$$X_1 = (x_1(1), x_1(2), x_1(3), x_1(4), x_1(5), x_1(7)) = (46, 58, 70, 77, 84, 98)$$

Step 2: Transform X_0 and X_1 into equal-time-interval sequences:

$$x_0(6) = \frac{1}{2}(x_0(5) + x_0(7)) = \frac{1}{2}(14 + 16) = 15$$

$$x_1(6) = \frac{1}{2}(x_1(5) + x_1(7)) = \frac{1}{2}(84 + 98) = 91$$

We have:

$$X_0 = (x_0(1), x_0(2), x_0(3), x_0(4), x_0(5), x_0(6), x_0(7))$$
$$= (10, 9, 15, 14, 14, 15, 16)$$
$$X_1 = (x_1(1), x_1(2), x_1(3), x_1(4), x_1(5), x_1(6), x_1(7))$$
$$= (46, 58, 70, 77, 84, 91, 98)$$

where X_0 and X_1 are 1-time-interval sequences.

Step 3: Compute the zero-starting point images of sequences X_0 and X_1.

$$X_0^0 = \left(x_0^0(1), x_0^0(2), x_0^0(3), x_0^0(4), x_0^0(5), x_0^0(6), x_0^0(7)\right)$$
$$= (0, -1, 5, 4, 4, 5, 6)$$
$$X_1^0 = \left(x_1^0(1), x_1^0(2), x_1^0(3), x_1^0(4), x_1^0(5), x_1^0(6), x_1^0(7)\right)$$
$$= (0, 12, 24, 31, 38, 45, 52)$$

Step 4: Calculate $|S_0|$, $\quad |S_1|$, $\quad |S_1 - S_0|$

$$|s_0| = \left| \sum_{k=2}^{6} x_0^0(k) + \frac{1}{2}x_0^0(7) \right| = 20$$

$$|s_1| = \left| \sum_{k=2}^{6} x_1^0(k) + \frac{1}{2}x_1^0(7) \right| = 176$$

$$|s_1 - s_0| = \left| \sum_{k=2}^{6} \left(x_1^0(k) - x_0^0(k)\right) + \frac{1}{2}\left(x_1^0(7) - x_0^0(7)\right) \right| = 156$$

Step 5: Compute the grey absolute relational degree ε_{01} of sequences X_0 and X_1.

$$\varepsilon_{01} = \frac{1 + |s_0| + |s_1|}{1 + |s_0| + |s_1| + |s_1 - s_0|} = \frac{197}{353} \approx 0.5581$$

Theorem 5.4.2 *The grey absolute relational degree ε_{ij} satisfies the following properties:*

(1) $0 < \varepsilon_{ij} \leq 1$;
(2) *ε_{ij} is only related to the geometric shapes of X_i and X_j, and has no relationship with the spatial positions of these sequences;*
(3) *Any two sequences are not absolutely unrelated. That is, ε_{ij} never equals zero;*
(4) *The more X_i and X_j are geometrically similar, the greater ε_{ij} is;*
(5) *If X_i and X_j are parallel or X_i^0 fluctuates around X_j^0, with the area of the parts of X_i^0 located above X_j^0 equal to that of the parts with X_i^0 located underneath X_j^0, then $\varepsilon_{ij} = 1$;*
(6) *When one of the observed values of X_i and X_j change, ε_{ij} also changes accordingly;*
(7) *When the lengths of X_i and X_j change, ε_{ij} also changes;*
(8) *$\varepsilon_{jj} = \varepsilon_{ii} = 1$; and*
(9) *$\varepsilon_{ij} = \varepsilon_{ji}$.*

5.5 Grey Relative and Synthetic Relational Degree

5.5.1 Relative Grey Relational Degree

Definition 5.5.1 Let X_i and X_j be sequences of the same length with non-zero initial values, and X_i' and X_j' the initial images of X_i and X_j, respectively. The grey absolute relational degree of X_i' and X_j' is referred to as the relative grey relational degree of X_i and X_j, denoted r_{ij} (Liu, 1991) This relative relational degree is a quantitative representation of the relationship between the rates of change of sequences X_i and X_j, relative to their initial values. The closer the rates of change of X_i and X_j are, the greater r_{ij} is, and vice versa.

Proposition 5.5.1 *Let X_i be a sequence with a non-zero initial value. If $X_j = cX_i$. If $c > 0$ is a constant, then $r_{ij} = 1$.*

Proof Assume that $X_i = (x_i(1), x_i(2), \ldots, x_i(n))$, then:

$$X_j = (x_j(1), x_j(2), \ldots, x_j(n)) = (cx_i(1), cx_i(2), \ldots, cx_i(n)).$$

The initial images of X_i and X_j are as follows:

$$X_i' = X_i/x_i(1) = (\frac{x_i(1)}{x_i(1)}, \frac{x_i(2)}{x_i(1)}, \ldots, \frac{x_i(n)}{x_i(1)})$$

$$X_j' = X_j/x_j(1) = (\frac{x_j(1)}{x_j(1)}, \frac{x_j(2)}{x_j(1)}, \ldots, \frac{x_j(n)}{x_j(1)})$$

$$= \frac{cx_i(1)}{cx_i(1)}, \frac{cx_i(2)}{cx_i(1)}, \ldots, \frac{cx_i(n)}{cx_i(1)} = (\frac{x_i(1)}{x_i(1)}, \frac{x_i(2)}{x_i(1)}, \ldots, \frac{x_i(n)}{x_i(1)}).$$

Therefore, $X_j' = X_i'$, so $r_{ij} = 1$.

Proposition 5.5.2 *Let X_i and X_j be two sequences of the same length with non-zero initial values. Additionally, the relative grey relational degree of r_{ij} and the grey absolute relational degree of ε_{ij} do not have any connections. When ε_{ij} is relatively large, r_{ij} can be very small; when ε_{ij} is very small, r_{ij} can also be very large.*

Proposition 5.5.3 *Let X_i and X_j be two sequences of the same length with non-zero initial values. Then, for any non-zero constants a and b, the relative grey relational degree r_{ij}' between aX_i and bX_j is the same as the r_{ij} of X_i and X_j.*

In fact, the initial images of a X_i and b X_j are equal to those of X_i and X_j, respectively. Thus, scalar multiplication does not act in any way under the function of initialing operators. Hence, $r_{ij}' = r_{ij}$.

Example 5.4.2 Calculate the relative grey relational degree r_{01} for sequences X_0 and X_1 of Example 5.4.1.

Solution
Step 1: Transform X_1 and X_0 into the same 1-time-interval sequences.

$$X_0 = (x_0(1), x_0(2), x_0(3), x_0(4), x_0(5), x_0(6), x_0(7))$$
$$= (10, 9, 15, 14, 14, 15, 16)$$
$$X_1 = (x_1(1), x_1(2), x_1(3), x_1(4), x_1(5), x_1(6), x_1(7))$$
$$= (46, 58, 70, 77, 84, 91, 98)$$

Step 2: Calculate the initial images of sequences X_0 and X_1.

$$X_0' = (1, 0.9, 1.5, 1.4, 1.4, 1.5, 1.6)$$
$$X_1' = (1, 1.26, 1.52, 1.67, 1.83, 1.98, 2.13)$$

Step 3: Compute the zero-starting point images of sequences X_0' and X_1'.

$$X_0'^0 = (x_0'^0(1), x_0'^0(2), x_0'^0(3), x_0'^0(4), x_0'^0(5), x_0'^0(6),$$

$$x_0^{'0}(7)) = (0, -0.1, 0.5, 0.4, 0.4, 0.5, 0.6)$$
$$X_1^{'0} = (x_1^{'0}(1), x_1^{'0}(2), x_1^{'0}(3), x_1^{'0}(4), x_1^{'0}(5), x_1^{'0}(6),$$
$$x_1^{'0}(7)) = (0, 0.26, 0.52, 0.67, 0.83, 0.98, 1.13)$$

Step 4: Calculate $|s_0'|, |s_1'|, |s_1' - s_0'|$.

$$|s_0'| = \left| \sum_{k=2}^{6} x_0^{'0}(k) + \frac{1}{2} x_0^{'0}(7) \right| = 2$$

$$|s_1'| = \left| \sum_{k=2}^{6} x_1^{'0}(k) + \frac{1}{2} x_1^{'0}(7) \right| = 3.828$$

$$|s_1' - s_0'| = \left| \sum_{k=2}^{6} (x_1^{'0}(k) - x_0^{'0}(k)) + \frac{1}{2}(x_1^{'0}(7) - x_0^{'0}(7)) \right| = 1.925$$

Step 5: Calculate the relative grey relational degree of r_{01}.

$$r_{01} = \frac{1 + |s_0'| + |s_1'|}{1 + |s_0'| + |s_1'| + |s_1' - s_0'|} = \frac{6.825}{8.75} \approx 0.78$$

Theorem 5.5.1 *The relative grey relational degree of r_{ij} satisfies the following properties:*

(1) $0 < r_{ij} \leq 1$;
(2) *The value of r_{ij} relates only the rates of change of the sequences X_i and X_j with respect to their individual initial values. It does not relate to the magnitudes of other entries. In other words, scalar multiplication does not change the value of relative grey relational degree;*
(3) *The rates of change of any two sequences are somehow related. That is, r_{ij} is never zero;*
(4) *The closer the individual rates of change of X_i and X_j with respect to their initial values, the greater the r_{ij};*
(5) *If $X_j = aX_i$, or when the images of zero initial points of the initial images of X_i and X_j satisfy that $X_i^{'0}$ fluctuates around $X_j^{'0}$, and if the area of the parts where $X_i^{'0}$ is located above $X_j^{'0}$ equals that of the parts where $X_i^{'0}$ is located underneath $X_j^{'0}$, then $r_{ij} = 1$;*
(6) *When an entry in X_i or X_j is changed, r_{ij} will change accordingly;*
(7) *When the length of X_i or X_j is changed , r_{ij} also changes;*
(8) $r_{jj} = r_{ii} = 1$; *and*
(9) $r_{ij} = r_{ji}$.

5.5.2 Grey Synthetic Relational Degree

Definition 5.5.2 Let X_i and X_j be sequences of the same length with non-zero initial entries, ε_{ij} and r_{ij} be respectively the absolute and relative relational degrees between X_i and X_j, and $\theta \in [0, 1]$. Then the following is referred to as the grey synthetic relational degree between X_i and X_j (Liu, 1991):

$$\rho_{ij} = \theta \varepsilon_{ij} + (1 - \theta) r_{ij} \tag{5.12}$$

The concept of grey synthetic relational degree reflects the degree of similarity between the zigzagged lines of X_i and X_j, and the closeness between the rates of change of X_i and X_j with respect to their individual initial values. It is an index that describes relatively completely the closeness relationship between sequences. In general, we take $\theta = 0.5$. If the focus of a study is the relationship between relevant absolute quantities, θ can take a greater value than 0.5. On the other hand, if the focus is more on comparison between rates of change, then θ can take a smaller value than 0.5.

Example 5.4.3 Calculate the synthetic grey relational degree of ρ_{01} for sequences X_0 and X_1 of Example 5.4.1.

Solution From Examples 5.4.1 and 5.4.2, we have $\varepsilon_{01} = 0.5581$ and $r_{01} = 0.78$. If $\theta = 0.5$:
   $\rho_{01} = \theta \varepsilon_{01} + (1 - \theta) r_{01} = 0.5 \times 0.5581 + 0.5 \times 0.78 \approx 0.669$.
   We can obtain different ρ_{01} values if we take $\theta = 0.2, 0.3, 0.4, 0.6, 0.8$, respectively (see Table 5.1).

Theorem 5.5.2 *The grey synthetic relational degree of ρ_{ij} satisfies the following properties:*

(1) $0 < \rho_{ij} \leq 1$;
(2) *The value of ρ_{ij} relates to the individual observed values of sequences X_i and X_j, as well as to the rates of change of these values with respect to their initial values;*
(3) ρ_{ij} *will never be zero;*
(4) ρ_{ij} *changes along with the values in X_i and X_j;*
(5) *When the lengths of X_i and X_j change, so does ρ_{ij};*
(6) *With different θ value, ρ_{ij} also varies;*
(7) *When $\theta = 1$, X_j; when $\theta = 0$, $\rho_{ij} = r_{ij}$;*

Table 5.1 The values of ρ_{01} with different θ

θ	0.2	0.3	0.4	0.6	0.8
ρ_{01}	0.73562	0.71343	0.69124	0.64686	0.60248

(8) $\rho_{jj} = \rho_{ii} = 1$; and
(9) $\rho_{ij} = \rho_{ji}$.

5.6 Grey Similarity, Closeness and Three-Dimensional Relational Degree

This section focuses on the new models which measure mutual influences and connections between sequences from two different angles: similarity and closeness. These new models are much easier to apply to practical problems than original model. Also, three-dimensional grey relational degree can be used to analyze the relationship among curved surfaces in three-dimensional space and this is discussed next.

5.6.1 Grey Relational Analysis Models Based on Similarity and Closeness

Definition 5.6.1 Let X_i and X_j be sequences of the same length, and $s_i - s_j$ the same as defined in Proposition 5.4.2. Then, the following Formula (5.13) is referred to as the grey similitude relational degree between X_i and X_j:

$$\varepsilon_{ij} = \frac{1}{1 + |s_i - s_j|} \tag{5.13}$$

The concept of similitude relational degree is employed to measure the geometric similarity of the shapes of sequences X_i and X_j. The more similar the geometric shapes of X_i and X_j, the greater the value of ε_{ij}, and vice versa.

Definition 5.6.2 Let X_i and X_j be sequences of the same length, and $S_i - S_j$ the same as defined in Proposition 5.4.2. Then, the following Formula (5.14) is referred to as the grey closeness relational degree between X_i and X_j:

$$\rho_{ij} = \frac{1}{1 + |S_i - S_j|} \tag{5.14}$$

The concept of the grey closeness relational degree is employed to measure the spatial closeness of sequences X_i and X_j. The closer the X_i and X_j sequences, the greater the value of ρ_{ij}, and vice versa.

Proposition 5.6.1 *Let X_i and X_j be sequences of 1-time-intervals with the same length. Then:*

$$\left| S_i - S_j \right| = \left| \frac{1}{2}[x_i(1) - x_j(1)] + \sum_{k=2}^{n-1} [x_i(k) - x_j(k)] + \frac{1}{2}[x_i(n) - x_j(n)] \right|$$

$$(5.15)$$

It should be noted that the concept of the grey closeness relational degree is only meaningful when sequences X_i and X_j possess similar meanings and identical units. Otherwise, it does not stand for any practical significance.

Theorem 5.6.1 *The grey similitude relational degree of ε_{ij} satisfies the following properties:*

(1) $0 < \varepsilon_{ij} \leq 1$;
(2) *The value of ε_{ij} is determined only by the geometric shape of sequences X_i and X_j without any relationship with their relative spatial positions. In other words, the transform translation of X_i and X_j will not change the value of ε_{ij};*
(3) *The more geometrically similar the sequences X_i and X_j, the greater the value of ε_{ij}, and vice versa;*
(4) *If X_i and X_j are parallel, or when X_i^0 fluctuates around X_j^0, and the area of the parts where X_i^0 is located above X_j^0 equals that of the parts where X_i^0 is located beneath X_j^0, then $\varepsilon_{ij} = 1$;*
(5) $\varepsilon_{ii} = 1, \varepsilon_{jj} = 1$; *and*
(6) $\varepsilon_{ij} = \varepsilon_{ji}$.

Theorem 5.6.2 *The grey closeness relational degree of ρ_{ij} satisfies the following properties:*

(1) $0 < \rho_{ij} \leq 1$;
(2) *The value of ρ_{ij} is determined not only by the geometric shape of sequences X_i and X_j, but also by their relative spatial positions. In other words, the transform translation of X_i and X_j will change the value of ρ_{ij};*
(3) *The closer the sequences X_i and X_j, the greater the ρ_{ij} value, and vice versa;*
(4) *If X_i and X_j coincide, or X_i fluctuates around X_j, and the area of the parts where X_i is located above X_j equals that of the parts where X_i is located beneath X_j, then $\rho_{ij} = 1$;*
(5) $\rho_{ii} = 1, \rho_{jj} = 1$; *and*
(6) $\rho_{ij} = \rho_{ji}$.

Example 5.6.1 Compute the grey similitude relational degrees of $\varepsilon_{12}, \varepsilon_{13}$ and the grey closeness relational degrees of ρ_{12}, ρ_{13} between X_1 and X_2, X_3, respectively, given the sequences below:

$$X_1 = (x_1(1), x_1(2), x_1(3), x_1(4), x_1(5),$$
$$x_1(7)) = (0.91, 0.97, 0.90, 0.93, 0.91, 0.95)$$
$$X_2 = (x_2(1), x_2(2), x_2(3), x_2(5),$$
$$x_2(7)) = (0.60, 0.68, 0.61, 0.63, 0.65)$$

$$X_3 = (x_3(1), x_3(3), x_3(7)) = (0.82, 0.90, 0.86)$$

Solution

Step 1: Let us translate both X_2 and X_3 into sequences with the same time intervals as X_1. To this end, consider the following:

$$x_2(4) = \frac{1}{2}(x_2(3) + x_2(5)) = \frac{1}{2}(0.61 + 0.63) = 0.62$$

$$x_3(2) = \frac{1}{2}(x_3(1) + x_3(3)) = \frac{1}{2}(0.82 + 0.90) = 0.86$$

$$x_3(5) = \frac{1}{2}(x_3(3) + x_3(7)) = \frac{1}{2}(0.90 + 0.86) = 0.88$$

$$x_3(4) = \frac{1}{2}(x_3(3) + x_3(5)) = \frac{1}{2}(0.90 + 0.88) = 0.89$$

Thus, we have:

$$X_2 = (x_2(1), x_2(2), x_2(3), x_2(4), x_2(5),$$
$$x_2(7)) = (0.60, 0.68, 0.61, 0.62, 0.63, 0.65)$$
$$X_3 = (x_3(1), x_3(2), x_3(3), x_3(4), x_3(5),$$
$$x_3(7)) = (0.82, 0.86, 0.90, 0.89, 0.88, 0.86)$$

Step 2: Let us translate X_1, X_2, and X_3 into sequences of equal time distance. To this end:

$$x_1(6) = \frac{1}{2}(x_1(5) + x_1(7)) = \frac{1}{2}(0.91 + 0.95) = 0.93$$

$$x_2(6) = \frac{1}{2}(x_2(5) + x_2(7)) = \frac{1}{2}(0.63 + 0.65) = 0.64$$

$$x_3(6) = \frac{1}{2}(x_3(5) + x_3(7)) = \frac{1}{2}(0.88 + 0.86) = 0.87$$

Therefore, the following sequences are all 1-time distance, which means that the time distances between consecutive entries are all 1.

$$X_1 = (x_1(1), x_1(2), x_1(3), x_1(4), x_1(5),$$
$$x_1(7)) = (0.91, 0.97, 0.90, 0.93, 0.91, 0.93, 0.95)$$
$$X_2 = (x_2(1), x_2(2), x_2(3), x_2(4), x_2(5),$$
$$x_2(7)) = (0.60, 0.68, 0.61, 0.62, 0.63, 0.64, 0.65)$$
$$X_3 = (x_3(1), x_3(2), x_3(3), x_3(4), x_3(5),$$
$$x_3(7)) = (0.82, 0.86, 0.90, 0.89, 0.88, 0.87, 0.86)$$

Step 3: Compute the images of zero-starting points provided below.

$$X_1^0 = (x_1^0(1), x_1^0(2), x_1^0(3), x_1^0(4), x_1^0(5), x_1^0(6),$$
$$x_1^0(7)) = (0, 0.06, -0.01, 0.02, 0, 0.02, 0.04)$$
$$X_2^0 = (x_2^0(1), x_2^0(2), x_2^0(3), x_2^0(4), x_2^0(5), x_2^0(6),$$
$$x_2^0(7)) = (0, 0.08, 0.01, 0.02, 0.03, 0.04, 0.05)$$
$$X_3^0 = (x_3^0(1), x_3^0(2), x_3^0(3), x_3^0(4), x_3^0(5), x_3^0(6),$$
$$x_3^0(7)) = (0, 0.04, 0.08, 0.07, 0.06, 0.05, 0.04)$$

Step 4: Compute $|s_1 - s_2|$, $|s_1 - s_3|$ and $|S_1 - S_2|$, $|S_1 - S_3|$ as follows.

$$|s_1 - s_2| = \left| \sum_{k=2}^{6} (x_1^0(k) - x_2^0(k)) + \frac{1}{2}(x_1^0(7) - x_2^0(7)) \right| = 0.095$$

$$|s_1 - s_3| = \left| \sum_{k=2}^{6} (x_1^0(k) - x_3^0(k)) + \frac{1}{2}(x_1^0(7) - x_3^0(7)) \right| = 0.21$$

$$|S_1 - S_2| = \left| \sum_{k=2}^{6} (x_1(k) - x_2(k)) + \frac{1}{2}(x_1(7) - x_2(7)) \right| = 1.91$$

$$|S_1 - S_3| = \left| \sum_{k=2}^{6} (x_1(k) - x_3(k)) + \frac{1}{2}(x_1(7) - x_3(7)) \right| = 0.375$$

Step 5: Calculate the similitude relational degrees of $\varepsilon_{12}, \varepsilon_{13}$ and closeness relational degrees of ρ_{12}, ρ_{13}.

$$\varepsilon_{12} = \frac{1}{1 + |s_1 - s_2|} = 0.91, \ \varepsilon_{13} = \frac{1}{1 + |s_1 - s_3|} = 0.83.$$

$$\rho_{12} = \frac{1}{1 + |S_1 - S_2|} = 0.34, \ \rho_{13} = \frac{1}{1 + |S_1 - S_3|} = 0.73.$$

Because $\varepsilon_{12} > \varepsilon_{13}$, it follows that X_2 is more similar to X_1 than X_3. Because $\rho_{12} < \rho_{13}$, it follows that X_3 is closer to X_1 than X_2.

Please note that the grey relational analysis focus on relevant order relationship and influence between sequences rather than the value of the grey relational degree. For instance, let us assume that one must compute the similitude relational degrees or closeness relational degrees based on Eqs. (5.13) or (5.14) When the absolute values of the sequence data are relatively large, the values of both $|s_i - s_j|$ and $|S_i - S_j|$ might be large, too, which in turn leads to the resultant similitude and closeness relational degrees being relatively small. This scenario does not have any substantial impact on the analysis of order relationships. If a particular problem demands relatively large numerical magnitudes in the value of grey relational degree, one can replace the number 1 appearing in the numerators or denominators of Eqs. (5.13) and (5.14) by a relevant constant, or use the grey absolute relational degree, or use other appropriate models.

5.6.2 Grey Three-Dimension Degree of Relational Degree

The above GRA models can be generalized to three-dimensional space based on geometric descriptions of a behavior matrix.

Definition 5.6.3 Assume that X is a two-dimensional system factor, and a_{ij} is an observation value of the system's behavior at two-dimensional point (i, j), where $1 \leq i \leq m;\ 1 \leq j \leq n$. Then the following expression is called the behavior matrix of system factor X:

$$A = (a_{ij})_{m \times n} = \begin{bmatrix} a_{11} & a_{12} & \cdots & a_{1n} \\ a_{21} & a_{22} & \cdots & a_{1n} \\ \cdots & \cdots & \cdots & \cdots \\ a_{m1} & a_{m2} & \cdots & a_{mn} \end{bmatrix}$$

For example, if the prices (e.g., opening prices, closing prices, maximum prices, or minimum prices) of a share have been recorded on different dates, we can obtain the behavior matrix of the different prices X of the share. The behavior matrix will reduce to a behavior sequence if only one share price has been recorded on different dates.

The scatter diagram in behavior matrix A and the corresponding behavioral curved surface in three-dimensional space are shown in Figs. 5.3 and 5.4.

Fig. 5.3 The scatter diagram as behavior matrix

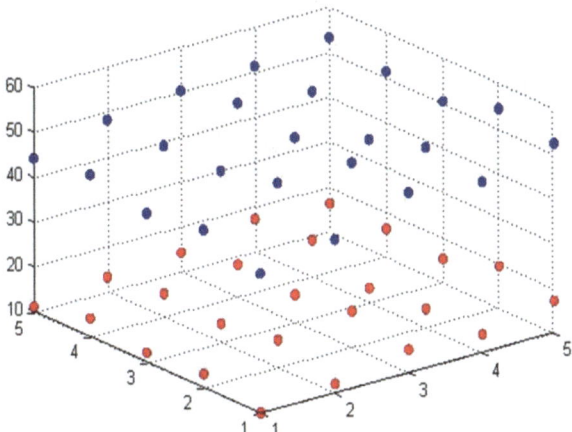

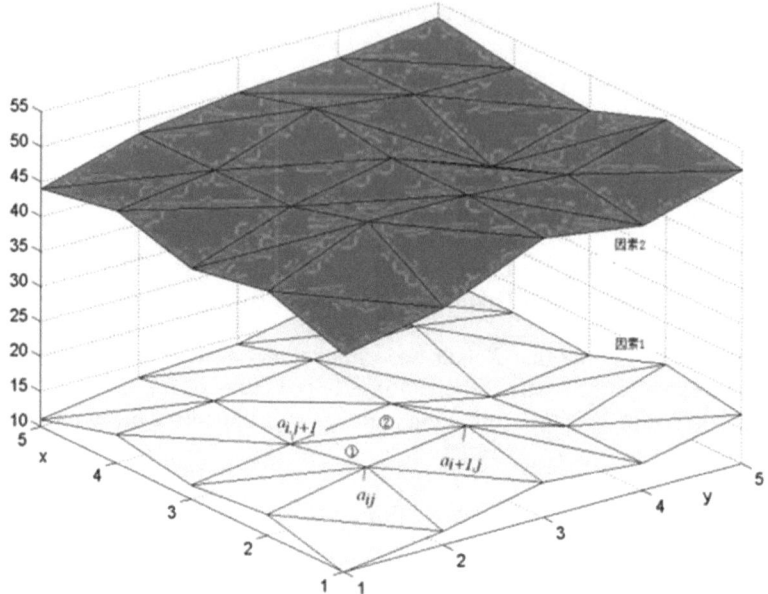

Fig. 5.4 The corresponding behavioral curved surface

Definition 5.6.4 Assume the following behavior matrix of system factor X.

$$A = (a_{ij})_{m \times n} = \begin{bmatrix} a_{11} & a_{12} & \cdots & a_{1n} \\ a_{21} & a_{22} & \cdots & a_{1n} \\ \cdots & \cdots & \cdots & \cdots \\ a_{m1} & a_{m2} & \cdots & a_{mn} \end{bmatrix}$$

$AD = (a_{ij}d)_{m \times n}$, where D is a matrix operator, $a_{ij}d = a_{ij} - a_{1j}$, then D is called a zero-starting edge operator, AD is called the zero-starting edge image of A, and they are denoted as $AD = A^0 = (a_{ij}^0)_{m \times n}$.

The zero-starting edge curved surface of A is shown in Fig. 5.5.

Definition 5.6.5 Assume that behavior matrices $A = (a_{ij})_{m \times n}$, $B = (b_{ij})_{m \times n}$ are matrices of the same type. Then the following formula is called the three-dimensional grey absolute relational degree between A and B:

$$\varepsilon_{ab} = \frac{1 + |s_a| + |s_b|}{1 + |s_a| + |s_b| + |s_a - s_b|} \tag{5.16}$$

This occurs when $s_a = \iint_{D_a} A^0 dx dy$, $s_b = \iint_{D_b} B^0 dx dy$, $s_a - s_b = \iint_{D_{ab}} (A^0 - B^0) dx dy$.

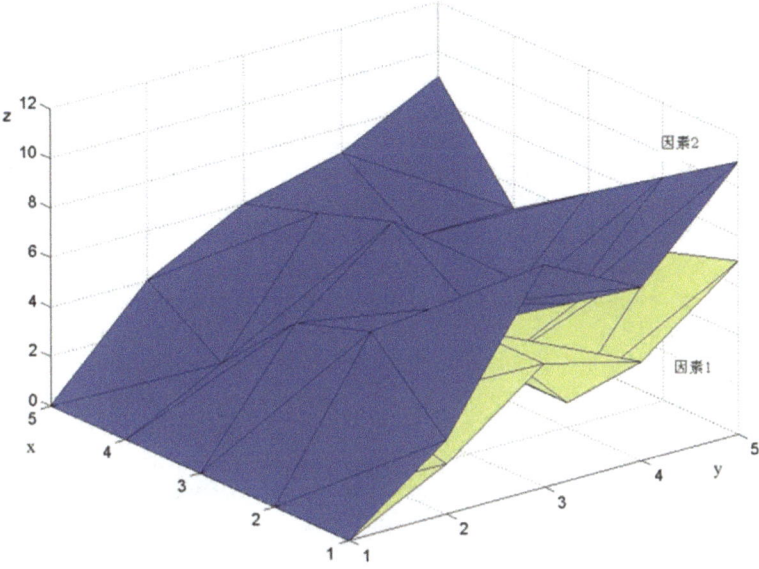

Fig. 5.5 The zero-starting edge image of a behavior curved surface

Formula (5.16) looks similar to the absolute GRA model shown in Formula (5.11) However, the meaning is different. The meaning of $|s_i|$, $|s_j|$, $|s_i - s_j|$ in Formula (5.11) is the area of curved edge trapezoids surrounded by axis X_i^0, X_j^0, the zero-starting point curves, and the area of curved edge trapezoid surrounded by X_i^0 and X_j^0. However, the meaning of $|s_p|$, $|s_q|$, $|s_p - s_q|$ in Formula (5.16) is the volume of curved roof cylinders surrounded by the axis plane and A^0, B^0, the curved surface of zero-starting edge, and the volume of curved roof cylinders surrounded by A^0 and B^0.

The three-dimensional grey relational analysis model can truly reflect the relational degree between system behavior matrices. The analysis results are objective, reliable and easy to implement on computer. The three-dimensional GRA model is seen to have expansive application prospects in many fields such as multi-criterion decision-making, panel data analysis, image processing, among others, which include matrices as objects of study.

5.7 Negative Grey Relational Analysis Models

In the past 40 years, driven by the realistic demand of measuring the relationship between the reverse sequences of the system, many scholars have made unremitting attempts and exploration around the construction of negative grey relational analysis model. In 2008, Shi Hongxing, Liu Sifeng, and Fang Zhigeng proposed a kind of grey relational analysis model referring to the grey absolute relational analysis model. In this paper, the positive and negative sign of the grey relational degree is determined according to the concave and convex direction of the periodic waveform to describe the inverse relationship between the periodic factors (Shi et al., 2008) In 2015, based on dissolved gas analysis (DGA), Song Bin et al. studied the latent fault diagnosis of power transformer. In order to correctly describe the reverse change relationship between different fault types, a calculation method of negative grey relational degree is proposed (Song et al., 2015) In 2019, Saad Ahmed Javed and Sifeng Liu proposed a bidirectional gabsolute GRA model for uncertain systems.. The proposed model can be used to evaluate both positive and negative relation of different sequences (Javed & Liu, 2019).

Firstly, the definition of inverse sequence will be given in this section. Then, several different negative grey relational analysis models, such as negative grey similarity relational analysis model, negative grey absolute relational analysis model, negative relative grey relational analysis model, negative grey synthetic relational analysis model, and negative Deng's grey relational analysis model will be put forward based on the corresponding common grey relational analysis models. The properties of the new models will be studied.

In order to build a negative grey relational model, it is necessary to give the definition of inverse sequence at first.

Definition 5.7.1 Assume that $X_i = (x_i(1), x_i(2), \ldots, x_i(n))$.

is a system's behavior data sequence,

(1) If $\forall k = 2, 3, \ldots, n$, $x_i(k) - x_i(k-1) > 0$, then X_i is referred to as a monotonic increasing sequence;
(2) If the inequality sign in (1)is inversed, then X_i is referred to as a monotonic decreasing sequence.

Monotonic increasing sequence and monotonic decreasing sequence are collectively referred to as monotone sequence. Please see Fig. 5.6 for the curves of monotonic increasing sequence and monotonic decreasing sequences.

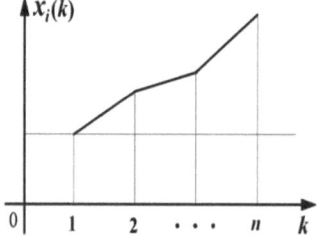 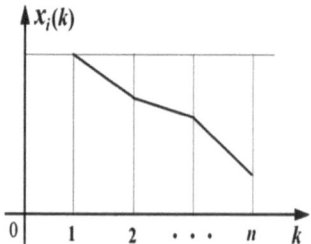

Fig. 5.6 Monotone sequence curves

Fig. 5.7 Oscillation
sequence curve

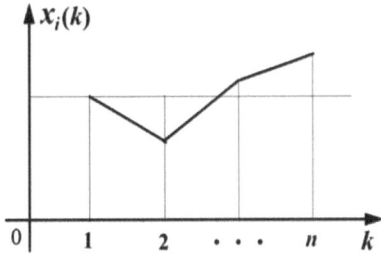

Definition 5.7.2 Assume that $X_i = (x_i(1), x_i(2), \ldots, x_i(n))$ is a system's behavior data sequence, if there are $k, k' \in \{2, 3, \ldots, n\}$ such that $x(k) - x(k-1) > 0$, $x(k') - x(k'-1) < 0$, then X is referred to as an oscillation sequence.

Figure 5.7 shows the case of a curve of oscillation sequence.

Definition 5.7.3 Assume that $X_i = (x_i(1), x_i(2), \ldots, x_i(n))$ is a system's behavior data sequence, $X_i^0 = X_i - x_i(1)$ is the zero-starting point sequence of X_i, let $s_i = \int_1^n (X_i - x_i(1))dt$, then

(1) If $s_i > 0$, then X_i is referred to as an increasing sequence;
(2) If $s_i < 0$, then X_i is referred to as a decreasing sequence;
(3) If $s_i = 0$, then X_i is referred to as a horizontal sequence.

Obviously, monotonic increasing sequence is a special case of increasing sequence and monotonic decreasing sequence is a special case of decreasing sequence. An oscillation sequence can be an increasing sequence, decreasing sequence, or a horizontal sequence. And stationary sequence is a special case of horizontal sequence.

Definition 5.7.4 Assume that

$$X_i = (x_i(1), x_i(2), \ldots, x_i(n))$$
$$X_j = (x_j(1), x_j(2), \ldots, x_j(n))$$

are two system's behavior data sequences.

(1) When both X_i, X_j are increasing sequences or decreasing sequences, then X_i and X_j are called sequences with the same direction;
(2) When one of X_i and X_j is an increasing sequence and the other is a decreasing sequence, then X_i and X_j are called reverse sequences (Liu, 2022).

The relationship between two sequences with the same direction can be measured by positive grey relational analysis model. The relationship between two reverse sequences needs to be measured by negative grey relational analysis model.

Proposition 5.7.1 *Assume that*

$$X_i = (x_i(1), x_i(2), \ldots, x_i(n))$$
$$X_j = (x_j(1), x_j(2), \ldots, x_j(n))$$

are two system's behavior data sequences. The zero-starting point sequences of X_i and X_j as follows,

$$X_i^0 = (x_i^0(1), x_i^0(2), \ldots, x_i^0(n))$$
$$X_j^0 = (x_j^0(1), x_j^0(2), \ldots, x_j^0(n))$$

Let

$$s_i = \int_1^n (X_i - x_i(1))dt \qquad (5.17)$$

$$s_i - s_j = \int_1^n (X_i^0 - X_j^0)dt \qquad (5.18)$$

then

$$|s_i| = \left| \sum_{k=2}^{n-1} x_i^0(k) + \frac{1}{2}x_i^0(n) \right| \qquad (5.19)$$

$$|s_i - s_j| = \left| \sum_{k=2}^{n-1} (x_i^0(k) - x_j^0(k)) + \frac{1}{2}(x_i^0(n) - x_j^0(n)) \right| \qquad (5.20)$$

Proof $|s_i|$ and $|s_i - s_j|$ are determined by areas of the following curved triangles, respectively.

$$X = 0, \ X = X_i^0, \text{ and } t = n$$
$$X = X_i^0, \ X = X_j^0, \ t = n$$

They are sums of little areas of n-1 small trapezoids of height 1. Note the length of the bottom edges of the small trapezoids, and it is easy to know that the conclusion is true (Liu, 2022).

Proposition 5.7.2 *Assume that* X_i, X_j, X_i^0, X_j^0, *and* $|s_i|, |s_i - s_j|$ *as shown in proposition 1 , then*

(1) *When* X_i *and* X_j *are with the same direction, and* X_i^0, X_j^0 *intersect only at the starting point, then* $|s_i - s_j| = ||s_i| - |s_j||$;

(2) *When X_i and X_j are two reverse sequences, then $|s_i - s_j| = |s_i| + |s_j|$;*
(3) *If X_i^0 fluctuates around X_j^0, then $|s_i - s_j|$ is the absolute value of algebraic sum of area enclosed by X_i^0 and X_j^0. The parts where X_i^0 are above X_j^0 take positive sign, and the parts where X_i^0 are underneath X_j^0 take negative sign.*

Positional relationship of X_i^0 and X_j^0 can be clearly seen from Fig. 5.8. Figure 5.8a shows the case where X_i and X_j are both increasing sequences, Fig. 5.8b shows the case where X_i and X_j are both increasing sequences, Fig. 5.8c shows the case where X_i and X_j are two reverse sequences, and Fig. 5.8d shows the case where X_i^0 fluctuates around X_j^0 (Liu, 2022).

It can be seen from Fig. 5.8c, when X_i and X_j are reverse sequences, the value of $|s_i - s_j|$ is large. At this time, the value of grey relational degree calculated by positive grey relational model will be very small. Before the negative grey relational analysis model was proposed, people usually convert the inverse sequence into the same direction sequence through inverse operator or reciprocal operator, then calculate

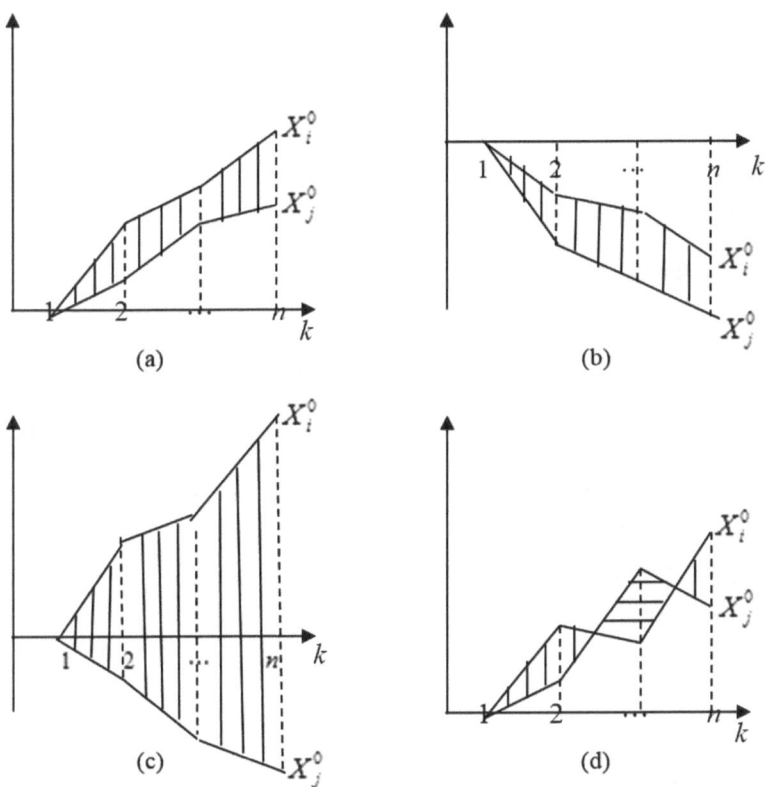

(a)   (b)

(c)   (d)

Fig. 5.8 Positional relationship of X_i^0 and X_j^0

the positive grey relational degree of the sequences with the same direction, but the results are not completely reasonable.

Therefore, for the measurement of the relationship between reverse sequences, the construction of negative grey relational analysis model has become an inevitable choice.

Corresponding to the normalization and proximity axioms of the grey relational analysis model, the negative grey relational degree ϕ_{ij}^N shall meet the following axioms.

Axiom 5.7.1 Normalization

$$-1 < \varphi_{ij}^N \leq 0, \varphi_{ij}^N = 0 \Leftarrow X_i = X_j.$$

The value of φ_{ij}^N is negative. The minimum value is -1 and the maximum value is 0 (Liu, 2022).

Axiom 5.7.2 Reversibility The stronger the inverse relation between X_i and X_j, the smaller the value of φ_{ij}^N.

Note that the value of negative grey relational degree belongs to interval $(-1,0]$, the smaller the value of φ_{ij}^N, the greater the absolute value of φ_{ij}^N (Liu, 2022).

Definition 5.7.5 Suppose the following system's behavior data sequences

$$X_i = (x_i(1), x_i(2), \ldots, x_i(n))$$
$$X_j = (x_j(1), x_j(2), \ldots, x_j(n))$$

are reverse sequences, then

$$\phi_{ij}^N = -\frac{|s_i - s_j|}{1 + |s_i - s_j|} \tag{5.21}$$

is called the negative grey similarity relational degree of X_i and X_j (Liu, 2022).

It can be easily proved that the negative grey similarity relational degree defined by Formula (5.21) satisfies the axioms of normalization and reversibility, and has the following properties:

Theorem 5.7.1 *The negative grey similarity relational degree ϕ_{ij}^N satisfies the following properties:*

(1) $-1 < \phi_{ij}^N < 0$.

(2) ϕ_{ij}^N *is only related to the geometric shapes of X_i and X_j, and has no relationship with the spatial positions of these sequences. In other words, the translation transformation does not change the value of negative grey similarity relational degree.*

(3) *The stronger the reverse relation between X_i and X_j, the closer ϕ_{ij}^N is to -1; The weaker the reverse relation between X_i and X_j, the closer ϕ_{ij}^N is to 0.*

(4) *If X_i and X_j are parallel or X_i^0 fluctuates around X_j^0, with the area of the parts of X_i^0 located above X_j^0 equal to that of the parts with X_i^0 located underneath X_j^0, then $\phi_{ij}^N = 0$.*

(5) $\phi_{ii}^N = \phi_{jj}^N = 0$.

(6) $\phi_{ij}^N = \phi_{ji}^N$.

(Liu, 2022)

Example 1 Let $X_1 = (x_1(1), x_1(2), x_1(3), x_1(4), x_1(5)) = (1, 2, 3, 4, 5)$ and $X_2 = (x_2(1), x_2(2), x_2(3), x_2(4), x_2(5)) = (5, 4, 2, 2, 1)$.

Then the zero-starting point sequences of X_1 and X_2 as follows

$$X_1^0 = (x_1^0(1), x_1^0(2), x_1^0(3), x_1^0(4), x_1^0(5)) = (0, 1, 2, 2, 4)$$
$$X_2^0 = (x_2^0(1), x_2^0(2), x_2^0(3), x_2^0(4), x_2^0(5)) = (0, -1, -3, -3, -4)$$

We have $s_1 = 7$, $s_2 = -9$, therefore, X_1 is an increasing sequence, and X_2 is a decreasing sequence. That is, X_1 and X_2 are reverse sequences. From Formula (5)

$$\phi_{12}^N = -\frac{|s_1 - s_2|}{1 + |s_1 - s_2|} = -\frac{16}{1 + 16} \approx -0.9412.$$

It shows that there is a strong inverse correlation between X_1 and X_2.

Similarly, the definitions of negative grey absolute relational degree, negative grey relative relational degree and negative grey comprehensive relational degree can be given as follows.

Definition 5.7.6 Assume that X_i and X_j are system's behavior data sequences,

(1) If X_i and X_j are reverse sequences, then

$$\varepsilon_{ij}^N = -\frac{|s_i - s_j|}{1 + |s_i| + |s_j| + |s_i - s_j|} \tag{5.22}$$

Is called negative grey absolute relational degree of X_i and X_j.

(2) If the initial valued sequences of X_i and X_j are reverse sequences, then

$$r_{ij}^N = -\frac{|s_i' - s_j'|}{1 + |s_i'| + |s_j'| + |s_i' - s_j'|} \tag{5.23}$$

Is called negative relative grey relational degree of X_i and X_j.

(3) If both of X_i and X_j, and the initial valued sequences of X_i and X_j are all reverse sequences, then

$$\rho_{ij}^N = \theta \varepsilon_{ij}^N + (1 - \theta) r_{ij}^N \tag{5.24}$$

is called negative grey synthetic relational degree of X_i and X_j. Where $\theta \in [0,1]$ (Liu, 2022).

It should be noted that the grey proximity relational analysis model have been constructed to measure the spatial relative position relationship of the sequences. The grey proximity relational degree does not consider the change direction of the sequences and does not pay attention to the same or reverse relationship between the two sequences. Therefore, it is not necessary to define the corresponding "negative grey proximity relational analysis model".

Definition 5.7.7 Let $X_0 = (x_0(1), x_0(2), \ldots, x_0(n))$ be a data sequence of a system's behavioral characteristic and the following are relevant factor sequences:

$$X_1 = (x_1(1), x_1(2), \ldots, x_1(n))$$

$$\ldots\ldots\ldots\ldots\ldots\ldots\ldots\ldots\ldots\ldots\ldots\ldots$$

$$X_i = (x_i(1), x_i(2), \ldots, x_i(n))$$

$$\ldots\ldots\ldots\ldots\ldots\ldots\ldots\ldots\ldots\ldots\ldots\ldots$$

$$X_m = (x_m(1), x_m(2), \ldots, x_m(n))$$

If X_i is a reverse sequence of X_0, for $\xi \in (0, 1)$, let

$$\gamma_{0i}^N(k) = \frac{\min\limits_{i} \min\limits_{k} |x_0(k) - x_i(k)| - |x_0(k) - x_i(k)|}{|x_0(k) - x_i(k)| + \xi \max\limits_{i} \max\limits_{k} |x_0(k) - x_i(k)|} \tag{5.25}$$

$$\gamma_{0i}^N = \frac{1}{n} \sum_{k=1}^{n} \gamma_{0i}^N(k) \tag{5.26}$$

Then γ_{0i}^N is called negative Deng's grey relational degree of X_i and X_0, and $\gamma_{0i}^N(k)$ is called the negative Deng's grey relational coefficient of relevant factor sequence X_i and the system's behavioral characteristic sequence X_0 at point k (Liu, 2022).

It is easy to show that the negative grey absolute relational degree, negative relative grey relational degree, negative grey synthetic relational degree, and negative Deng's grey relational degree are all satisfy the axioms of normalization and Reversibility.

5.8 Superiority Analysis

Definition 5.8.1 Assume that $Y_1, Y_2, \ldots, Y_s$ are a system's characteristic behavioral sequences, and $X_1, X_2, \ldots, X_m$ are behavioral sequences of relevant factors with the same length. Let γ_{ij} be the grey relational degree between Y_i and X_j, $i = 1, 2, \ldots, s$, and $j = 1, 2, \ldots, m$. Then:

$$\Gamma = \left(\gamma_{ij}\right)_{s \times m} = \begin{bmatrix} \gamma_{11} & \gamma_{12} & \cdots & \gamma_{1m} \\ \gamma_{21} & \gamma_{22} & \cdots & \gamma_{2m} \\ \cdots & \cdots & \cdots & \cdots \\ \gamma_{s1} & \gamma_{s2} & \cdots & \gamma_{sm} \end{bmatrix}.$$

This formula is referred to as the grey relational matrix of the system, where the ith row is made up of the grey relational degree between the characteristic sequence $Y_i(i = 1, 2, \ldots, s)$ and each of the factor sequences $X_1, X_2, \ldots, X_m$; and the jth column consists of the grey relational degree between each of the characteristic sequences $Y_1, Y_2, \ldots, Y_s$ and $X_j(j = 1, 2, \ldots, m)$. We can analyze both the superiority of a system's characteristic behavioral variables or the behavioral variables of relevant factors.

Definition 5.8.2 Assume that $Y_1, Y_2, \ldots, Y_s$ are a system's characteristic behavioral sequences, $X_1, X_2 \ldots, X_m$ are behavioral sequences of relevant factors, and $\Gamma = \left(\gamma_{ij}\right)_{s \times m}$ is the grey relational matrix. If there are $k, i \in \{1, 2, \ldots, s\}$ such that $\gamma_{kj} \geq \gamma_{ij}$, $j = 1, 2, \ldots, m$, then the system's characteristic variable Y_k is said to be more favorable than the system's characteristic variable Y_i, written as $Y_k \succ Y_i$.

If $\forall i = 1, 2, \ldots, s$, $i \neq k$, $Y_k \succ Y_i$ always holds true, then Y_k is said to be the most favorable characteristic variable.

Definition 5.8.3 Assume that $Y_1, Y_2, \ldots, Y_s$ are a system's characteristic behavioral sequences, $X_1, X_2, \ldots, X_m$ are behavioral sequences of relevant factors, and $\Gamma = \left(\gamma_{ij}\right)_{s \times m}$ is the grey relational matrix. If there are $l, j \in \{1, 2, \ldots, m\}$ such that $\gamma_{il} \geq \gamma_{ij}$, $i = 1, 2, \ldots, s$, then we say that the system's factor X_l is more favorable than factor X_j, written as $X_l \succ X_j$.

If $\forall j = 1, 2, \ldots, m$, $j \neq l$, $X_l \succ X_j$ always holds true, then $Y_1 = (170, 174, 197, 216.4, 235.8)$ is said to be the most favorable factor.

Definition 5.8.4 Assume that $Y_1, Y_2, \ldots, Y_s$ are a system's characteristic behavioral sequences, $X_1, X_2, \ldots, X_m$ are behavioral sequences of relevant factors, and $\Gamma = \left(\gamma_{ij}\right)_{s \times m}$ is the grey relational matrix.

(1) If there are $k, i \in \{1, 2, \ldots, s\}$ satisfying $\sum\limits_{j=1}^{m} \gamma_{kj} \geq \sum\limits_{j=1}^{m} \gamma_{ij}$, then the system's characteristic variable Y_k is said to be more quasi-favorable than Y_i, which is denoted as $Y_k \geq Y_i$.

(2) If there are $l, j \in \{1, 2, \ldots, m\}$ satisfying $\sum\limits_{i=1}^{m} \gamma_{il} \geq \sum\limits_{i=1}^{m} \gamma_{ij}$, then the system's factor X_l is more quasi-favorable than X_j, which is denoted as $X_l \geq X_j$.

Definition 5.8.5 Assume that $Y_1, Y_2, \ldots, Y_s$ are a system's characteristic behavioral sequences, $X_1, X_2, \ldots, X_m$ are behavioral sequences of relevant factors, and $\Gamma = (\gamma_{ij})_{s \times m}$ is the grey relational matrix.

(1) If there is $k \in \{1, 2, \ldots, s\}$ such that $\forall i = 1, 2, \ldots, s, \ i \neq k, Y_k \geq Y_i$, then the system's characteristic variable Y_k is said to be quasi-preferred.
(2) If there is $l \in \{1, 2, \ldots, m\}$ such that $\forall j = 1, 2, \ldots, m, j \neq l, X_l \geq X_j$, then the system's factor X_l is said to be quasi-preferred.

Proposition 5.8.1 *In a system of s characteristic variables and m relevant factors, there may not be a most favorable characteristic variable and a most favorable factor. However, there must be quasi-preferred characteristic variable and factor.*

Example 5.8.1 The formulas below are system's characteristic behavioral sequences.

$$Y_1 = (170, 174, 197, 216.4, 235.8)$$
$$Y_2 = (57.55, 70.74, 76.8, 80.7, 89.85)$$
$$Y_3 = (68.56, 70, 85.38, 99.83, 103.4)$$

The formulas below are behavioral sequences of relevant factors.

$$X_1 = (308.58, 310, 295, 346, 367)$$
$$X_2 = (195.4, 189.9, 189.2, 205, 222.7)$$
$$X_3 = (24.6, 21, 12.2, 15.1, 14.57)$$
$$X_4 = (20, 25.6, 23.3, 29.2, 30)$$
$$X_5 = (18.98, 19, 22.3, 23.5, 27.655)$$

Try and analyze the superiority of the system's characteristic behavioral variables and the superiority of the behavioral variables of relevant factors.

Solution
We analyze the superiority of the system's characteristic behavioral sequences and the behavioral sequences of relevant factors by absolute degree of GRA model.

(1) Find the matrix of the grey absolute relational degree. Calculate the images of zero-starting point for all the system's characteristic behavioral sequences as well as the behavioral sequences of relevant factors as follows:

$$Y_1^0 = (0, 4, 27, 46.4, 65.8)$$
$$Y_2^0 = (0, 13.19, 19.25, 23.15, 32.3)$$
$$Y_3^0 = (0, 1.44, 16.82, 31.27, 34.84)$$
$$X_1^0 = (0, 1.42, -13.58, 37.42, 58.42)$$
$$X_2^0 = (0, -5.5, -8.2, 9.6, 27.3)$$
$$X_3^0 = (0, -3.6, -12.4, , -9.5, -10.03)$$
$$X_4^0 = (0, 5.6, 3.3, 9.2, 10)$$
$$X_5^0 = (0, 0.02, 3.32, 4.52, 8.675)$$

For the system's characteristic behavioral variable Y_1, we have:

$$\left| s_{y_1} \right| = \left| \sum_{k=2}^{4} y_1^0(k) + \frac{1}{2} y_1^0(5) \right| = \left| 4 + 27 + 46.4 + \frac{1}{2} \times 65.8 \right| = 110.3$$

$$\left| s_{x_1} \right| = \left| \sum_{k=2}^{4} x_1^0(k) + \frac{1}{2} x_1^0(5) \right| = \left| 1.42 + (-13.58) + 37.42 + \frac{1}{2} \times 58.42 \right| = 54.47$$

$$\left| s_{y_1} - s_{x_1} \right| = \left| \sum_{k=2}^{4} (y_1^0(k) - x_1^0(k)) + \frac{1}{2} (y_1^0(5) - x_1^0(5)) \right| = 55.9$$

$$\varepsilon_{11} = \frac{1 + \left| s_{y_1} \right| + \left| s_{x_1} \right|}{1 + \left| s_{y_1} \right| + \left| s_{x_1} \right| + \left| s_{y_1} - s_{x_1} \right|} = \frac{1 + 110.3 + 54.47}{1 + 110.3 + 54.47 + 55.9} = 0.748$$

$$\left| s_{x_2} \right| = \left| \sum_{k=2}^{4} x_2^0(k) + \frac{1}{2} x_2^0(5) \right| = \left| (-5.5) + (-8.2) + 9.6 + \frac{1}{2} \times 27.3 \right| = 9.55$$

$$\left| s_{y_1} - s_{x_2} \right| = \left| \sum_{k=2}^{4} (y_1^0(k) - x_2^0(k)) + \frac{1}{2} (y_1^0(5) - x_2^0(5)) \right| = 100.75$$

$$\varepsilon_{12} = \frac{1 + \left| s_{y_1} \right| + \left| s_{x_2} \right|}{1 + \left| s_{y_1} \right| + \left| s_{x_2} \right| + \left| s_{y_1} - s_{x_2} \right|} = \frac{1 + 110.3 + 9.55}{1 + 110.3 + 9.55 + 100.75} = 0.545$$

Similarly:

$$\varepsilon_{13} = \frac{1 + \left| s_{y_1} \right| + \left| s_{x_3} \right|}{1 + \left| s_{y_1} \right| + \left| s_{x_3} \right| + \left| s_{y_1} - s_{x_3} \right|} = 0.502$$

$$\varepsilon_{14} = \frac{1 + \left| s_{y_1} \right| + \left| s_{x_4} \right|}{1 + \left| s_{y_1} \right| + \left| s_{x_4} \right| + \left| s_{y_1} - s_{x_4} \right|} = 0.606$$

$$\varepsilon_{15} = \frac{1 + |s_{y_1}| + |s_{x_5}|}{1 + |s_{y_1}| + |s_{x_5}| + |s_{y_1} - s_{x_5}|} = 0.557$$

For the system's characteristic behavioral variable Y_2, Y_3, we have:

$$\varepsilon_{21} = 0.880, \varepsilon_{22} = 0.570, \varepsilon_{23} = 0.502, \varepsilon_{24} = 0.663, \varepsilon_{25} = 0.588$$
$$\varepsilon_{31} = 0.907, \varepsilon_{32} = 0.574, \varepsilon_{33} = 0.503, \varepsilon_{34} = 0.675, \varepsilon_{35} = 0.594$$

Therefore, we have the grey absolute relational degree matrix as follows:

$$A = (\varepsilon_{ij}) = \begin{bmatrix} \varepsilon_{11} & \varepsilon_{12} & \varepsilon_{13} & \varepsilon_{14} & \varepsilon_{15} \\ \varepsilon_{21} & \varepsilon_{22} & \varepsilon_{23} & \varepsilon_{24} & \varepsilon_{25} \\ \varepsilon_{31} & \varepsilon_{32} & \varepsilon_{33} & \varepsilon_{34} & \varepsilon_{35} \end{bmatrix} = \begin{bmatrix} 0.748 & 0.545 & 0.502 & 0.606 & 0.557 \\ 0.880 & 0.570 & 0.502 & 0.663 & 0.588 \\ 0.907 & 0.574 & 0.503 & 0.675 & 0.594 \end{bmatrix}$$

(2) Calculate the relative grey relational degree matrix. Calculate the initial images: $Y_i'(i = 1, 2, 3)$ and $X_j'(j = 1, 2, 3, 4, 5)$ of $Y_i(i = 1, 2, 3)$ and $X_j(j = 1, 2, 3, 4, 5)$.

Then find the images of zero-starting point for all system's characteristic behavioral sequences $Y_i(i = 1, 2, 3)$ and the behavioral sequences of relevant factors $X_j(j = 1, 2, 3, 4, 5)$.

$Y_i'^0(i = 1, 2, 3)$ and $X_j'^0(j = 1, 2, 3, 4, 5)$ of $Y_i'(i = 1, 2, 3)$ and $X_j'(j = 1, 2, 3, 4, 5)$.

From:

$$\left| s_{y_i}' \right| = \left| \sum_{k=2}^{4} y_i'^0(k) + \frac{1}{2} y_i'^0(5) \right|; \quad i = 1, 2, 3$$

$$\left| s_{x_j}' \right| = \left| \sum_{k=2}^{4} x_j'^0(k) + \frac{1}{2} x_j'^0(5) \right|; \quad j = 1, 2, 3, 4, 5$$

$$\left| s_{y_i}' - s_{x_j}' \right|$$
$$= \left| \sum_{k=2}^{4} (y_i'^0(k) - x_j'^0(k)) + \frac{1}{2} (y_i'^0(5) - x_j'^0(5)) \right|; \quad i = 1, 2, 3, ; j = 1, 2, 3, 4, 5$$

$$r_{ij} = \frac{1 + |s_{y_i}'| + |s_{x_j}'|}{1 + |s_{y_i}'| + |s_{x_j}'| + |s_{y_i}' - s_{x_j}'|}; i = 1, 2, 3; j = 1, 2, 3, 4, 5,$$

we have:

$$r_{11} = 0.7945, r_{12} = 0.7389, r_{13} = 0.6046, r_{14} = 0.8471, r_{15} = 0.9973$$
$$r_{21} = 0.6937, r_{22} = 0.6571, r_{23} = 0.5837, r_{24} = 0.9738, r_{25} = 0.8271$$

$$r_{31} = 0.7300, r_{32} = 0.6866, r_{33} = 0.6101, r_{34} = 0.9444, r_{35} = 0.8884$$

Therefore, we have the relative grey relational degree matrix as follows:

$$B = \begin{bmatrix} r_{11} \ r_{12} \ r_{13} \ r_{14} \ r_{15} \\ r_{21} \ r_{22} \ r_{23} \ r_{24} \ r_{25} \\ r_{31} \ r_{32} \ r_{33} \ r_{34} \ r_{35} \end{bmatrix} = \begin{bmatrix} 0.7945 \ 0.7389 \ 0.6046 \ 0.8471 \ 0.9973 \\ 0.6937 \ 0.6571 \ 0.5837 \ 0.9738 \ 0.8271 \\ 0.7300 \ 0.6866 \ 0.6101 \ 0.9444 \ 0.8884 \end{bmatrix}$$

(3) Compute the grey synthetic relational degree matrix. If $\theta = 0.5$, we have:

$$C = \theta A + (1 - \theta) B = (\theta \varepsilon_{ij} + (1 - \theta) r_{ij}) = (\rho_{ij})$$

$$= \begin{bmatrix} \rho_{11} \ \rho_{12} \ \rho_{13} \ \rho_{14} \ \rho_{15} \\ \rho_{21} \ \rho_{22} \ \rho_{23} \ \rho_{24} \ \rho_{25} \\ \rho_{31} \ \rho_{32} \ \rho_{33} \ \rho_{34} \ \rho_{35} \end{bmatrix}$$

$$= \begin{bmatrix} 0.7713 \ 0.6420 \ 0.5533 \ 0.7266 \ 0.7772 \\ 0.7869 \ 0.6136 \ 0.5429 \ 0.8184 \ 0.7076 \\ 0.8185 \ 0.6303 \ 0.5566 \ 0.8097 \ 0.7412 \end{bmatrix}$$

(4) Analysis and discussion. In matrix A of the grey absolute relational degree, the rows of A satisfy the following formula:

$$\varepsilon_{3j} > \varepsilon_{2j} \geq \varepsilon_{1j}; \quad j = 1, 2, 3, 4, 5.$$

Therefore, we have $Y_3 \succ Y_2 \succ Y_1$. That is, Y_3 is the most favorable characteristic variable, Y_2 is the second, and Y_1 the least favorable characteristic variable. All columns of A satisfy:

$$\varepsilon_{i1} > \varepsilon_{i4} > \varepsilon_{i5} > \varepsilon_{i2} > \varepsilon_{i3}; \quad i = 1, 2, 3.$$

Therefore, we have:
$X_1 \succ X_4 \succ X_5 \succ X_2 \succ X_3$.
That is, X_1 is the most favorable factor, X_4 the second, X_5 the third, X_2 the fourth, and X_3 the least.

From the matrix B of relative degree of relational, it can be seen that because the elements of B satisfy

$$r_{i4} > r_{i1} > r_{i2} > r_{i3}; \quad i = 1, 2, 3$$
$$r_{i5} > r_{i1} > r_{i2} > r_{i3}; \quad i = 1, 2, 3$$

Thus, we can conclude that:

$$X_4 \succ X_1 \succ X_2 \succ X_3, X_5 \succ X_1 \succ X_2 \succ X_3.$$

Hence, X_3 is the most unfavorable factor of the system. Further, let us consider the following:

$$\sum_{j=1}^{5} r_{1j} = 3.9824 > \sum_{j=1}^{5} r_{3j} = 3.8595 > \sum_{j=1}^{5} r_{2j} = 3.7354.$$

Thus, we can conclude that $Y_1 \succeq Y_3 \succeq Y_2$, that is, Y_1 is the quasi-preferred characteristic. Also, given that:

$$\sum_{i=1}^{3} r_{i4} = 2.7653 > \sum_{i=1}^{3} r_{i5} = 2.7128 > \sum_{i=1}^{3} r_{i1} = 2.2182$$

$$> \sum_{i=1}^{3} r_{i2} = 2.0826 > \sum_{i=1}^{3} r_{i3} = 1.7984,$$

we have:

$$X_4 \succeq X_5 \succeq X_1 \succeq X_2 \succeq X_3.$$

That is, X_4 is the quasi-preferred factor, X_5 the next, and X_3 the most unfavorable factor.

On matrix C of the grey synthetic relational degree, it can be seen that the elements of C satisfy:

$$\rho_{i1} > \rho_{i2} > \rho_{i3}, \ \rho_{i4} > \rho_{i2} > \rho_{i3}, \ \rho_{i5} > \rho_{i2} > \rho_{i3}, \ i = 1, 2, 3.$$

Therefore, we have:

$$X_1 \succeq X_2 \succeq X_3, \ X_4 \succeq X_2 \succeq X_3, \ X_5 \succeq X_2 \succeq X_3.$$

That is, X_3 is the least preferred factor. We further consider the following:

$$\sum_{j=1}^{5} \rho_{3j} = 3.5563 > \sum_{j=1}^{5} \rho_{1j} = 3.4704 > \sum_{j=1}^{5} \rho_{2j} = 3.4694.$$

Thus,

$$Y_3 \succeq Y_1 \succeq Y_2.$$

That is, Y_3 is the quasi-preferred characteristic variable. Also, based on:

$$\sum_{i=1}^{3} \rho_{i1} = 2.3767 > \sum_{i=1}^{3} \rho_{i4} = 2.3547 > \sum_{i=1}^{3} \rho_{i5} = 2.226$$

$$> \sum_{i=1}^{3} \rho_{i2} = 1.8859 > \sum_{i=1}^{3} \rho_{i3} = 1.6528,$$

it follows that:

$$X_1 \succeq X_4 \succeq X_5 \succeq X_2 \succeq X_3.$$

Therefore, X_1 is the quasi-preferred factor, X_4 the next, X_5 is more favorable than X_2, and X_3 is the most unfavorable factor.

When investigating practical problems, the analyses of the three relational orders may not provide cohesive conclusions. This is because the absolute relational order looks at the relationship between absolute quantities, the relative relational order focuses on the rates of change with respect to the initial values of the observed sequences, while the synthetic relational order combines both the relationships between absolute quantities and rates of change. When considering the background of the problem of concern, we can choose one of the relational orders. For parsimony purposes, after a particular grey relational operator is applied to the system's characteristic behavioral sequences and relevant factor sequences, one only needs to employ the absolute relational order to the processed data.

5.9 Practical Application

Through the example below, we look at how to apply GRA models to analyze the time difference of economic indices.

Example 5.9.1 In order to effectively monitor the performance of macro-economic systems and provide timely warnings, there is a need to investigate the time relationship of various economic indices with respect to economic cycles in terms of their peaks and valleys. In order to do so, questions such as the following must be addressed: Which indices can provide warning ahead of time? Which indices would be synchronic with the evolution of economic systems? And which indices tend to lag behind economic development? In other words, there is a need to divide economic indices into three classes: leading indicators, synchronic indices, and stagnant representations. To this end, grey relational analysis is an effective method for classifying economic indices (Chen & Liu, 2005).

Through careful research and analysis, we selected the following 8 major classes and 17 criteria as indices for economic performance:

(1) The Energy and raw materials class: the total production of energy;
(2) The investments class: the total investment in real estate;

(3) The production class: increase in industry output, increase in light industry output, increase in heavy industry output;

(4) The revenue class: national income, national expenditure;

(5) The currency and credit class: currency in circulation, savings at various financial institutions, amount of loans issued by financial institutions, cash payout in the form of salary and wages, net amount of currency in circulation;

(6) The consumption class: the gross retail amount of the society;

(7) The foreign trade class: gross amount of imports, gross amount of exports, direct investments by foreign entities; and

(8) The commodity prices class: the consumer price index.

By applying the following standards, we classify the previous criteria into three classes: leading indicators, synchronic indices, and stagnant representations. The standards for determining leading indicators are as follows:

(1) The indicated appearance of economic cyclic peaks needs to be at least three months ahead of their actual occurrence. Such leading relationship must be relatively stable with few exceptions;

(2) Indicated cycles and historical cycles are nearly one-to-one corresponded to each other. Also, for the most recent three economic cycles, the indicated cycles must be at least two times ahead of the actual occurrences with at least 3 months of lead time; and

(3) The economic characteristics of the indices provide relatively definite and clear leading relationships with respect to the background economic cycles.

The standards for determining both synchronic indices and stagnant representations are similar to those outlined above. However, for synchronic indices the time differences between the indicated appearances and the actual occurrences of economic cycles must be within plus and minus 3 months, while for stagnant representations the indicated appearances of economic cycles are behind the actual occurrences by at least 3 months.

In practice, it is almost impossible to find an index that meets all the stated standards. Therefore, based on the recorded reference cycles, we look for the statistical indices that meet the previously stated standards as closely as possible. In reality, a leading indicator can sometimes lag behind actual economic development, while an identified stagnant representation can also provide good lead-time in its forecast of a specific economic evolution. Similar scenarios also occur with regard to synchronic indices. However, theoretically, if the index is leading the actual occurrences among the one-to-one correspondences between an index and the actually recorded cycles over 2/3 of times, then we treat such an index as leading. Similar treatments are applied to synchronic indices and stagnant representations.

Given that the increase in industry output has played a significant role in the Chinese economy, as a synchronic index it has high quality. Therefore, it can be employed as the basic index in our grey relational analysis. We will compute not only the grey absolute relational degree between each criterion and the increase in industry output, but also the grey absolute relational degree of the other 16 criteria

Table 5.2 The absolute degrees of grey relational of the criteria when $L = 0$

Index	Absolute degree of grey relational	Index	Absolute degree of grey relational
Increase in heavy industry output	0.979810	National income	0.559540
Increase in light industry output	0.972655	Gross amount of exports	0.544870
Gross retail amount	0.862105	Total production of energy	0.541044
Cash payout as salaries	0.789278	Net amount of currency in circulation	0.525936
Currency in circulation	0.753681	Loans issued by financial institutions	0.507958
Total investment in real estate	0.726366	Savings at financial institutions	0.505226
Gross amount of imports	0.598248	Consumer price index	0.500173
National expenditure	0.566914	Direct investments by foreign entities	0.500002

with their data translated 1–12 months along the time axis either left or right. When data are translated to the left, the months will take negative values; when translated to the right, the months will take positive values. The amount of horizontal translation is denoted by L. That is, we compute the grey absolute relational degree between all 16 individual criteria, excluding that of increase in industry output, and that of increase in industry output for $L = -12, \ldots, 12$. For each L-value, we order the obtained the grey absolute relational degree from the smallest to the largest, with the criterion listed in the front chosen as candidate criterion for that specific L-value. For instance, when $L = 0$, the grey absolute relational degree of the criteria are listed in Table 5.2.

Synchronic indices should be selected from those with large grey absolute relational degree, because large degrees of relational indicate that these criteria have greater similarities in comparison with that of increase in industry output, which we employ as the basic standard of the Chinese economic cycles. However, we still do not have theoretical evidence to support that an index with large grey absolute relational degree must be synchronic. To this end, we also need to consider whether or not the related grey absolute relational degree will be even greater when $L \neq 0$. If when $L = 0$ the value of the grey absolute relational degree of a certain index is ranked in the front, and if when $L = -4$ its value is even greater, it means that after this index is translated to four months earlier, it is more similar to the pattern of the increase in industry output. Thus, in this case, this specific index can be seen as one leading the economic cycle by as much as about four months. By using these two standards, we can not only classify indices as synchronic, leading, or stagnant, but also specify the amount of leading or staggering time.

Table 5.3 The grey absolute relational degrees of "cash payout as salaries" when $L \neq 0$

L	The grey absolute relational degree	L	The grey absolute relational degree
−12	0.664615	1	0.877090
−11	0.705983	2	0.867859
−10	0.733564	3	0.857366
−9	0.752740	4	0.832260
−8	0.753598	5	0.825027
−7	0.732221	6	0.806787
−6	0.723942	7	0.806782
−5	0.731232	8	0.820384
−4	0.742249	9	0.803771
−3	0.752628	10	0.806649
−2	0.770216	11	0.805679
−1	0.800838	12	0.836308
0	0.789278		

When $L = 0$, the index of "cash payout as salaries" is ranked relatively in the front. Therefore, it is a natural candidate for being a synchronic indicator. When the L–value changes, the relevant changes in its absolute degrees of grey relational are given in Table 5.3.

From Table 5.3, it follows that when $L = 1$, the grey absolute relational degree reaches its maximum. Therefore, this specific index should be seen as one that is lagging the economic cycle by as much as one month. An index which is leading or lagging no more than two months is usually seen as synchronic. However, if it exceeds this range of time it will be treated as either a leading or staggering index.

As a second example, the computational results for the index of "gross retail amount" are provided in Table 5.4.

From Table 5.4, it can be seen that when $L = -6$ the grey absolute relational degree of the particular index reaches its maximum. Therefore, it can be seen as a leading indicator. By using this method, we can compute the L–values corresponding to the maximum grey absolute relational degree of each of the indices of our interest. The results are listed in Table 5.5.

Table 5.5 indicates that we can classify the 16 indices of the eight major classes into three classes as leading, synchronic, and stagnant indices, as shown in Table 5.6.

***Example 5.9.2* Measurement of reverse incentive effect of Fields Medal** (Liu, 2022).

Most people agree that knowledge production can promote long-term economic growth. Yet little is known about how knowledge is produced (Borjas & Doran, 2015) For example, it is difficult for the author to explain clearly how the models of the negative grey similarity relational degree, the negative grey absolute relational degree, negative grey relative relational degree, negative grey comprehensive relational degree, and negative Deng's grey relational degree are finally proposed in

Table 5.4 The grey absolute relational degrees of "gross retail amount"

L	The grey absolute relational degree	L	The grey absolute relational degree
−12	0.914466	1	0.856944
−11	0.915117	2	0.866789
−10	0.918527	3	0.876758
−9	0.887243	4	0.882430
−8	0.888258	5	0.889590
−7	0.928151	6	0.895899
−6	0.948684	7	0.900899
−5	0.939351	8	0.900130
−4	0.923900	9	0.895977
−3	0.909621	10	0.894374
−2	0.884610	11	0.892662
−1	0.846814	12	0.889532
0	0.862105		

Table 5.5 L–values corresponding to maximum grey absolute relational degree of the indices of our interest

Index	L	Absolute degree	Index	L	Absolute degree
Currency in circulation	− 6	0.983452	National income	+ 12	0.718998
Increase in heavy industry output	0	0.979810	Gross amount of imports	− 9	0.606556
Increase in light industry output	0	0.972655	Gross amount of exports	+ 10	0.560054
Gross retail amount	− 6	0.948684	Total production of energy	− 6	0.555035
Cash payout as salaries	+ 1	0.877090	Direct investments by foreign entities	− 11	0.510016
National expenditure	+ 12	0.800533	Loans issued by financial institutions	− 5	0.508375
Net amount of currency in circulation	+ 8	0.796688	Savings at financial institutions	− 6	0.505588
Total investment in real estate	− 11	0.769778	Consumer price index	+ 11	0.503235

Table 5.6 Classifications of leading, synchronic, and stagnant indices

	Leading index	Synchronic index	Stagnant index
Energy and raw materials	Total production of energy (−6)[a]		
Investment	Total investment in real estate (−11)		
Production		Increase in light industry output (0) Increase in heavy industry output (0)	
Finance			National income (+12) National expenditure (+12)
Currency and credit	Currency in circulation (−6) Savings at financial institutions (−6) Loans issued by financial institutions (−5)	Cash payout as salaries (+1)	Net amount of currency in circulation (+8)
Consumption	Gross retail amount (−6)		
Foreign trade	Gross amount of imports (−9) Direct investments by foreign entities (−11)		Gross amount of exports (+10)
Commodity price			Consumer price index (+11)

[a] Numbers in parentheses stand for the time difference between indicated cycles and reference cycles

this paper after 40 years of thinking. People try to motivate knowledge producers through awards. Hundreds of scientific prizes are awarded throughout the world and across all scientific disciplines. Although these prizes are frequently awarded with the explicit goal of inspiring more and better scientific work (Scotchmer, 2006) But a question remains: what kind of incentive effect does these prizes have produced (Rosen, 1986)?

Fields Medal is recognized internationally as the highest academic award project in the field of mathematics. Mathematicians all over the world are proud to win the Fields Medal. Because there is no mathematics award in the Nobel Prize for natural science, Fields Medal is also known as the "Nobel Prize in mathematics".

In 1932, according to the proposal of the Canadian mathematician John Charles Fields, the 9th International Conference of mathematicians held in Zurich decided to establish an international mathematics award named after his surname—Fields Medal.

Fields Medal is awarded every four years. The award ceremony is held at the Quadrennial International Conference of mathematicians hosted by the International Mathematical Federation. Each time, it is awarded to 2–4 young mathematicians with outstanding contributions. Winners will receive a bonus of 15,000 Canadian dollars and a gold medal. According to the award rules, Fields Medal is only awarded to mathematicians under the age of 40 on January 1, the year of award.

Fields Medal was first awarded in 1936. By 2018, a total of 60 mathematicians in the world had won Fields Medal.

As a prestigious World Award, Fields Medal has played an important role in attracting a large number of talented young scholars to participate in mathematical research and solve the world's mathematical problems.

Unlike the Nobel Prize, Fields Medal is awarded only to mathematicians under the age of 40. Mathematicians over the age of 40, no matter how much academic achievements they have made, are not eligible for Fields Medal. If there are a large number of mathematicians who have made greater contributions than the winners and can not win the prize only because of their age, the fairness of such a "grand prize" is obviously debatable.

For those scholars who won Fields Medal, what effect does the award have on their research work?

In 2015, George J. Borjas and Kirk B. Doran with University of Notre Dame conducted an in-depth study on the effect of Fields Medal. They selected 142 mathematicians at first, including all 56 Fields Medal winners (Medalists) at that time and 86 mathematicians in the control group (Contenders). Then collected the data of the published academic papers and other relevant data every year from the beginning of their academic career to the age of 60. Trying to analyze the impact of Fields Medal on the research output of the winners according to the actual data (Borjas & Doran, 2015).

The 86 mathematicians of contenders are all the winners of other prestigious mathematics awards. Such as the Abel Prize and the Wolf Prize. Other important awards are issued by the American Mathematical Society which including the Cole Prize for algebra, the Bôcher Prize for mathematical analysis, the Veblen Prize for Geometry, and Salem Prize for Fourier Series. Most of the winners of Fields Medal were won the above awards at first, and then won their Fields Medal. Therefore, it can be said that the 86 mathematicians in the control group are scholars who have the strength to participate competition for Fields Medal and finally fail to win Fields Medal.

We divide the sequences of annual average number of papers published by the Medalists and the Contenders into two parts: 16 years before the award and 20 years after the award. The data sequences of annual average number of papers published by Medalists and Contenders for 16 years before the award are denoted by X_M, X_C respectively. And the data sequences of annual average number of papers published by the Medalists and the Contenders for 20 years after the award are denoted by Y_M, Y_C respectively.

Calculate the three term center moving average smoothing sequence of X_M, X_C; Y_M, Y_C. Still denoted by X_M, X_C; Y_M, Y_C as before.

$$X_M = (x_M(1), x_M(2), \ldots, x_M(16))$$
$$= (2.64, 2.63, 2.70, 2.69, 2.80, 2.91, 3.01, 3.09, 3.38, 3.63$$
$$, 3.81, 3.83, 3.95, 3.75, 3.85, 3.62)$$

$$X_C = (x_C(1), x_C(2), \ldots, x_C(16))$$
$$= (1.95, 2.11, 2.32, 2.72, 3.03, 3.48, 3.51, 3.61, 3.60, 3.50$$
$$, 3.59, 3.75, 4.02, 4.10, 4.02, 4.00)$$

$$Y_M = (y_M(1), y_M(2), \ldots, y_M(20))$$
$$= (3.72, 3.50, 3.51, 3.35, 3.15, 2.90, 2.95, 2.92, 3.02, 3.01, 3.02, 3.10, 3.25,$$
$$3.30, 3.40, 3.35, 3.33, 2.96, 2.72, 2.60)$$

$$Y_C = (y_C(1), y_C(2), \ldots, y_C(20))$$
$$= (3.95, 3.90, 4.20, 4.40, 4.50, 4.53, 4.48, 4.46, 4.01, 4.54, 4.75, 4.72, 4.49,$$
$$4.23, 4.50, 4.62, 4.91, 4.95, 5.24, 5.49).$$

Calculate the zero-starting point sequences of $X_M, X_C; Y_M, Y_C,$

$$X_M^0 = (0, -0.01, 0.06, 0.05, 0.16, 0.27, 0.37, 0.45, 0.74, 0.99,$$
$$1.17, 1.19, 1.31, 1.11, 1.21, 0.98)$$
$$X_C^0 = (0, 0.16, 0.37, 0.77, 1.08, 1.53, 1.56, 1.66, 1.65, 1.55,$$
$$1.64, 1.80, 2.07, 2.15, 2.07, 2.05)$$
$$Y_M^0 = (0, -0.22, -0.21, -0.37, -0.57, -0.82, -0.77, -0.80,$$
$$- 0.70, -0.71, -0.70, -0.62,$$
$$- 0.47, -0.42, -0.32, -0.37, -0.39, -0.76, -1.00, -1.12)$$

$$Y_C^0 = (0, -0.05, 0.25, 0.45, 0.55, 0.58, 0.53, 0.51, 0.06,$$
$$0.59, 0.80, 0.77, 0.54, 0.28, 0.55, 0.67, 0.96, 1.00, 1.29, 1.07)$$

From Definition 2, we have.

$$s_{X_M} = 9.56, \; s_{X_C} = 21.085.$$

and

$$s_{Y_M} = -10.78, \; s_{Y_C} = 10.865.$$

Therefore, both X_M and X_C are all increasing sequences. Y_M and Y_C are reverse sequences.

By Formula (5), we have

$$\phi_{Y_M Y_C}^N = -\frac{\left|s_{Y_M} - s_{Y_C}\right|}{1 + \left|s_{Y_M} - s_{Y_C}\right|} \approx -0.96$$

Before the award, the average annual number of papers published by both the Medalists and the Contenders are all showed increasing trend, and the number was roughly the same. After winning the award, the average annual number of papers published by the Contenders is still an increase sequence, while the average annual number of papers published by the Medalists is an attenuation sequence, and the two show a strong inverse relation. The results clearly reveals that the "highest award" won by researchers in their prime of life has a significant reverse incentive effect on their research output.

At the same time, George J. Borjas and Kirk B. Doran are collected and analyzed the relevant data that can reflect the research "quality" of the Medalists and the Contenders. They found that from the data such as the citation of the papers, the quality of research work of the winners of Fields Medal were also significantly reduced (Borjas & Doran, 2015).

Borjas and Doran's research further shows that, compared with the Contenders, more the Medalists changed their research direction after winning the prize. The Medalists are usually not as worried about the "failure" of the research as before. Therefore, the proportion of those who change the research direction in the Medalists is significantly higher than that in the Contenders (Borjas & Doran, 2015).

The Fields Medal not only won the winners social reputation and respect, but also produced a huge wealth effect. Many academic institutions have hired or hope to hire the winners of Fields Medal with high salaries, giving them more opportunities and choices. After winning the prize, some people began to "Revel in being sought after" and "To play the game of life", giving up their previous academic pursuit (Borjas & Doran, 2015) This may be one of the reasons for the reverse incentive effect.

Some people say, "small awards inspire people to forge ahead, and big awards stop people" maybe it's not unreasonable.

References

Borjas, G. J., & Doran, K. B. (2015). Prizes and productivity: How winning the fields medal affects scientific output. *Journal of Human Resources, 50*(3), 728–758.

Chen, K. J., & Liu, S. F. (2005). Time difference analysis of economic indicators based on grey relational model. In *Proceedings of IEEE International Conference on Networking, Sensing and Control*. 311–342.

Dang, Y. G., & Liu, S. F. (2004). The GM models that x(n) is taken as initial value. *Kybernetes: The International Journal of Systems & Cybernetics, 33* (2), 247–254.

Deng, J. L. (1985). *Grey control systems*. Press of Huazhong University of Science and Technology.

Liu, S. F. (1991). The three axioms of buffer operator and their application. *Journal of Grey System, 3*(1), 39–48.

Liu, S. F. (2021). *Grey system theory and its application*, (9th ed.). Beijing: Science Press.

Liu, S. F. (2022). Negative grey relational model and measurement of the reverse incentive effect of Fields Medal. Grey systems: Theory and application. Published online.

Liu, S. F., Yang, Y. J., & Cao, Y., et al. (2013). A summary on the research of GRA models. *Grey systems: theory and application, 3*(1), 7–15.

Liu, S. F., Yang, Y. J., Xie, N. M., & Forrest, J. (2016). New progress of grey system theory in the new millennium. *Grey Systems theory and application, 6*(1), 2–31.

Rosen, S. (1986). Prizes and incentives in elimination tournaments. *American Economic Review, 76*, 701–715.

Scotchmer, S. (2006). *Innovation and incentives. Cambridge.* MA: MIT Press.

Shi, B. Z. (1995). Presentation of range incidence quantity. *Journal of the University of Petroleum, China, 19*(5), 114–116.

Zhao, Y. L., Wei, S. Y., Mei, Z. X. (1998b) A new theoretical model for analysis of grey relation. *Systems Engineering and Electronics, 9*(10), 34–36.

Zhao, H., Ma, L. Y., & Jia, Q. (2007). The theory and application of grey coincidence analysis model based on variant coefficient. *Heilongjiang Science and Technology of Water Conservancy, 35*(2), 26–27.

Zhang, Q. S., Guo, X. J., & Deng, J. L. (1996). The entropy analysis method of grey incidences. *Systems Engineering: Theory and Practice, 16*(8), 7–11.

Zhang, Q. Y., Zhou, X. H., & Wang, W. T. (2007). Evaluation of defending efficiency in engineering based on improved grey interrelated analysis method. *Journal of PLA University of Science and Technology (Natural Science Edition), 8*(3), 283–287.

Zhou, T., & Peng, C. H. (2008). Application of grey model on analyzing the passive natural circulation residual heat removal system of HTR-10. *Nuclear Science and Techniques, 19*(5), 308–313.

Chapter 6
Grey Clustering Evaluation Models

6.1 Introduction

There are two kinds grey clustering models. One is based on grey relational degree, mainly clustering indicators. The other is based on possibility functions, mainly used to classify objects. When investigating practical problems, it is often the case that each observational object possesses quite a few characteristic indices, which are difficult to accurately classify. Depending on the objects to be clustered, grey clustering can be based on two methods: clustering using GRA models, and clustering using grey possibility functions. The first method is mainly applied to group the same kinds of factors into their individual categories, so that a complicated system can be simplified. By using the clustering method of grey relational analysis, we can examine whether or not some of the factors under consideration really belong to the same kind. This allows a synthetic index of these factors, or one of these factors, to be used to represent all factors without losing any part of the available information carried by such factors. This problem regards the selection of variables to be used in the study of a system. Before conducting a large-scale survey, which generally costs a lot of money and man power, by using the clustering method of grey relational analysis on a typical sample data, one can reduce the amount of data collection to a minimal level by eliminating the unnecessary variables so that tangible savings can be achieved.

The clustering method based on grey possibility functions is mainly used for checking whether or not the observational objects belong to pre-determined classes so that they can be treated differently. In practice, we need to set the possibility functions and the weights for different criterion according to the corresponding clustering index and the grey classes we intend to partition if using the clustering method based on grey possibility functions.

Grey clustering evaluation models using possibility functions are used widely for uncertain systems analysis. For the past four decades, much research on modeling techniques has been done, and new research results emerge constantly. For example, Professor Julong Deng has proposed the variable weight grey clustering model (Deng,

S. Liu, *Grey Systems Analysis*, Series on Grey System,
https://doi.org/10.1007/978-981-19-6160-1_6

1985), while Professor Sifeng Liu et al. has proposed the fixed weight grey clustering evaluation model (Liu, 1993), the grey clustering evaluation model using end-point triangular possibility functions (Liu, 1991; Liu & Zhu, 1993), the grey cluster evaluation model using center-point triangular possibility functions (Liu & Xie, 2011), among others. These models are all used widely. Grey variable weight clustering model is applicable to the problems with criteria that have the same meanings and dimensions. When the criteria for clustering involve different meanings and dimensions, the fixed weight grey clustering evaluation model and grey clustering evaluation model using triangular possibility functions are suitable. In particular, compared with the variable weight grey clustering and fixed weight grey clustering models, the grey clustering evaluation model using triangular possibility functions is more suitable for problems of poor information clustering evaluation. The grey clustering evaluation model using mixed end-point triangular possibility functions is suitable for situations where all grey boundaries are clear, but where the most likely points belonging to each grey class are unknown. Conversely, the grey clustering evaluation model using mixed center-point triangular possibility functions is suitable for problems where it is easy to judge the most likely points belonging to each grey class, but where the grey boundaries are unclear (Liu et al., 2015a). Additionally, both of the last two grey clustering evaluation models based on mixed possibility function which composed by the possibility function of moderate measure, the possibility function of lower measure, and the possibility function of upper measure (Liu et al., 2015a, 2017; Liu, 2021).

Further, Dong et al. (2010), Pei et al. (2012), Xiao (1997), Xiong & Chen (1999), Xu et al. (2006), and others are improved and optimized grey clustering evaluation models from different perspectives. Furthermore, Zhang (2002) has studied the measurement problem of Grey Characteristics of Grey Clustering Result. The author has investigated the relation between a grey clustering analysis result and the entropy of the weight sequence, and proposed a measure method for the grey characteristics of a grey clustering analysis result.

In this chapter, two novel grey cluster evaluation models based on mixed center-point triangular possibility functions and mixed end-point triangular possibility functions are put forward. These new grey clustering models based on mixed possibility functions are especially applicable to evaluation and classification of poor information objects, and have broad application prospects.

6.2   Grey Relational Clustering Model

Definition 6.2.1 Assume that there are n observational objects. For each object the data of m attribute indexes are collected, producing the following sequences:

$$X_1 = (x_1(1), x_1(2), \ldots, x_1(n))$$
$$X_2 = (x_2(1), x_2(2), \ldots, x_2(n))$$

$$\cdots\cdots\cdots\cdots\cdots\cdots\cdots\cdots\cdots\cdots\cdots$$

$$X_m = (x_m(1), x_m(2), \ldots, x_m(n))$$

Then, for all $i < j, i, j = 1, 2, \ldots, m$, calculate ε_{ij}, the absolute grey relational degree between X_i and X_j, so that we have the following upper triangular matrix A:

$$A = \begin{pmatrix} \varepsilon_{11} & \varepsilon_{12} & \cdots & \varepsilon_{11} \\ & \varepsilon_{22} & \cdots & \varepsilon_{2m} \\ & & \ddots & \vdots \\ & & & \varepsilon_{mm} \end{pmatrix}$$

A is referred to as the grey relational matrix of the attribute indexes, where $\varepsilon_{ii} = 1$, $i = 1, 2, \ldots, m$. For a chosen threshold value $r \in [0, 1]$, which in general satisfies $r > 0.5$, if $\varepsilon_{ij} \geq r, i \neq j$, the variables X_j and X_i are seen as the same attribute.

Definition 6.2.2 The classification of the attribute indexes with the chosen value r is referred to as the r- classification by grey relational degree.

When studying a specific problem, the particular value r is determined based on the circumstances involved. The closer the r is to 1, the finer the classification and the fewer the variables in each class. Conversely, the smaller the r, the coarser the classification and the greater the number of variables in each class.

Example 6.2.1 The talent search committee of a firm has proposed 15 candidate recruitment criteria as follows:

1. Impression of overall application package;
2. Academic abilities;
3. Likability by others;
4. Level of self-confidence;
5. Intelligence;
6. Honesty;
7. Ability to sell;
8. Experience;
9. Motivation;
10. Ambition;
11. Presentation skills;
12. Ability to comprehend instructions;
13. Potential for future growth;
14. Interpersonal skills; and
15. Adaptability.

Members of the committee admit that some of these 15 criteria can overlap and hope that through the study of a sample of a few data points, these 15 criteria can be classified into fewer categories. By using the scoring method to quantify the criteria,

Table 6.1 The scores of 9 observational objects

Attributes	Objects								
	1	2	3	4	5	6	7	8	9
X_1	6	9	7	5	6	7	9	9	9
X_2	2	5	3	8	8	7	8	9	7
X_3	5	8	6	5	8	6	8	8	8
X_4	8	10	9	6	4	8	8	9	8
X_5	7	9	8	5	4	7	8	9	8
X_6	8	9	9	9	9	10	8	8	8
X_7	8	10	7	2	2	5	8	8	5
X_8	3	5	4	8	8	9	10	10	9
X_9	8	9	9	4	5	6	8	9	8
X_{10}	9	9	9	5	5	5	10	10	9
X_{11}	7	10	8	6	8	7	9	9	9
X_{12}	7	8	8	8	8	8	8	9	8
X_{13}	5	8	6	7	8	6	9	9	8
X_{14}	7	8	8	6	7	6	8	9	8
X_{15}	10	10	10	5	7	6	10	10	10

9 observational objects have been scored according to each of the criteria. Table 6.1 gives the scores, where Oi stands for the ith object, $i = 1, 2, ..., 9$.

To calculate the absolute grey relational degree of ε_{ij} of X_i and X_j for all $i \leq j$, $i, j = 1, 2, ... 15$, we obtained the upper triangular matrix A as shown in Table 6.2.

We divided the 15 criteria into different classes based on Table 6.2, where the value of threshold r can be different based on the requirements involved. For example, if we take $r = 1$, all 15 criteria above belong to their own classes with each in its own class. If we take $r = 0.80$, then we check the values in Table 6.2, row by row, and pick out all the values of ε_{ij} which are greater than 0.80. Thus, we have:

$$\varepsilon_{1,3} = 0.88, \quad \varepsilon_{1,11} = 0.90, \quad \varepsilon_{1,12} = 0.88, \quad \varepsilon_{1,13} = 0.80, \quad \varepsilon_{2,8} = 0.99$$
$$\varepsilon_{3,11} = 0.80, \quad \varepsilon_{3,13} = 0.90, \quad \varepsilon_{6,11} = 0.84, \quad \varepsilon_{6,12} = 0.86, \quad \varepsilon_{6,14} = 0.81$$
$$\varepsilon_{7,10} = 0.83, \quad \varepsilon_{7,15} = 0.89, \quad \varepsilon_{9,10} = 0.81, \quad \varepsilon_{10,15} = 0.92, \quad \varepsilon_{11,12} = 0.97$$

Therefore, we know that X_3, X_{11}, X_{12}, and X_{13} belong to the same class as X_1; X_8 belong to the same class as X_2; X_{11} and X_{13} belong to the same class as X_3; X_{11}, X_{12}, and X_{14} belong to the same class as X_6; X_{10} and X_{15} belong to the same class as X_7; X_{10} belong to the same class as X_9; X_{15} belong to the same class as X_{10}; and X_{12} belong to the same class as X_{11}.

Let each class be represented with the criterion with the minimum index contained in the class, and combine the classes containing X_6 and X_{11}, respectively, with the

Table 6.2 The grey relational matrix of attribute indexes

	X_1	X_2	X_3	X_4	X_5	X_6	X_7	X_8	X_9	X_{10}	X_{11}	X_{12}	X_{13}	X_{14}	X_{15}
X_1	1	0.66	0.88	0.52	0.58	0.77	0.51	0.66	0.51	0.51	0.9	0.88	0.8	0.67	0.51
X_2		1	0.72	0.51	0.53	0.59	0.5	0.99	0.51	0.51	0.63	0.62	0.77	0.55	0.51
X_3			1	0.56	0.7	0.51	0.72	0.51	0.51	0.51	0.8	0.78	0.9	0.63	0.51
X_4				1	0.56	0.53	0.58	0.51	0.69	0.62	0.52	0.52	0.51	0.54	0.6
X_5					1	0.65	0.51	0.53	0.53	0.52	0.61	0.61	0.55	0.75	0.52
X_6						1	0.51	0.59	0.52	0.52	0.84	0.86	0.66	0.81	0.51
X_7							1	0.5	0.7	0.83	0.51	0.51	0.51	0.51	0.89
X_8								1	0.51	0.51	0.63	0.62	0.77	0.55	0.51
X_9									1	0.81	0.0.52	0.52	0.51	0.53	0.76
X_{10}										1	0.51	0.51	0.51	0.52	0.92
X_{11}											1	0.97	0.74	0.71	0.51
X_{12}												1	0.73	0.72	0.51
X_{13}													1	0.6	0.51
X_{14}														1	0.52
X_{15}															1

class containing X_1. Put X_9 and X_{10} into the class containing X_7, and treat X_4 and X_5 as individual classes. Then, we have obtained a classification of the 15 attribute criteria for our shortened list as follows:

$$\{X_1, X_3, X_6, X_{11}, X_{12}, X_{13}, X_{14}\}, \{X_2, X_8\}, \{X_4\}, \{X_5\},$$
$$\{X_7, X_9, X_{10}, X_{15}\}$$

Here, the class of $\{X_1, X_3, X_6, X_{11}, X_{12}, X_{13}, X_{14}\}$ including the attribute criteria such as impression of overall application package, likability by others, honesty, presentation skills, ability to comprehend instructions, potential for future growth, and interpersonal skills, all of which direct impression, can be obtained through the application form or interviews. These attribute criteria can be replaced by one synthetic impression attribute criterion because all these attribute criteria correlate and it is difficult to be separate them completely. The class of $\{X_2, X_8\}$ includes two attribute criteria, namely academic abilities and experience, which can be evaluated through investigation and understanding of the academic research and practical work accomplished by the candidate. The class of $\{X_7, X_9, X_{10}, X_{15}\}$ includes four attribute criteria, namely ability to sell, motivation, ambition, and adaptability, which can be judged synthetically by investigating the learning and working background of the candidate. Special investigation is required for assessment of the attribute criterion level of self-confidence of $\{X_4\}$, and the attribute criterion intelligence of $\{X_5\}$.

6.3 Common Possibility Functions

The variable weight grey clustering model, the fixed weight grey clustering evaluation model, the grey clustering evaluation model using end-point and center-point triangular possibility functions, and the grey clustering evaluation model based on mixed possibility functions are all grey clustering evaluation models based on different possibility functions. Therefore, the four kinds of common possibility functions are explained in this section (Deng, 1985; Liu, 2021; Liu et al., 2017).

The possibility function of the jth criterion about the kth class is denoted by $f_j^k(\bullet)$, $j = 1, 2, \ldots m, \quad k = 1, 2, \ldots s$.

Definition 6.3.1 Assume that the possibility function $f_j^k(\bullet)$ of the jth criterion about kth class is a trapezoidal function shown in Fig. 6.1. Then $f_j^k(\bullet)$ is referred to as possibility function of typical form, and $x_j^k(1)$, $x_j^k(2)$, $x_j^k(3)$, and $x_j^k(4)$ are referred to as turning points of $f_j^k(\bullet)$.

The possibility function of typical form is denoted by $f_j^{\,k}[x_j^k(1), x_j^k(2), x_j^k(3), x_j^k(4)]$.

Definition 6.3.2 Assume that the possibility function $f_j^k(\bullet)$ of the jth criterion about kth class does not have the first and second turning points $x_j^k(1)$ and $x_j^k(2)$, as shown in Fig. 6.2. Then $f_j^k(\bullet)$ is referred to as the possibility function of lower measure.

The possibility function of lower measure is denoted by $f_j^{\,k}[-, -, x_j^k(3), x_j^k(4)]$.

Fig. 6.1 The possibility function of typical form

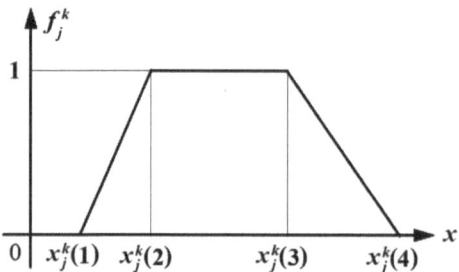

Fig. 6.2 The possibility function of lower measure

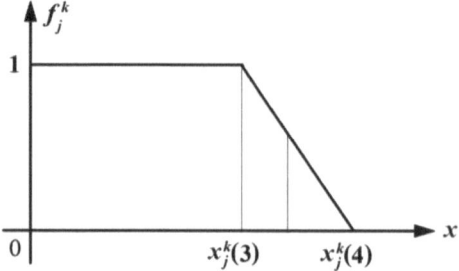

Fig. 6.3 The possibility function of moderate measure

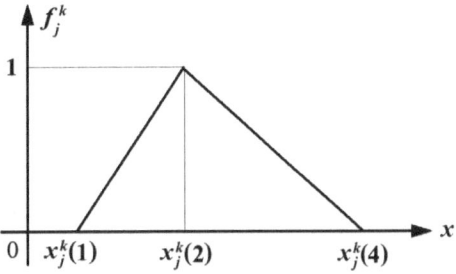

Definition 6.3.3 Assume that the possibility function $f_j^k(\bullet)$ of the jth criterion about kth class does not have the third turning point $x_j^k(3)$, or that the second and third turning points $x_j^k(2)$ and $x_j^k(3)$ of $f_j^k(\bullet)$ coincide, as shown in Fig. 6.3. In this case, $f_j^k(\bullet)$ is referred to as a possibility function of moderate measure, or a triangular possibility function. The possibility function of moderate measure, or triangular possibility function, is denoted by $f\,_j^k[x_j^k(1), x_j^k(2), -, x_j^k(4)]$.

Definition 6.3.4 Assume that the possibility function $f_j^k(\bullet)$ of the jth criterion about kth class does not have turning points $x_j^k(3)$ and $x_j^k(4)$, as shown in Fig. 6.4. Function $f_j^k(\bullet)$ is then referred to as a possibility function of upper measure. The possibility function of upper measure is denoted by $f\,_j^k[x_j^k(1), x_j^k(2), -, -]$.

Proposition 6.3.1

(1) *For the possibility function of typical form as shown in* Fig. 6.1, *we have:*

$$
f_j^k(x) = \begin{cases}
0 & x \notin [x_j^k(1), x_j^k(4)] \\
\frac{x-x_j^k(1)}{x_j^k(2)-x_j^k(1)} & x \in [x_j^k(1), x_j^k(2)] \\
1 & x \in [x_j^k(2), x_j^k(3)] \\
\frac{x_j^k(4)-x}{x_j^k(4)-x_j^k(3)} & x \in [x_j^k(3), x_j^k(4)]
\end{cases}
\tag{6.1}
$$

(2) *For the possibility function of lower measure as shown in* Fig. 6.2, *we have:*

Fig. 6.4 The possibility function of upper measure

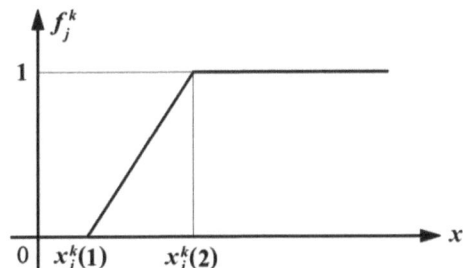

$$f_j^k(x) = \begin{cases} 0 & x \notin [0, x_j^k(4)] \\ 1 & x \in [0, x_j^k(3)] \\ \frac{x_j^k(4)-x}{x_j^k(4)-x_j^k(3)} & x \in [x_j^k(3), x_j^k(4)] \end{cases} \tag{6.2}$$

(3) *For the possibility function of moderate measure as shown in* Fig. 6.3, *we have:*

$$f_j^k(x) = \begin{cases} 0 & x \notin [x_j^k(1), x_j^k(4)] \\ \frac{x-x_j^k(1)}{x_j^k(2)-x_j^k(1)} & x \in [x_j^k(1), x_j^k(2)] \\ \frac{x_j^k(4)-x}{x_j^k(4)-x_j^k(2)} & x \in [x_j^k(2), x_j^k(4)] \end{cases} \tag{6.3}$$

(4) *For the possibility function of upper measure as shown in* Fig. 6.4, *we have:*

$$f_j^k(x) = \begin{cases} 0, & x < x_j^k(1) \\ \frac{x-x_j^k(1)}{x_j^k(2)-x_j^k(1)}, & x \in [x_j^k(1), x_j^k(2)] \\ 1, & x \geq x_j^k(2) \end{cases} \tag{6.4}$$

6.4 Variable Weight Grey Clustering Model

Definition 6.4.1 Assume that there are n objects to be classified according to m criteria into s different grey classes. Classifying the ith object into the kth grey class according to the observed value of the ith object judged against the jth criterion, x_{ij}, $i = 1, 2, \ldots n$, $j = 1, 2, \ldots m$, is called grey clustering (Deng, 1985).

Definition 6.4.2

(1) For the possibility function of typical form as shown in Fig. 6.1, let $\lambda_j^k = \frac{1}{2}(x_j^k(2) + x_j^k(3))$.
(2) For the possibility function of lower measure as shown in Fig. 6.2, let $\lambda_j^k = x_j^k(3)$.
(3) For the possibility function of moderate measure as shown in Fig. 6.3 and the possibility function of upper measure as shown in Fig. 6.4, let $\lambda_j^k = x_j^k(2)$.

Then λ_j^k is referred to as the basic value of the jth criterion about the kth class.

Definition 6.4.3 Assume that λ_j^k is the basic value of the jth criterion about the kth class. Then the following formula is referred to as the weight of the jth criterion about kth class (Deng, 1985):

$$\eta_j^k = \lambda_j^k / \sum_{j=1}^{m} \lambda_j^k \tag{6.5}$$

Definition 6.4.4 Assume that x_{ij}, $i = 1, 2, \ldots n$, $\quad j = 1, 2, \ldots m$ is the observed value of object i with regard to the jth criterion, $f_j^k(\bullet)$ the possibility function and η_j^k the weight of the jth criterion about the kth class, with $j = 1, 2, \ldots m$, $\quad k = 1, 2, \ldots s$. Then the following Formula (6.6) is referred to as the grey clustering coefficient of variable weight for object i to belong to the kth grey class (Deng, 1985):

$$\sigma_i^k = \sum_{j=1}^{m} f_j^k(x_{ij}) \cdot \eta_j^k \tag{6.6}$$

Definition 6.4.5

(1) The following formula is referred to as the clustering coefficient vector of object i:

$$\sigma_i = (\sigma_i^1, \sigma_i^2, \ldots, \sigma_i^s) = \left(\sum_{j=1}^{m} f_j^1(x_{ij}) \cdot \eta_j^1, \sum_{j=1}^{m} f_j^2(x_{ij}) \cdot \eta_j^2, \ldots, \sum_{j=1}^{m} f_j^s(x_{ij}) \cdot \eta_j^s \right)$$

(2) The following matrix is referred to as the cluster coefficient matrix:

$$\sum = (\sigma_i^k) = \begin{bmatrix} \sigma_1^1 & \sigma_1^2 & \cdots & \sigma_1^s \\ \sigma_2^1 & \sigma_2^2 & \cdots & \sigma_2^s \\ \vdots & \vdots & \ddots & \vdots \\ \sigma_n^1 & \sigma_n^2 & \cdots & \sigma_n^s \end{bmatrix}$$

Definition 6.4.6 If $\max_{1 \le k \le s} \{\sigma_i^k\} = \sigma_i^{k^*}$, then it is called object i belongs to grey class k^*.

The variable weight clustering method is used to study problems with criteria that have the same meanings and units. Otherwise, it is not appropriate to employ this method. Also, if the numbers of observed values of individual criteria are greatly different from each other, this clustering method should not be applied.

Example 6.4.1 Assume that we are interested in the study of three economic districts with the added value by the primary, secondary and tertiary industries as the three cluster criteria. The observational values x_{ij}, $i = 1, 2, 3; j = 1, 2, 3$, of the ith economic district with respect to the jth criterion is given in the following matrix A, where the unit of the three criteria is same as a hundred million RMB:

$$A = (x_{ij}) = \begin{bmatrix} x_{11} \ x_{12} \ x_{13} \\ x_{21} \ x_{22} \ x_{23} \\ x_{31} \ x_{32} \ x_{33} \end{bmatrix} = \begin{bmatrix} 80 \ 20 \ 100 \\ 40 \ 30 \ 30 \\ 10 \ 90 \ 60 \end{bmatrix}$$

Please try to perform a synthetic clustering based on high, medium, and low added values.

Solution Assume that the possibility functions $f_j^k(\bullet)$ for the jth criterion about the kth class are as follows:

$$f_1^1[30, 80, -, -], \quad f_1^2[10, 40, -, 70], \quad f_1^3[-, -, 10, 30]$$
$$f_2^1[30, 90, -, -], \quad f_2^2[20, 50, -, 90], \quad f_2^3[-, -, 20, 40]$$
$$f_2^1[40, 100, -, -], \quad f_3^2[30, 60, -, 90], \quad f_3^3[-, -, 30, 50]$$

It follows that:

$$f_1^1(x) = \begin{cases} 0, & x < 30 \\ \frac{x-30}{80-30}, & 30 \le x < 80 \; ; \\ 1, & x > 80 \end{cases} \quad f_1^2(x) = \begin{cases} 0, & x \notin [10, 70] \\ \frac{x-10}{40-10}, & 10 \le x < 40 \\ \frac{70-x}{70-40}, & 40 \le x < 70 \end{cases}$$

$$f_1^3(x) = \begin{cases} 0, & x \notin [0, 30] \\ 1, & 0 \le x < 10 \; ; \\ \frac{30-x}{30-10}, & 10 \le x < 30 \end{cases} \quad f_2^1(x) = \begin{cases} 0, & x < 30 \\ \frac{x-30}{90-30}, & 30 \le x < 90 \\ 1, & x > 90 \end{cases}$$

$$f_2^2(x) = \begin{cases} 0, & x \notin [20, 90] \\ \frac{x-20}{50-20}, & 20 \le x < 50 \; ; \\ \frac{90-x}{90-50}, & 50 \le x < 90 \end{cases} \quad f_2^3(x) = \begin{cases} 0, & x \notin [0, 40] \\ 1, & 0 \le x < 20 \\ \frac{40-x}{40-20}, & 20 \le x < 40 \end{cases}$$

$$f_3^1(x) = \begin{cases} 0, & x < 40 \\ \frac{x-40}{100-40}, & 40 \le x < 100 \; ; \\ 1, & x > 100 \end{cases} \quad f_3^2(x) = \begin{cases} 0, & x \notin [30, 90] \\ \frac{x-30}{50-30}, & 30 \le x < 50 \\ \frac{90-x}{90-50}, & 50 \le x < 90 \end{cases}$$

$$f_3^3(x) = \begin{cases} 0, & x \notin [0, 50] \\ 1, & 0 \le x < 30 \\ \frac{50-x}{50-30}, & 30 \le x < 50 \end{cases}$$

Therefore:

$$\lambda_1^1 = 80, \lambda_2^1 = 90, \lambda_3^1 = 100, \lambda_1^2 = 40,$$
$$\lambda_2^2 = 50, \lambda_3^2 = 60, \lambda_1^3 = 10, \lambda_2^3 = 20, \lambda_3^3 = 30$$

By $\eta_j^k = \frac{\lambda_j^k}{\sum_{j=1}^{3} \lambda_j^k}$ we have:

$$\eta_1^1 = \frac{80}{270}, \eta_2^1 = \frac{90}{270}, \eta_3^1 = \frac{100}{270}, \eta_1^2 = \frac{40}{150},$$

$$\eta_2^2 = \frac{50}{150}, \eta_3^2 = \frac{60}{150}, \eta_1^3 = \frac{10}{60}, \eta_2^3 = \frac{20}{60}, \eta_3^3 = \frac{30}{60}$$

Thus, from $\sigma_i^k = \sum_{j=1}^m f_j^k(x_{ij}) \cdot \eta_j^k$, when $i = 1$ for economic district 1, we have:

$$\sigma_1^1 = \sum_{j=1}^3 f_j^1(x_{1j}) \cdot \eta_j^1 = f_1^1(80) \times \frac{80}{270}$$

$$+ f_2^1(20) \times \frac{90}{270} + f_3^1(100) \times \frac{100}{270} = 0.6667$$

Similarly, we obtained the following:

$$\sigma_1^2 = 0, \sigma_1^3 = 0.3333$$

Therefore, $\sigma_1 = (\sigma_1^1, \sigma_1^2, \sigma_1^3) = (0.6667, 0, 0.3333)$.

Similarly, we can calculate the clustering coefficient vector for economic districts 2 and 3 as done for economic district 1.

When $i = 2$, $\sigma_2 = (\sigma_2^1, \sigma_2^2, \sigma_2^3) = (0.0593, 0.3778, 0.6667)$.

When $i = 3$, $\sigma_3 = (\sigma_3^1, \sigma_3^2, \sigma_3^3) = (0.4667, 0.4, 0.1667)$.

The clustering coefficient matrix is as follows:

$$\sum = (\sigma_i^k) = \begin{bmatrix} \sigma_1^1 & \sigma_1^2 & \sigma_1^3 \\ \sigma_2^1 & \sigma_2^2 & \sigma_2^3 \\ \sigma_3^1 & \sigma_3^2 & \sigma_3^3 \end{bmatrix} = \begin{bmatrix} 0.6667 & 0 & 0.3333 \\ 0.0593 & 0.3778 & 0.6667 \\ 0.4667 & 0.4 & 0.1667 \end{bmatrix}$$

From $\max_{1 \le k \le 3} \{\sigma_1^k\} = \sigma_1^1 = 0.6667$, $\max_{1 \le k \le 3} \{\sigma_2^k\} = \sigma_2^3 = 0.6667$, $\max_{1 \le k \le 3} \{\sigma_3^k\} = \sigma_3^1 = 0.4667$, it follows that the second economic district belongs to the low grey class of added value, and the first and third economic districts belong to the high grey class of added value. Furthermore, from the cluster coefficients $\sigma_1^1 = 0.6667$ and $\sigma_3^1 = 0.4667$, it follows that there still exists some differences between the first and third districts, even though both belong to the high grey class of added value. If the grey classes of added value are refined, that is, if we use five grey classes such as high, mid-high, medium, mid-low, and low added value, then different results can be obtained.

Furthermore, to determine the possibility function for the jth criterion about the kth class, it is generally possible to use the background information of the problem at hand. When resolving practical problems, one can determine the possibility functions from either the angle of the objects that are to be clustered or by looking at all the same type objects in the whole system, not just the ones involved in the clustering. For example, in Example 6.4.1, we could determine the possibility functions not only from the three economic districts in question, but also from the same level of economic districts in a city, a province, or from around the nation. Therefore, the results of grey clustering evaluation can only be applied to a certain range, which is the same as the one used in the determination of relevant possibility functions.

6.5 Fixed Weight Grey Clustering Model

When the criteria for clustering have different meanings, dimensions (units), and drastically different numbers of observed data points, the variable weight clustering method will fail. Because the indexes with different meanings and dimensions do not meet the additivity. There are two ways to get around this problem. The first is to transform the sample of data values of all the criteria into non-dimensional values by applying either the initiating operator or the averaging operator, and then clustering the transformed data. When employing this method, all the criteria are treated equally so that no difference played by the criteria in the process of clustering is reflected. The second way to solve non additive problem is to assign each clustering criterion a weight ahead of the clustering process. In this section, we address this second method.

Definition 6.5.1 Assume that x_{ij} is the observed value of object i with regard to criterion j, $i = 1, 2, \ldots n$, $j = 1, 2, \ldots m$, and the possibility function $f_j^k(\bullet)$ of the jth criterion about kth class, $j = 1, 2, \ldots m$, $k = 1, 2, \ldots, s$. If the weight η_j^k of the jth criterion about the kth class is not a function of k, $j = 1, 2, \ldots m$, $k = 1, 2, \ldots, s$. That is, if for any $k_1, k_2 \in \{1, 2, \ldots, s\}$ we always have $\eta_j^{k_1} = \eta_j^{k_2}$, then the symbol η_j^k can be written as η_j, $j = 1, 2, \ldots m$, with the superscript k removed. In this case, the following Formula (6.7) is referred to as the fixed weight clustering coefficient for object i to belong to the kth grey class (Liu, 1993).

$$\sigma_i^k = \sum_{j=1}^m f_j^k(x_{ij})\eta_j \qquad (6.7)$$

Definition 6.5.2 In Formula (6.7), if $\eta_j = \frac{1}{m}$, for $j = 1, 2, \ldots, m$, then the following formula is referred to as the equal weight clustering coefficient for object i to belong to the kth grey class:

$$\sigma_i^k = \sum_{j=1}^m f_j^k(x_{ij}) \cdot \eta_j = \frac{1}{m} \sum_{j=1}^m f_j^k(x_{ij})$$

The method of clustering objects by using grey fixed weight clustering coefficients is known as grey fixed weight clustering. The method which uses grey equal weight clustering coefficients is known as grey equal weight clustering.

Grey fixed weight clustering can be carried out according to the following steps:

Step 1: Determine the possibility function $f_j^k(\bullet)$ for the jth criterion about the kth class, $j = 1, 2, \ldots m, k = 1, 2, \ldots, s$.

Step 2: Determine a clustering weight η_j for each criterion $j = 1, 2, \ldots m$.

Step 3: Based on the possibility functions $f_j^k(\bullet)$ obtained in step 1, the clustering weights η_j obtained in step 2, and the observed data value x_{ij} of object i with respect to criterion j, calculate the fixed weight clustering coefficients $\sigma_i^k = \sum_{j=1}^m f_j^k(x_{ij})$, $i = 1, 2, \ldots n, j = 1, 2, \ldots m, k = 1, 2, \ldots, s$.

Step 4: If $\max\limits_{1 \le k \le s} \{\sigma_i^k\} = \sigma_i^{k^*}$, then it is called object i belongs to grey class k^*.

Example 6.5.1 Let us perform a grey clustering for the ecological adaptation of major strains of trees commercially used in China (Li et al., 1994). China is a huge country with a very diverse ecological environment, and different strains of trees obviously require different growing conditions. The area where a certain strain of trees has been growing reflects the adaptability of the strain to that particular ecological environment. We now classify ecological environmental conditions into four main quantification criteria:

(1) Geographical measure;
(2) Temperature measure;
(3) Precipitation measure; and
(4) Arid measure.

Here, geographical measure is an index representing the geographical width of the region in which the strain of trees grows. The numerical value of this measure is given by the product of differences of longitudes in the directions of east and west, and latitudes in the directions of south and north. The temperature measure indicates the adaptability of the strain of trees to various temperatures. Its numerical value is computed by using the difference of annual average temperatures of the southern and the northern bounds of the growing region. The precipitation measure is the adaptability of the trees to precipitation conditions. Its numerical value is recorded as the difference between the maximum and minimum annual average precipitation in all areas of the growing region. The arid measure is selected to describe a strain's adaptability to arid conditions in the atmosphere. Its value is the difference between the maximum and minimum annual average aridities in different areas of the growing region.

Statistics regarding the four measures for the 17 main strains of trees planted in China are given in Table 6.3.

With such data it is possible to carry out grey clustering based on wide adaptability, medium adaptability, and narrow adaptability.

Solution Because the meanings of the criteria are different and there exists much difference among the values observed, we must apply the fixed weight clustering method.

Table 6.3 The four measures of the 17 main strains of trees in China

Trees	Measure			
	Geo. eco measure	Temp. eco measure	Prec. eco measure	Arid eco measure
1 Camphor pine	22.50	4	0	0
2 Korean pine	79.37	6	600	0.75
3 Northeast China ash	144.00	7	300	0.75
4 Diversiform-leaved poplar	300.00	6.1	189	12.00
5 Sacsaoul	456.00	12	250	12.00
6 Chinese pine	189.00	8	700	1.5
7 Oriental arborvitae	369.00	8	1300	2.25
8 White elm	1127.11	16.2	550	3.00
9 Dryland willow	260.00	11	600	1.00
10 Chinese white poplar	200.00	8	600	1.25
11 Oak	475.00	10	1000	0.75
12 Huashan pine	314.10	8	900	0.75
13 Masson pine	282.80	7.4	1300	0.5
14 China fir	240.00	8	1200	0.5
15 Bamboo	160.00	5	1000	0.25
16 Camphor tree	270.00	8	1200	0.25
17 Southern Asian pine	9.00	1	200	0

Step 1: Assume that the possibility functions $f_j^k(\bullet)(j = 1, 2, 3, 4; \quad k = 1, 2, 3)$ for the jth criterion about the kth class are as follows:

$$f_1^1[100, 300, -, -], f_1^2[50, 150, -, 250], f_1^3[-, -, 50, 100]$$
$$f_2^1[3, 10, -, -], f_2^2[2, 6, -, 10], f_2^3[-, -, 15, 30]$$
$$f_3^1[200, 1000, -, -], f_3^2[100, 600, -, 1100], f_3^3[-, -, 300, 600]$$
$$f_4^1[0.25, 1.25, -, -], f_4^2[0, 0.5, -, 1], f_4^3[-, -, 0.25, 0.5]$$

Step 2: Let the weights for the geographical, temperature, precipitation, and aridity measures be:

$$\eta_1 = 0.3, \quad \eta_2 = 0.25, \quad \eta_3 = 0.25, \quad \eta_4 = 0.2$$

Step 3: Based on $\sigma_i^k = \sum_{j=1}^m f_j^k(x_{ij}) \cdot \eta_j; i = 1, 2, \ldots, 17; \quad k = 1, 2, 3$ and Table 6.3, when i = 1,

$$\sigma_1^1 = \sum_{j=1}^{4} f_j^1(x_{1j}) \cdot \eta_j = f_1^1(22.5) \times 0.3 + f_2^1(4) \times 0.25;$$

$$+ f_3^1(0) \times 0.25 + f_4^1(0) \times 0.2 = 0.0357$$

and $\sigma_1^2 = \sum_{j=1}^{m} f_j^2(x_{1j}) \cdot \eta_j = 0.125, \sigma_1^3 = \sum_{j=1}^{m} f_j^3(x_{1j}) \cdot \eta_j = 1$

Therefore,

$$\sigma_1 = (\sigma_1^1, \sigma_1^2, \sigma_1^3) = (0.0357, 0.125, 1)$$

Similarly, we can calculate and obtain:

$$\sigma_2 = (\sigma_2^1, \sigma_2^2, \sigma_2^3) = (0.3321, 0.6881, 0.2488)$$
$$\sigma_3 = (\sigma_3^1, \sigma_3^2, \sigma_3^3) = (0.3401, 0.6695, 0.3125)$$
$$\sigma_4 = (\sigma_4^1, \sigma_4^2, \sigma_4^3) = (0.6107, 0.2883, 0.3688)$$
$$\sigma_5 = (\sigma_5^1, \sigma_5^2, \sigma_5^3) = (0.7656, 0.075, 0.25)$$
$$\sigma_6 = (\sigma_6^1, \sigma_6^2, \sigma_6^3) = (0.6683, 0.508, 0)$$
$$\sigma_7 = (\sigma_7^1, \sigma_7^2, \sigma_7^3) = (0.9286, 0.125, 0)$$
$$\sigma_8 = (\sigma_8^1, \sigma_8^2, \sigma_8^3) = (0.8594, 0.225, 0.0417)$$
$$\sigma_9 = (\sigma_9^1, \sigma_9^2, \sigma_9^3) = (0.765, 0.25, 0)$$
$$\sigma_{10} = (\sigma_{10}^1, \sigma_{10}^2, \sigma_{10}^3) = (0.6536, 0.525, 0)$$
$$\sigma_{11} = (\sigma_{11}^1, \sigma_{11}^2, \sigma_{11}^3) = (0.9, 0.15, 0)$$
$$\sigma_{12} = (\sigma_{12}^1, \sigma_{12}^2, \sigma_{12}^3) = (0.7973, 0.325, 0)$$
$$\sigma_{13} = (\sigma_{13}^1, \sigma_{13}^2, \sigma_3^3) = (0.7313, 0.3625, 0.0375)$$
$$\sigma_{14} = (\sigma_{14}^1, \sigma_{14}^2, \sigma_4^3) = (0.6886, 0.355, 0)$$
$$\sigma_{15} = (\sigma_{15}^1, \sigma_5^2, \sigma_{15}^3) = (0.4114, 0.6075, 0.3875)$$
$$\sigma_{16} = (\sigma_{16}^1, \sigma_{16}^2, \sigma_{16}^3) = (0.6836, 0.225, 0.2)$$

Furthermore,

$$\sigma_{17} = (\sigma_{17}^1, \sigma_{17}^2, \sigma_7^3) = (0, 0.05, 1)$$

Step 4: Based on the following facts, it follows that trees with numberings 4, 5, 6, 7, 8, 9, 10, 11, 12, 13, 14, 16, are strains with wide adaptability:

$$\max_{1 \leq k \leq 3} \{\sigma_1^k\} = \sigma_1^3 = 1, \quad \max_{1 \leq k \leq 3} \{\sigma_2^k\} = \sigma_2^2 = 0.6881, \quad \max_{1 \leq k \leq 3} \{\sigma_3^k\} = \sigma_3^2 = 0.6695$$

$$\max_{1\leq k\leq 3}\{\sigma_4^k\} = \sigma_4^1 = 0.6107, \quad \max_{1\leq k\leq 3}\{\sigma_5^k\} = \sigma_5^1 = 0.7656, \quad \max_{1\leq k\leq 3}\{\sigma_6^k\} = \sigma_6^1 = 0.6683$$

$$\max_{1\leq k\leq 3}\{\sigma_7^k\} = \sigma_7^1 = 0.9286, \quad \max_{1\leq k\leq 3}\{\sigma_8^k\} = \sigma_8^1 = 0.8594, \quad \max_{1\leq k\leq 3}\{\sigma_9^k\} = \sigma_9^1 = 0.765$$

$$\max_{1\leq k\leq 3}\{\sigma_{10}^k\} = \sigma_{10}^1 = 0.6536, \quad \max_{1\leq k\leq 3}\{\sigma_{11}^k\} = \sigma_{11}^1 = 0.9, \quad \max_{1\leq k\leq 3}\{\sigma_{12}^k\} = \sigma_{12}^1 = 0.91$$

$$\max_{1\leq k\leq 3}\{\sigma_{13}^k\} = \sigma_{13}^1 = 0.82, \quad \max_{1\leq k\leq 3}\{\sigma_{14}^k\} = \sigma_{14}^1 = 0.6886, \quad \max_{1\leq k\leq 3}\{\sigma_{15}^k\} = \sigma_{15}^2 = 0.6075$$

$$\max_{1\leq k\leq 3}\{\sigma_{16}^k\} = \sigma_{16}^1 = 0.6836, \quad \max_{1\leq k\leq 3}\{\sigma_{17}^k\} = \sigma_{17}^3 = 1$$

Such strains are diversiform-leaved poplars, sacsaouls, Chinese pines, oriental arborvitaes, white elms, dryland willows, Chinese white poplars, oaks, Huashan pines, masson pines, China firs, and camphor trees. These trees have an extremely strong ability to adapt themselves to natural ecological environments, can grow well in most parts of China, and should be widely introduced. The trees named Korean pine, Northeast China Ash, and bamboo with numberings 2, 3, and 15, respectively, belong to the grey class of medium adaptability, and can be introduced to a relatively large area in China. Finally, trees with the names camphor pine and South Asian pine, and numberings 1 and 17, respectively, belong to the grey class of narrow adaptability, where camphor pines are found near the Northern border of China and South Asian pines are mainly located near the Southern border of China.

6.6 Grey Clustering Evaluation Models Based on Mixed Possibility Functions

6.6.1 Grey Clustering Evaluation Model Based on End-Point Mixed Possibility Functions

The grey clustering evaluation model based on mixed end-point triangular possibility functions is a new model. Compared with end-point triangular possibility functions, the new model has changed the possibility function for grey class 1 to the possibility function of lower measure, and the possibility function for grey class s to the possibility function of upper measure. Additionally, the new model has avoided the problem of extension of the bound of value of each clustering index. The Grey clustering evaluation model based on mixed end-point triangular possibility function is suitable for situations where all grey boundaries are clear, but the most likely points belonging to each grey class are unknown. The modeling steps are explained below (Liu et al., 2015a, 2015b).

Step1 Assume that according to the assessment requirements, the number of grey classes to be divided is s. Then the value range of each index is also divided into s classes. For example, the value range $[a_1, a_{s+1}]$ of index j can be divided into s small intervals:

$$[a_1, a_2], \ldots, [a_{k-1}, a_k], \ldots, [a_{s-1}, a_s], [a_s, a_{s+1}]$$

The value of $a_k(k = 2, \ldots, s)$ can be determined by the actual assessment requirements or the qualitative research results.

Step 2: Determine the turning point λ_j^1 and λ_j^s of $[a_1, a_2]$ and $[a_s, a_{s+1}]$ that correspond to grey classes 1 and s. At the same time, calculate the geometric center-point $\lambda_k = (a_k + a_{k+1})/2$ for each small interval $[a_k, a_{k+1}], k = 2, \ldots, s - 1$.

Step 3: For grey class 1 and grey class s, construct the corresponding possibility function of lower measure $f_j^1[-, -, \lambda_j^1, \lambda_j^2]$ and the possibility function of upper measure $f_j^s[\lambda_j^{s-1}, \lambda_j^s, -, -]$.

Assume that x is an observation of index j, when $x \in [a_1, \lambda_j^2]$ or $x \in [\lambda_j^{s-1}, a_{s+1}]$, using Formulas (6.8) or (6.9), respectively:

$$f_j^1(x) = \begin{cases} 0 & x \notin [a_1, \lambda_j^2] \\ 1 & x \in [a_1, \lambda_j^1] \; ; \\ \frac{\lambda_j^2 - x}{\lambda_j^2 - \lambda_j^1} & x \in [\lambda_j^1, \lambda_j^2] \end{cases} \tag{6.8}$$

or

$$f_j^s(x) = \begin{cases} 0 & x \notin [\lambda_j^{s-1}, a_{s+1}] \\ \frac{x - \lambda_j^{s-1}}{\lambda_j^s - \lambda_j^{s-1}} & x \in [\lambda_j^{s-1}, \lambda_j^s] \; . \\ 1 & x \in [\lambda_j^s, a_{s+1}] \end{cases} \tag{6.9}$$

By using these formulas, the possibility degree of $f_j^1(x)$ or $f_j^s(x)$ regarding grey class 1 and grey class s can be calculated.

Step 4: For grey class $k(k \in \{2, 3, \ldots, s - 1\})$, connecting point $(\lambda_j^k, 1)$ with center-point $(\lambda_j^{k-1}, 0)$ of grey class $k - 1$ (or turning point$(\lambda_j^1, 0)$ of grey class 1), and connecting $(\lambda_j^k, 1)$ with center-point $(\lambda_j^{k+1}, 0)$ of grey class $k + 1$ (or turning point $(\lambda_j^s, 0)$ of grey class s), we can get the triangular possibility function $f_j^k[\lambda_j^{k-1}, \lambda_j^k, -, \lambda_j^{k+1}], j = 1, 2, \ldots; m; \quad k = 2, 3, \ldots, s - 1$ of index j regarding grey class k (shown in Fig. 6.5).

For index j, x is an observation of it when $k = 2, 3, \ldots, s - 1$, according to Formula (6.10):

$$f_j^k(x) = \begin{cases} 0 & x \notin [\lambda_j^{k-1}, \lambda_j^{k+1}] \\ \frac{x - \lambda_j^{k-1}}{\lambda_j^k - \lambda_j^{k-1}} & x \in [\lambda_j^{k-1}, \lambda_j^k] \; . \\ \lambda_j^{k+1} - \lambda_j^k & x \in [\lambda_j^k, \lambda_j^{k+1}] \end{cases} \tag{6.10}$$

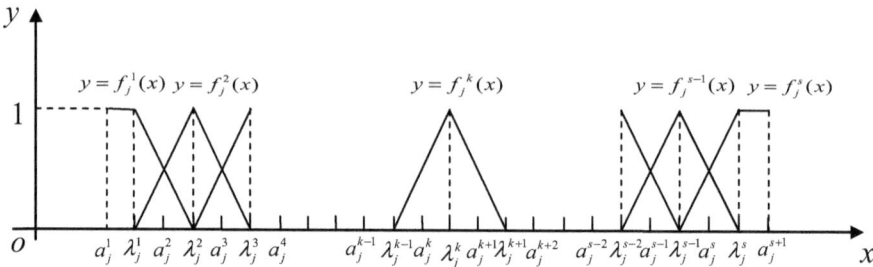

Fig. 6.5 The end-point mixed possibility function

This formula allows the possibility degree of $f_j^k(x)$ regarding grey class $k(k \in \{2, 3, \ldots, s-1\})$ to be calculated.

Step 5: Determine the weight w_j, $j = 1, 2, \ldots, m$ of each index.
Step 6: Calculate the clustering coefficient σ_i^k of object $i(i = 1, 2, \ldots, n)$ regarding grey class $k(k = 1, 2, \ldots, s)$:

$$\sigma_i^k = \sum_{j=1}^{m} f_j^k(x_{ij}) \cdot w_j. \tag{6.11}$$

$f_j^k(x_{ij})$ is the possibility function of index j about grey class k, w_j is the weight of index j among comprehensive clustering.

Step 7: By $\max_{1 \leq k \leq s} \{\sigma_i^k\} = \sigma_i^{k*}$, determine that object i belongs to grey class k^*. When there are multiple objects belonging to the same grey class k^*, we can further determine individual objects' precedence in grey class k^* on the basis of the size of integrate clustering coefficients.

6.6.2 Grey Clustering Evaluation Model Based on Center-Point Mixed Possibility Functions

This section addresses an improvement in the triangular possibility function. Such an improvement entails changing the center-point triangular possibility function which corresponds to class 1 to a possibility function of lower measure, and changing the triangular possibility function which corresponds to class s to a possibility function of upper measure. This improvement allows us to avoid having to extend the bound of value of each clustering index.

Definition 6.6.1 For grey class $k(k \in \{2, 3, \ldots, s-1\})$, the point which most likely belongs to grey class k is called the center-point of grey class k.

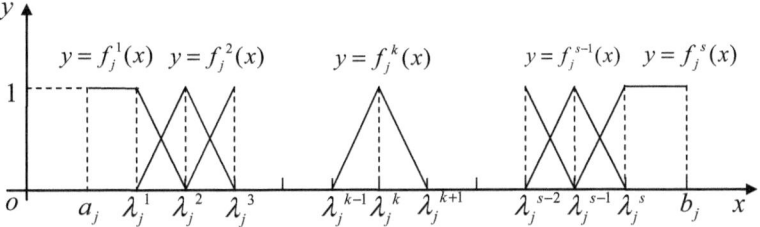

Fig. 6.6 Center-point mixed possibility function

The center-point may or may not be the midpoint. This is determined by the maximum likelihood of such a point to belong to the grey class.

The modeling steps of the grey cluster evaluation model using center-point triangular possibility functions are as follows.

Step 1: Assume that $[a_j, b_j]$ is the range of index j. According to the evaluation requirements, we divide $[a_j, b_j]$ into s small intervals. Then we determine the turning point λ_j^1, λ_j^s of grey classes 1 and s, and the center-point $\lambda_j^2, \lambda_j^3, \ldots, \lambda_j^{s-1}$ of grey class $k(k \in \{2, 3, \ldots, s - 1\})$, respectively.

Step 2: Construct the corresponding lower measure possibility function $f_j^1[-, -, \lambda_j^1, \lambda_j^2]$, and the upper measure possibility function $f_j^s[\lambda_j^{s-1}, \lambda_j^s, -, -]$ for grey classes 1 and s (see Fig. 6.6).

Assume x is an observation value of index j. When $x \in [a_j, \lambda_j^2]$ or $x \in [\lambda_j^{s-1}, b_j]$, the possibility degree of $f_j^1(x)$ or $f_j^s(x)$ regarding grey classes 1 and s can be calculated by using Formulas (6.12) and (6.13), respectively.

$$
f_j^1(x) = \begin{cases} 0 & x \notin [a_j, \lambda_j^2] \\ 1 & x \in [a_j, \lambda_j^1] \\ \frac{\lambda_j^2 - x}{\lambda_j^2 - \lambda_j^1} & x \in [\lambda_j^1, \lambda_j^2] \end{cases} \tag{6.12}
$$

$$
f_j^s(x) = \begin{cases} 0 & x \notin [\lambda_j^{s-1}, b_j] \\ \frac{x - \lambda_j^{s-1}}{\lambda_j^s - \lambda_j^{s-1}} & x \in [\lambda_j^{s-1}, \lambda_j^s] \\ 1 & x \in [\lambda_j^s, b_j] \end{cases} \tag{6.13}
$$

Step 3: For grey class $k(k \in \{2, 3, \ldots, s - 1\})$, by connecting point $(\lambda_j^k, 1)$ with center-point $(\lambda_j^{k-1}, 0)$ of grey class $k - 1$ (or turning point $(\lambda_j^1, 0)$ of grey class 1), and by connecting $(\lambda_j^k, 1)$ with center-point $(\lambda_j^{k+1}, 0)$ of grey class $k + 1$ (or turning point $(\lambda_j^s, 0)$ of grey class s), we get triangular possibility function $f_j^k[\lambda_j^{k-1}, \lambda_j^k, -, \lambda_j^{k+1}]$, $j = 1, 2, \ldots; m$; $k = 2, 3, \ldots, s - 1$ of index j regarding grey class k (see Fig. 6.6).

Assume that x is an observation value of index j. The degree of membership $f_j^k(x)$ regarding grey class $k(k \in \{2, 3, \ldots, s - 1\})$ can be calculated by using Formula (6.14).

$$f_j^k(x) = \begin{cases} 0 & x \notin [\lambda_j^{k-1}, \lambda_j^{k+1}] \\ \frac{x - \lambda_j^{k-1}}{\lambda_j^k - \lambda_j^{k-1}} & x \in [\lambda_j^{k-1}, \lambda_j^k] \\ \frac{\lambda_j^{k+1} - x}{\lambda_j^{k+1} - \lambda_j^k} & x \in [\lambda_j^k, \lambda_j^{k+1}] \end{cases} \tag{6.14}$$

Step 4: Determine the weight w_j, $j = 1, 2, \ldots, m$ of each index.
Step 5: Compute clustering coefficient σ_i^k of object $i(i = 1, 2, \ldots, n)$ regarding grey class $k(k = 1, 2, \ldots, s)$, as seen in Eq. (6.15).

$$\sigma_i^k = \sum_{j=1}^m f_j^k(x_{ij}) \cdot w_j \tag{6.15}$$

$f_j^k(x_{ij})$ is the possibility function of index j about class k, while w_j is the weight of comprehensive clustering of index j.

Step 6: By $\max_{1 \le k \le s} \{\sigma_i^k\} = \sigma_i^{k*}$, determine that object i belongs to grey class k^*; when there are multiple objects that belong to the same grey class k^*, we can further determine the precedence of individual objects in grey class k^* on the basis of the size of integrate clustering coefficients.

6.7 Practical Applications

Example 6.7.1 Five suppliers A, B, C, D, E who undertake the development of the C919 body component for Commercial Aircraft Corporation of China Ltd. (COMAC) are evaluated on their performance and are divided into four classes including "excellent", "good", "medium" and "poor" (Liu et al., 2015b).

Step 1: Set the evaluation index system for supplier performance.

The evaluation index system for supplier performance reflects the specific requirements of the main manufacturers to supplier. It is an important basis for the main manufacturers to comprehensively evaluate the supplier and make final management decisions.

Factors that affect supplier performance are very complex. Main manufactures' foci on supplier performance are also not the same across different stages in C919 development. After four rounds of expert investigation, six first-grade evaluation indexes are determined, including quality, cost, delivery, cooperation, technology and service.

At development stage, the four second-grade indexes of quality are pass rate of product, quality control system, airworthiness certification ability and control of sub-supplier. The three second-grade indexes of cost are price, logistics costs and price stability. The two second-grade indexes of delivery are punctuality and flexibility. The three second-grade indexes of cooperation are credit, information communication and cooperation intention. The five second-grade indexes of technology are professional R&D staff, R&D investment, number of invention patents, market share and technology level. The four second-grade indexes of service are quick response, spare part support, training and technology support.

Among those indexes, pass rate and market share are shown in percentage. Price and logistics costs are quantitative indexes and the unit is ten thousand yuan. The smaller the indexes, the better. The unit of professional R&D staff is person, R&D investment is ten thousand yuan, and the unit of patent number is an item. The bigger the indexes, the better. Other indexes like quality control system, price stability, delivery punctuality, flexibility, credit, information communication, technology level, quick response, spare part support, training and technology support are all qualitative indexes. They are usually quantified by expert grade. Here, the grade is a 10-point scale score and decimal points are allowed.

If supplier who have different tasks are evaluated together, most quantitative indexes such as price and logistics costs cannot be compared. Therefore, at this point we need to invite experienced experts to make qualitative assessments of quantitative indexes by grading them as a 10-point scale score. The evaluation index system for supplier performance and its weight at development stage of C919 are shown in Table 6.4.

For the evaluation of supplier performance at development stage, we use the index system shown in Table 6.4.

Step 2: According to the evaluation results, the value range of each index is divided into four grey classes. The value of second-grade indexes are usually divided into four small sections based on the sample value. Considering the opinion of COMAC, the effect sample matrix of second-grade index is omitted. Here are the actual values of six first-grade indexes that are obtained by weighted integration of the second-grade indexes as 10-point scale scores. The values are $y_{ij}, (i = 1, 2 \ldots, 5; \quad j = 1, 2, \ldots 6)$ as shown in Table 6.5.

The six first-grade indexes are all in 10-point scores, and the value range is [0,10]. Interval [0,10] is sub-divided into 4 small intervals as [0,6), [6,7.5), [7.5,9), [9,10], which correspond to "poor", "medium", "good" and "excellent".

Step 3 : Determine the turning point $\lambda_j^1 = 5$, $\lambda_j^4 = 9.5$ of [0,6) and [9,10] that correspond to grey class 1 and grey class 4. At the same time, calculate the center-point of [6, 7.5) and [7.5, 9), $\lambda_j^2 = 6.75$, $\lambda_j^3 = 8.25$.

Step 4 : By using Formulas (6.8), (6.9), and (6.10), the possibility functions of index j regarding grey class k(k = 1, 2, 3, 4) can be obtained as follows:

Table 6.4 The evaluation index system for supplier performance and its weight at development stage

First-grade index and its weight	Second-grade index	Code	Unit	Weight
Quality (22%)	Pass rate	x_1	%	6
	Quality control system	x_2	Qualitative	6
	Airworthiness certification ability	x_3	Qualitative	5
	Control of sub-supplier	x_4	Qualitative	5
Cost (18%)	Price	x_5	Ten thousand yuan	8
	Logistic cost	x_6	Ten thousand yuan	4
	Price stability	x_7	Qualitative	6
Delivery (17%)	Punctuality	x_8	Qualitative	12
	Flexibility	x_9	Qualitative	5
Cooperation (13%)	Credit	x_{10}	Qualitative	6
	Information communication	x_{11}	Qualitative	4
	Cooperation intention	x_{12}	Qualitative	3
Technology (16%)	Professional R&D staff	x_{13}	Person	3
	R&D investment	x_{14}	Ten thousand yuan	3
	Number of invention patent	x_{15}	Item	3
	Market share	x_{16}	%	3
	Technology level	x_{17}	Qualitative	4
Service (14%)	Quick response	x_{18}	Qualitative	4
	Spare part support	x_{19}	Qualitative	4
	Training	x_{20}	Qualitative	3
	Technology support	x_{21}	Qualitative	3

Table 6.5 The actual values of first-grade index of five suppliers

Supplier	Actual value y_{ij}					
	Quality	Cost	Delivery	Technology	Cooperation	Service
A	9.1	7.8	8.4	9	9.5	9.3
B	9.3	7.5	9	9.2	9	9
C	9	8.6	8.7	9	9.1	9.1
D	8.9	8.5	9	9.1	9.6	9.2
E	8.6	9	8.6	9	9.7	9.5

Table 6.6 The clustering coefficient regarding to each grey class of five suppliers

The clustering objects	The clustering coefficient			
	σ_i^1	σ_i^2	σ_i^3	σ_i^4
A	0	5.4	42.04	52.56
B	0	9	34.44	56.56
C	0	0	47.44	52.56
D	0	0	39.28	60.72
E	0	0	40.48	59.52

$$f_j^1(x) = \begin{cases} 0 & x \notin [0, 6.75] \\ 1 & x \in [0, 5) \\ \frac{6.75-x}{1.75} & x \in [5, 6.75] \end{cases} \qquad f_j^2(x) = \begin{cases} 0 & x \notin [5, 8.25] \\ \frac{x-5}{1.75} & x \in [5, 6.75) \\ \frac{8.25-x}{1.5} & x \in [6.75, 8.25] \end{cases}$$

$$f_j^3(x) = \begin{cases} 0 & x \notin [6.75, 9.5] \\ \frac{x-6.75}{1.5} & x \in [6.75, 8.25) \\ \frac{9.5-x}{1.25} & x \in [8.25, 9.5] \end{cases} \qquad f_j^4(x) = \begin{cases} 0 & x \notin [8.25, 10] \\ \frac{x-8.25}{1.25} & x \in [8.25, 9.5) \\ 1 & x \in [9.5, 10] \end{cases}$$

where $j = 1, 2, \ldots 6$.

Step 5: The weight of each index w_j, $j = 1, 2, 3, 4, 5, 6$ is shown in Table 6.4.

Step 6: According to Formula (6.11), the clustering coefficient regarding the grey class of five suppliers can be calculated ($i = 1, 2, 3, 4, 5$; $k = 1, 2, 3, 4$), as shown in Table 6.6.

Step 7: As can be seen from the results of $\max_{1\le k\le 4}\{\sigma_A^k\} = 52.56 = \sigma_A^4$, $\max_{1\le k\le 4}\{\sigma_B^k\} = 56.56 = \sigma_B^4$, $\max_{1\le k\le 4}\{\sigma_C^k\} = 52.56 = \sigma_C^4$, $\max_{1\le k\le 4}\{\sigma_D^k\} = 60.72 = \sigma_D^4$ $\max_{1\le k\le 4}\{\sigma_E^k\} = 59.52 = \sigma_E^4$, the performance of five suppliers A, B, C, D, E at development stage all reach the level of "excellent". Among those supplier, the clustering coefficient of supplier D regarding grey class "excellent" is the highest and supplier E takes the second place. However, the difference between D and E is very small, so the two supplier belong to the same level. Then comes supplier B. The coefficient of supplier A and C regarding grey class "excellent" is the smallest.

Further investigation reveals that the indexes belonging to class "excellent" of supplier D and E are technology, cooperation and service. There is much room for improvement in terms of quality and cost for D, and in terms of quality and delivery for E.. The main problem for supplier B is its high cost. Although the evaluation on cooperation and service is quite good, the value is still on the low side compared with other supplier. For supplier A and C, the main problems are cost and delivery. The management department of COMAC can focus on each supplier according to their own problems and improve their whole performance level promptly.

In this example, the value range of each index as well as its turning point and center-point $\lambda_j^1,\quad \lambda_j^2, \lambda_j^3, \lambda_j^4, j = 1, 2, \ldots, 6$ regarding different grey classes are determined according to the expert evaluation results of supplier A, B, C, D, and E. Also, the conclusion only applies to the current situation of those supplier. The results of grey clustering evaluation can be used with a certain scope: the scope used when determining the possibility function is the one that can be used in the evaluation results. The so called "excellent", "good", "medium" and "poor" classes are also relative. Supplier A, B, C, D, and E are all prominent enterprises in China. Although they are very strong, there is a big gap between their performance and that of similar manufacturers around the world.

Example 6.7.2 The evaluation of a project for discipline development at a university will be used to illustrate the application of the grey clustering evaluation models, which are based on mixed center-point triangular possibility functions.

Based on extensive surveys, there are 6 primary indicators to reflect the performance of a discipline development project, including faculty, scientific research, student cultivation, discipline platform development, conditions for development and academic communication. The corresponding weights are 0.21, 0.24, 0.23, 0.14, 0.1, and 0.08, respectively (see Fig. 6.7) (Jian et al., 2007; Liu, 2021).

We convert the evaluation scores of each indicator to centesimal system for convenience. The evaluation results are divided into four grey classes including "excellent", "good", "medium" and "poor", according to requirements of the university authorities. 41 projects for discipline development have been conducted from 2016 to 2020. All the evaluation scores of the 6 indexes of these 41 projects for discipline development are laid in the interval of [40, 100]. We set up the turning point $\lambda_j^4 = 90$ for grey class "excellent" and the turning point $\lambda_j^1 = 60$ for grey class "poor", as well as the most likely points $\lambda_j^3 = 80$, $\lambda_j^2 = 70$, which belong to grey classes "good" and "medium".

Fig. 6.7 Evaluation indicator system of project for discipline construction

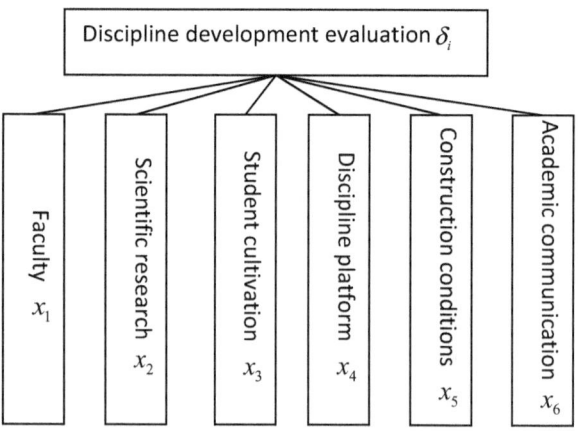

Table 6.7 The values of 6 indicators of a project for discipline development

Indicator	Faculty	Scientific research	Student cultivation	Discipline platform	Development conditions	Academic communication
Value	81	87	92	78	74	53

Since the evaluation scores of each indicator are converted to centesimal system, the possibility function of all 6 indicators on four grey classes of "poor", "medium", "good", and "excellent" are the same:

$$f_j^1(x) = \begin{cases} 0 & x \notin [40, 70] \\ 1 & x \in [40, 60] \\ \frac{70-x}{70-60} & x \in [60, 70] \end{cases}, \qquad f_j^2(x) = \begin{cases} 0 & x \notin [60, 80] \\ \frac{x-60}{70-60} & x \in [60, 70] \\ \frac{80-x}{80-70} & x \in [70, 80] \end{cases}$$

$$f_j^3(x) = \begin{cases} 0 & x \notin [70, 90] \\ \frac{x-70}{80-70} & x \in [70, 80] \\ \frac{90-x}{90-80} & x \in [80, 90] \end{cases}, \qquad f_j^4(x) = \begin{cases} 0 & x \notin [80, 100] \\ \frac{x-80}{90-80} & x \in [80, 90] \\ 1 & x \in [90, 100] \end{cases}$$

where the possibility function of each indicator for grey class "poor" is a possibility function of lower measure, each indicator for grey class "excellent" is a possibility function of upper measure, and each indicator for grey classes "medium" and "good" are triangular possibility functions. The values of the 6 indicators for a university's discipline development project are shown in Table 6.7.

The values of possibility functions for the different grey classes of each indicator can be calculated by using $f_j^1(x)\ f_j^2(x)f_j^3(x)f_j^4(x), j = 1, 2, \ldots, 6$. The grey clustering coefficient δ_i can be calculated by using Formula (6.7). The outcomes are shown in Table 6.8.

Based on the results in Table 6.8, we can confirm that the project belongs to grey class "excellent" according to $\max_{1 \leq k \leq 4}\{\delta_i^k\} = \delta_i^4 = 0.419$. Therefore, the effect of the project for discipline development is remarkable. But the grey clustering coefficient which suggests that the project belongs to grey class "good" is $\delta_i^3 = 0.413$. This result is very close to δ_i^4. It also shows that the execution effect of the

Table 6.8 Grey clustering coefficients of each indicator for different grey classes

Grey class	x_1	x_2	x_3	x_4	x_5	x_6	δ_i
Excellent	0.1	0.7	1.0	0	0	0	0.419
Good	0.9	0.3	0	0.8	0.4	0	0.413
Medium	0	0	0	0.2	0.6	0	0.088
Poor	0	0	0	0	0	1.0	0.080

project for discipline development is situated between grey classes "excellent" and "good". As for the sub-indicators, the indicator on student cultivation belongs to grey class "excellent", and reached a high level. The indicator on scientific research is situated between grey classes "good" and "excellent", but close to grey class "excellent". The indicators on faculty and discipline platform development basically belong to grey class "good", which indicates that the implementation effect of these two indicators are satisfactory. The indicator on development conditions is situated between grey classes "good" and "medium", but closer to grey class "medium". The indicator on academic communication belongs to grey class "poor ", which suggests that there are still significant shortcomings in development conditions and academic communication that require further strengthening.

The grey cluster evaluation model based on mixed possibility function is more suitable to solve problems of poor information clustering evaluation. On the other hand, grey cluster evaluation model using center-point triangular possibility functions is suitable for problems where it is relatively easy to judge the most likely points belonging to each grey class, but the grey boundaries are not clear. Finally, grey cluster evaluation model using end-point triangular possibility functions is suitable for situations where all grey boundaries are clear, but the most likely points belonging to each grey class are unknown.

References

Deng, J. L. (1985). *Grey control systems*. Press of Huazhong University of Science and Technology.

Dong, F. Y., Liu, J. J., Liu, B., & Li, B. J. (2010). Improved grey integrated clustering method and its application in the evaluation to rural economic development of Henan province. *System Sciences and Comprehensive Studies in agriculture, 26*(4), 478–483.

Jian, L. R., Liu, S. F., & Fang, Z.G. (2007). Performance evaluation of discipline construction projects in research universities. *Journal of Management Engineering, 21*(3), 132–137.

Li, S. R., Zhao, Y., Liu, S. F., & Wang, T. C. (1994). Classification and stability analysis of forest ecosystem in Henan Province. *Journal of Henan Agricultural University, 28*(2), 111–118.

Liu, S. F. (1991). The three axioms of buffer operator and their application. *Journal of Grey System, 3*(1), 39–48.

Liu, S. F. (1993). Fixed weight grey clustering evaluation analysis. In *New methods of grey systems* (pp. 178–184). Agriculture Press.

Liu, S. F. (2021). *Grey system theory and its application*. (9th ed.). Beijing: Science Press.

Liu, S. F., & Zhu, Y. D. (1993). Models based on criteria's triangular membership functions for the evaluation of regional economics. *Journal of Agricultural Engineering, 9*(2), 8–13.

Liu, S. F., & Xie, N. M. (2011). A new grey evaluation method based on reformative triangular whitenization weight function. *Journal of Systems Engineering, 26*(2), 244–250.

Liu, S. F., Yang, Y. J., & Fang, Z. G., et al. (2015a). Grey cluster evaluation models based on mixed triangular whitenization weight functions. *Grey systems: theory and application, 5*(3), 410–418.

Liu, S. F., Zhang, S. G., Jian, L. R., et al. (2015b). Performance evaluation of large commercial aircraft vendors. *Journal of Grey System, 27*(1), 1–11.

Liu, S. F., Yang Y. J., & Forrest, J. (2017). *Grey data analysis*. Singapore: Springer-Verlag.

Pei, L. L., Chen, W. M., & Shen, C. G. (2012). Study on optimization of grey clustering evaluation model. *Journal of Inner Mongolia Normal University, 41*(5), 462–466.

Xiao, X. P. (1997). Theoretical research and remarks on quantified models of grey degrees of incidence. *Systems Engineering: Theory and Practice, 17*(8), 76–81.

Xiong, H. J., & Chen, M. Y. (1999). *Some problems on grey clustering systems engineering and electronics, 21*(5), 6–91.

Xu, W. G., Zhang, Q. Y., Guo, H., & He, A. B. (2006). Improvement and application of grey clustering model in atmospheric quality comprehensive evaluation. *Mathematics in Practice and Theory, 36*(6), 200–205.

Zhang, Q. S. (2002). Measure of grey characteristics of grey clustering result. *Chinese Journal of Management Science, 10*(1), 54–561.

Chapter 7
Series of GM Models

7.1 Introduction

Model GM(1,1) is the basic model of grey prediction theory and has been used widely since its development in the early 1980s. Grey system theory is a new methodology that focuses on uncertain problems involving small data and poor information. Incomplete and inaccurate information is the basic characteristic of uncertainty systems. In the case of incomplete information and inaccurate data, it is impossible to pursue a refined model (Liu et al., 2012). Professor Zadeh's incompatibility principle also clearly states that, when the complexity of the system grows, our ability to make an accurate and significant description of a system's characteristics decreases until it reaches a threshold value that, if it exceeded, accuracy and significance will become mutually exclusive characteristics (Zadeh, 1994). The incompatibility principle tells us that pursuing a refinement model one-sidedly will reduce the feasibility and significance of the results. A refined model is not an effective means address complex systems.

In the last 40 years, much research has been carried out on the practical applications of Model GM(1,1), and new research results emerge continuously. Recent studies on model GM(1,1) are focusing on how to further optimize the model and improve its simulated and predictive results. Such research can be roughly divided into the following areas: (1) research on the nature and characteristics of model GM(1,1) (Ji et al., 2001); (2) studies about initial value selection (Dang et al., 2005); (3) research on the optimization of model parameters (Xiao, 2000); (4) attempts to improve the simulation accuracy of the model by recreating the background value (Li & Dai, 2004; Tan, 2005); (5) attempts to optimize the model through different modeling methods (Song et al., 2002; Wang, 2003); (6) research on discrete model GM(1,1) (Xie & Liu, 2009; Xie & Liu, 2005); (7) modeling for non-equidistant sequence and model optimization (Luo 2010; Wang et al., 2008); (8) research on the application bound of different models (Liu & Deng, 2000); (9) research on fractional order (Wu et al., 2013; Mao et al., 2016) or self-memory grey model (Guo

© The Author(s), under exclusive license to Springer Nature Singapore Pte Ltd. 2022
S. Liu, *Grey Systems Analysis*, Series on Grey System,
https://doi.org/10.1007/978-981-19-6160-1_7

et al., 2015) and research combining a grey system model with other soft computing methods to improve the accuracy of the model (Salmeron, 2010; Zhang, 2007).

The buffer operator proposed in 1991 (Liu, 1991) has attracted much research attention in recent years. Buffer operator is essentially a method for processing raw data, rather than a technique to improve the degree of accuracy of the GM(1,1) model simulation and prediction. During the period of raw data collection, a system is likely to suffer interference from external shocks, which means that such data will be distorted and unlikely to reflect the operation of the system's behavior. In such case, researchers can choose or construct a suitable buffer operator by following a qualitative analysis of the data to eliminate the impact of the distorted data sequence and keep the true nature of the data.

The above-mentioned studies have played a positive role in improving the degree of accuracy of the simulation and prediction of model GM(1, 1), and in helping scholars engaged in applied research make appropriate selection and use of grey prediction models.

7.2 The Four Basic Forms of GM(1,1)

In this section we present definitions of four basic forms of model GM(1,1), including Even Grey Model (EGM), Original Difference Grey Model (ODGM), Even Difference Grey Model (EDGM) and Discrete Grey Model (DGM). The properties and characteristics of different models are discussed in-depth (Liu et al., 2015a, 2017; Liu, 2021;).

7.2.1 The Basic Forms of Model GM(1,1)

Definition 7.2.1 Let $X^{(0)} = (x^{(0)}(1), x^{(0)}(2), \ldots, x^{(0)}(n)), x^{(0)}(k) \geq 0, X^{(1)}$ be the 1-AGO sequence of $X^{(0)}$; that is

$$X^{(1)} = [x^{(1)}(1), x^{(1)}(2), \ldots, x^{(1)}(n)]$$

where $x^{(1)}(k) = \sum_{i=1}^{k} x^{(0)}(i), k = 1, 2, \ldots n$. Then

$$x^{(0)}(k) + ax^{(1)}(k) = b \tag{7.1}$$

is referred to as the original form of model GM(1,1), which is a difference equation.

The parameter vector $\hat{a} = [a, b]^T$ of formula (7.1) can be estimated using the least square method, which satisfies

$$\hat{a} = (B^T B)^{-1} B^T Y \tag{7.2}$$

where

$$B = \begin{bmatrix} -x^{(1)}(2) & 1 \\ -x^{(1)}(3) & 1 \\ \vdots & \vdots \\ -x^{(1)}(n) & 1 \end{bmatrix}, Y = \begin{bmatrix} x^{(0)}(2) \\ x^{(0)}(3) \\ \vdots \\ x^{(0)}(n) \end{bmatrix} \tag{7.3}$$

Definition 7.2.2 Based on the original form of model GM(1,1) and formula (7.2), which is used to estimate the model's parameters, then the model that takes the solution of the original difference Eq. (7.1) as the time response formula is called the original difference form of model GM(1,1), and is referred to as Original Difference Grey Model(ODGM) for short (Liu et al., 2015b).

Definition 7.2.3 Let $X^{(0)}$, $X^{(1)}$ and, just like Definition 7.2.1, let

$$Z^{(1)} = (z^{(1)}(2), z^{(1)}(3), \ldots, z^{(1)}(n)),$$

where $z^{(1)}(k) = \frac{1}{2}(x^{(1)}(k) + x^{(1)}(k-1))$, then

$$x^{(0)}(k) + az^{(1)}(k) = b \tag{7.4}$$

is referred to as the even form of model GM(1,1).

The even form of model GM(1,1) is also essentially a difference equation. The parameter vector of formula (7.4) can also be estimated with formula (7.2), but it should be noted that the elements of matrix B are different from those in formula (7.3), which is

$$B = \begin{bmatrix} -z^{(1)}(2) & 1 \\ -z^{(1)}(3) & 1 \\ \vdots & \vdots \\ -z^{(1)}(n) & 1 \end{bmatrix} \tag{7.5}$$

Definition 7.2.4 The following differential equation

$$\frac{dx^{(1)}}{dt} + ax^{(1)} = b \tag{7.6}$$

is called a shadow equation of the even form $x^{(0)}(k) + az^{(1)}(k) = b$ of model GM(1,1).

Definition 7.2.5 Replace matrix B of formula (7.2) with (7.5), according to parameter vector $\hat{a} = [a, b]^T$ of the least squares estimator of (7.6) and the solution of whitenization Eq. (7.6), and model the difference, differential hybrid model of the time response formula of GM(1,1). This is called the even hybrid form of model

GM(1,1), and is referred to as Even Grey Model (EGM) for short (Deng, 1982, 1985).

Definition 7.2.6 The parameter $-a$ of Even GM(1,1) is called development index and b is called grey actuating quantity. The development index reflects the trend of $\hat{x}^{(1)}$ and $\hat{x}^{(0)}$.

Even Model GM(1,1) is the grey prediction model proposed firstly by Professor Deng Julong, and is currently the most influential, widely used form. When researchers mention model GM(1,1) they are often referring to EGM.

Definition 7.2.7 Based on the even form of model GM(1,1) and the estimated model parameters, then the model that takes the solution of the even difference Eq. (7.4) as the time response formula is called the even difference form of model GM(1,1), and is referred to as Even Difference Grey Model (EDGM) for short (Liu et al., 2015b).

Definition 7.2.8 The difference equation as follows

$$x^{(1)}(k+1) = \beta_1 x^{(1)}(k) + \beta_2 \tag{7.7}$$

is called a discrete form of model GM(1,1), and is referred to as Discrete Grey Model (DGM) for short (Xie & Liu, 2005).

The parameter vector $\hat{\beta} = [\beta_1, \beta_2]^T$ in Eq. (7.7) is similar to formula (7.2), where

$$B = \begin{bmatrix} x^{(1)}(1) & 1 \\ x^{(1)}(2) & 1 \\ \vdots & \vdots \\ x^{(1)}(n-1) & 1 \end{bmatrix}, Y = \begin{bmatrix} x^{(1)}(2) \\ x^{(1)}(3) \\ \vdots \\ x^{(1)}(n) \end{bmatrix}.$$

The four different models of GM(1,1) use only the system's behavior data sequence to model the predictive models and belong to the simple and practical modeling method with a single sequence. In the case of time series data, only a regular time variables are involved; In the case of horizontal sequence data, only a regular object sequence number variables are involved, and other explanatory variables are not involved. GM(1,1) model is a modeling method which is relatively simple to apply and can mine valuable development and change information, so it is widely used.

7.2.2   Properties and Characteristics of the Basic Model

Theorem 7.2.1 *The time response sequence of the Even Model GM(1,1) is as follows:*

$$\hat{x}^{(1)}(k) = \left(x^{(0)}(1) - \frac{b}{a} \right) e^{-a(k-1)} + \frac{b}{a}, k = 1, 2, \ldots n \tag{7.8}$$

Proof The solution of whitenization or shadow equation $\frac{dx^{(1)}}{dt} + ax^{(1)} = b$ is

$$x^{(1)}(t) = C e^{-at} + \frac{b}{a}. \tag{7.9}$$

When $t = 1$, we let $x^{(1)}(1) = x^{(0)}(1)$, and feed into Eq. (7.9); we can obtain $C = \left[x^{(0)}(1) - \frac{b}{a}\right] e^a$. After that we take C into Eq. (7.9) and can get

$$\hat{x}^{(1)}(t) = \left(x^{(0)}(1) - \frac{b}{a}\right) e^{-a(t-1)} + \frac{b}{a}. \tag{7.10}$$

Equation (7.8) is the discrete form of Eq. (7.10). From Eq. (7.8)'s regressive reduction formula

$$\hat{x}^{(0)}(k) = \alpha^{(1)} \hat{x}^{(1)}(k) = \hat{x}^{(1)}(k) - \hat{x}^{(1)}(k-1), k = 1, 2, \cdots n,$$

we can obtain the time response formula of $X^{(0)}$, that is

$$\hat{x}^{(0)}(k) = (1 - e^a)\left(x^{(0)}(1) - \frac{b}{a}\right) e^{-a(k-1)}, k = 1, 2, \cdots n \tag{7.11}$$

Theorem 7.2.2 *The time response formula of formula (7.7) of the Discrete Model GM(1,1) is*

$$\hat{x}^{(1)}(k) = \left[x^{(0)}(1) - \frac{\beta_2}{1-\beta_1}\right]\beta_1^k + \frac{\beta_2}{1 - \beta_1} \tag{7.12}$$

Proof The general solution of the difference Eq. (7.13)

$$x^{(1)}(k+1) = Ax^{(1)}(k) + B \tag{7.13}$$

is

$$x^{(1)}(k) = CA^k + \frac{B}{1 - A}, \tag{7.14}$$

where C is an arbitrary constant and can be defined by the initial conditions.

Formula (7.7) and (7.14) are exactly the same difference equation. Let $A = \beta_1$, $B = \beta_2$, then

$$x^{(1)}(k) = C\beta_1^k + \frac{\beta_2}{1 - \beta_1}. \tag{7.15}$$

When k = 0, let $x^{(1)}(0) = x^{(0)}(1)$, and feed into formula (7.15), we can get $C = \left[x^{(0)}(1) - \frac{\beta_2}{1-\beta_1}\right]$. Then take C into formula (7.15) and we can obtain formula (7.12).

From formula (7.12)'s regressive reduction formula

$$\hat{x}^{(0)}(k) = \alpha^{(1)}\,\hat{x}^{(1)}(k) = \hat{x}^{(1)}(k) - \hat{x}^{(1)}(k-1), k = 1, 2, \ldots n,$$

we can obtain the time response formula of $X^{(0)}$, that is

$$\hat{x}^{(0)}(k) = (\beta_1 - 1)\left[x^{(0)}(1) - \frac{\beta_2}{1-\beta_1}\right]\beta_1^{k-1}. \tag{7.16}$$

Theorem 7.2.3 *The time response formula of Original Difference Model GM(1,1) is*

$$\hat{x}^{(1)}(k) = \left(x^{(0)}(1) - \frac{b}{a}\right)\left(\frac{1}{1+a}\right)^k + \frac{b}{a} \tag{7.17}$$

Proof From the original form (7.1) of model GM(1,1) we can get

$$x^{(1)}(k+1) - x^{(1)}(k) + ax^{(1)}(k+1) = b. \tag{7.18}$$

After transposition, we obtain

$$x^{(1)}(k+1) = \left(\frac{1}{1+a}\right)x^{(1)}(k) + \frac{b}{1+a}.$$

Contrast with the difference Eq. (7.13), when we feed $A = \frac{1}{1+a}$, $B = \frac{b}{1+a}$ into Eq. (7.14), we can obtain

$$x^{(1)}(k) = C\left(\frac{1}{1+a}\right)^k + \frac{b}{a} \tag{7.19}$$

When k = 0, let $x^{(1)}(0) = x^{(0)}(1)$, feed into formula (7.18) and get $C = \left[x^{(0)}(1) - \frac{b}{a}\right]$. Then we feed C into formula (7.19) and can obtain formula (7.17).

From formula (7.17)'s regressive reduction formula

$$\hat{x}^{(0)}(k) = \alpha^{(1)}\,\hat{x}^{(1)}(k) = \hat{x}^{(1)}(k) - \hat{x}^{(1)}(k-1), k = 1, 2, \cdots n,$$

we can obtain the time response formula of $X^{(0)}$, which is

$$\hat{x}^{(0)}(k) = \left(x^{(0)}(1) - \frac{b}{a}\right)\left(\frac{1}{1+a}\right)^k + \frac{b}{a} - \left[\left(x^{(0)}(1) - \frac{b}{a}\right)\left(\frac{1}{1+a}\right)^{k-1} + \frac{b}{a}\right].$$

That is

$$\hat{x}^{(0)}(k) = (-a)\left(x^{(0)}(1) - \frac{b}{a}\right)\left(\frac{1}{1+a}\right)^k \tag{7.20}$$

Theorem 7.2.4 *The time response formula of Even Difference Model GM(1,1) is*

$$x^{(1)}(k) = \left(x^{(0)}(1) - \frac{b}{a}\right)\left(\frac{1 - 0.5a}{1 + 0.5a}\right)^k + \frac{b}{a} \tag{7.21}$$

Proof From the even form (7.4) of model GM(1,1) we can get

$$x^{(1)}(k+1) - x^{(1)}(k) + a\left(\frac{x^{(1)}(k+1) + x^{(1)}(k)}{2}\right) = b.$$

After transposition, we obtain

$$x^{(1)}(k+1) = \left(\frac{1 - 0.5a}{1 + 0.5a}\right)x^{(1)}(k) + \frac{b}{1 + 0.5a}.$$

Contrast with the difference Eq. (7.13), and feed $A = \frac{1-0.5a}{1+0.5a}$, $B = \frac{b}{1+0.5a}$ into formula (7.14). We can obtain

$$x^{(1)}(k) = C\left(\frac{2 - a}{2 + a}\right)^k + \frac{b}{a} \tag{7.22}$$

When $k = 0$, let $x^{(1)}(0) = x^{(0)}(1)$, feed it into formula (7.22) and get $C = \left[x^{(0)}(1) - \frac{b}{a}\right]$. Then feed C into formula (7.22) and we can obtain formula (7.21).
From formula (7.21)'s regressive reduction formula

$$\hat{x}^{(0)}(k) = \alpha^{(1)}\,\hat{x}^{(1)}(k) = \hat{x}^{(1)}(k) - \hat{x}^{(1)}(k-1), k = 1, 2, \ldots n,$$

we can obtain the time response formula of $X^{(0)}$, which is

$$\begin{aligned}
\hat{x}^{(0)}(k) = {}&\left(x^{(0)}(1) - \frac{b}{a}\right)\left(\frac{1 - 0.5a}{1 + 0.5a}\right)^k + \frac{b}{a} \\
&- \left[\left(x^{(0)}(1) - \frac{b}{a}\right)\left(\frac{1 - 0.5a}{1 + 0.5a}\right)^{k-1} + \frac{b}{a}\right].
\end{aligned}$$

That is

$$\hat{x}^{(0)}(k) = \left(\frac{-a}{1 - 0.5a}\right)\left(x^{(0)}(1) - \frac{b}{a}\right)\left(\frac{1 - 0.5a}{1 + 0.5a}\right)^k \tag{7.23}$$

Lemma 7.2.1 *When* $-a \to 0^+$, $\frac{1-0.5a}{1+0.5a} \approx e^{-a}$.

Proof The Maclaurin expansions of e^{-a} and $\frac{1-0.5a}{1+0.5a}$ are as follows:

$$e^{-a} = 1 - a + \frac{a^2}{2!} - \frac{a^3}{3!} + \cdots + (-1)^n \frac{a^n}{n!} + o(a^n)$$

$$\frac{1-0.5a}{1+0.5a} = 1 - a + \frac{a^2}{2} - \frac{a^3}{2^2} + \cdots + (-1)^{n+1} \frac{a^{n+1}}{2^n} + o(a^{n+1})$$

As n = 3, then there is $\Delta = e^{-a} - \frac{1-0.5a}{1+0.5a} = -\frac{a^3}{6} + \frac{a^3}{4} = \frac{a^3}{12}$, therefore, when $-a \to 0^+$, $\frac{1-0.5a}{1+0.5a} \approx e^{-a}$.

Theorem 7.2.5 *When* $-a \to 0^+$, *Even Model GM(1,1) and Discrete Model GM(1,1) are equivalent.*

Proof From the even form (7.4) of Model GM(1,1)

$$x^{(1)}(k+1) = \left(\frac{1-0.5a}{1+0.5a}\right)x^{(1)}(k) + \frac{b}{1+0.5a},$$

and contrast with the discrete form (7.7), we can obtain $\beta_1 = \frac{1-0.5a}{1+0.5a}$, $\beta_2 = \frac{b}{1+0.5a}$ and

$$a = \frac{2(1-\beta_1)}{1+\beta_1}, b = \frac{2\beta_2}{1+\beta_1}, \frac{b}{a} = \frac{\beta_2}{1-\beta_1}. \tag{7.24}$$

Take $\frac{b}{a} = \frac{\beta_2}{1-\beta_1}$ into formula (7.8), we can get

$$\hat{x}^{(1)}(k) = \left[x^{(0)}(1) - \frac{\beta_2}{1-\beta_1}\right]e^{-a(k-1)} + \frac{\beta_2}{1-\beta_1}, k = 1, 2, \ldots n \tag{7.25}$$

It is known from Lemma 1 that when $-a \to 0^+$, therefore, Even Model GM(1,1) and Discrete Model GM(1,1) are equivalent.

Analogously, we can prove that when $-a \to 0^+$, the four basic forms of model GM(1,1), namely Even Model GM(1,1) (EGM), Original Difference Model GM(1,1)(ODGM), Even Difference Model GM(1,1)(EDGM) and Discrete Model GM(1,1)(DGM) are pairwise equivalent. However, the degree of approximation between different forms is a difference. This difference leads to different forms of Model GM(1,1) being suitable for different situations, and it also offers a variety of possible options for the actual modeling process.

Theorem 7.2.6 *Original Difference Model GM(1,1)(ODGM), Even Difference Model GM(1,1) (EDGM) and Discrete Model GM(1,1) (DGM) can all accurately simulate homogeneous exponential sequences.*

Since the time response formulas of Original Difference Model GM(1,1)(ODGM), Even Difference GM(1,1) model (EDGM) and Discrete GM(1,1) model (DGM) are all geometric sequences, they can accurately simulate homogeneous exponential sequences.

In the basic forms of GM(1,1), the development coefficient $(-a)$ reflects the development states of $\hat{x}^{(1)}$ and $\hat{x}^{(0)}$. In general, the variables that act upon the system of interest should be external or pre-defined. Because GM(1,1) is a kind of model constructed on a single sequence, it uses only the behavioral sequence (also referred to as output sequence or background values) of the system without considering any externally acting sequences (also referred to as input sequences, or driving quantities). The grey action quantity b in the basic forms of GM(1,1) is a value derived from the background values. It reflects changes contained in the data and its exact intension is grey. This quantity realizes the extension of the relevant intension. Its existence distinguishes grey systems modeling from the general input–output (or black-box) modeling. It is also an important test stone of separating the thoughts of grey systems and those of grey boxes.

7.3 Suitable Ranges of Different GM(1,1)

The suitable sequences of different basic models of GM(1,1) (Liu et al., 2015a, 2015b) and the applicable ranges of EGM (Liu & Deng, 2000) are studied by simulation and analysis with homogeneous exponential sequences, non-exponential increasing sequences, and vibration sequences. It can provide reference and a basis for people to choose the correct model in the actual modeling process.

7.3.1 Suitable Sequences of Different GM(1,1)

For further study of the suitable sequences of four basic forms of model GM(1,1), we let

$$-a = 0.01, 0.02, 0.03, 0.04, 0.05, 0.1, 0.15, 0.2, 0.25,$$
$$0.3, 0.35, 0.4, 0.45, 0.5, 0.55, 0.6, 0.65, 0.7, 0.8, 0.9$$
$$1.0, 1.1, 1.2, 1.5, 1.8$$

and conduct simulation analysis, respectively. Let $k = 1, 2, 3, 4, 5$, with the homogeneous exponential function $x_i^{(0)}(k) = e^{-ak}$, and accurate to six decimal places. Then we can get the corresponding sequences as follows:

$$-a = 0.01, X_1^{(0)} = (x_1^{(0)}(1), x_1^{(0)}(2), x_1^{(0)}(3), x_1^{(0)}(4), x_1^{(0)}(5))$$
$$= (1.010050, 1.020201, 1.030455, 1.040811, 1.051271)$$

$$- a = 0.02, X_2^{(0)} = (x_2^{(0)}(1), x_2^{(0)}(2), x_2^{(0)}(3), x_2^{(0)}(4), x_2^{(0)}(5))$$
$$= (1.020201, 1.040811, 1.061837, 1.083287, 1.105171)$$

$$\cdots \quad \cdots \quad \cdots \quad \cdots \quad \cdots \quad \cdots \quad \cdots \quad \cdots \quad \cdots \quad \cdots \quad \cdots \quad \cdots$$

$$\cdots \quad \cdots \quad \cdots \quad \cdots \quad \cdots$$

$$- a = 1.8, X_{25}^{(0)} = (x_{25}^{(0)}(1), x_{25}^{(0)}(2), x_{25}^{(0)}(3), x_{25}^{(0)}(4), x_{25}^{(0)}(5))$$
$$= (6.049647, 36.59823, 221.4064, 1339.431, 8103.084).$$

We use $X_1^{(0)}$, $X_2^{(0)}$, ..., $X_{25}^{(0)}$ as the original data to establish Even Model GM(1,1)(EGM), Original Difference Model GM(1,1)(ODGM), Even Difference Model GM(1,1)(EDGM) and Discrete Model GM(1,1)(DGM), respectively. We can find that Original Difference Model GM(1,1)(ODGM), Even Difference Model GM(1,1)(EDGM) and Discrete Model GM(1,1)(DGM) can accurately simulate homogeneous exponential sequence, which confirms the conclusions of Theorem 7.2.7 once again. Using Even Model GM(1,1)(EGM) to simulate $X_1^{(0)}$, $X_2^{(0)}$, ..., $X_{25}^{(0)}$, it is fond that with the increasing of $-a$, the error will also increase. Table 7.1 shows the average relative error using four kinds of model GM(1,1) to simulate the homogeneous exponential sequence $X_1^{(0)}$, $X_2^{(0)}$, ..., $X_{25}^{(0)}$.

In Table 7.1, we can see that the small errors of Original Difference Model GM(1,1)(ODGM), Even Difference Model GM(1,1)(EDGM) and Discrete Model GM(1,1)(DGM) which simulate the homogeneous exponential sequence are all caused by round-off errors. In fact, the three models can all accurately simulate the homogeneous exponential sequence.

Then, we limited the range of random numbers at first, and got the non-exponential increasing sequence $Y_1^{(0)}$, $Y_2^{(0)}$, ..., $Y_{25}^{(0)}$ randomly generated by the homogeneous exponential sequence $X_1^{(0)}$, $X_2^{(0)}$, ..., $X_{25}^{(0)}$, along with the vibration sequence $Z_1^{(0)}$, $Z_2^{(0)}$, ..., $Z_{25}^{(0)}$. With that, when $k = 2, 3, \ldots, 5, z_i^{(0)}(k) < z_i^{(0)}(k-1), i = 1, 2, \ldots, 25$ will arise in the sequence data but there is a growth trend as a whole, both are equally accurate to six decimal places. Then we build Even Model GM(1,1)(EGM), Original Difference Model GM(1,1)(ODGM), Even Difference Model GM(1,1)(EDGM) and Discrete Model GM(1,1)(DGM) using sequences $Y_1^{(0)}$, $Y_2^{(0)}$, ..., $Y_{25}^{(0)}$ and $Z_1^{(0)}$, $Z_2^{(0)}$, ..., $Z_{25}^{(0)}$ respectively. The errors we can see are in Tables 7.2 and 7.4. Due to limited space, the generating data are not shown here.

From Table 7.2 we can see that four kinds of model GM(1,1) can all simulate the non-exponential increasing sequence to a certain degree. Generally speaking, the simulation error will increase with the increasing of the development index. In most cases, the simulation error of the difference, differential hybrid form of Even Model GM(1,1)(EGM), is smaller than that of the three discrete forms of Original Difference Model GM(1,1)(ODGM), Even Difference Model GM(1,1)(EDGM) and Discrete Model GM(1,1)(DGM). As the non-exponential increasing sequence is closer to the homogeneous exponential sequence, the simulation accuracy of the three discrete models is higher than. When the non-exponential increasing sequence is close to the homogeneous exponential sequence to a certain extent, the simulation accuracy of

Table 7.1 The simulation errors of the homogeneous exponential sequence of four kinds of GM(1,1) (%)

Code	$-a$	EGM	DGM	ODGM	EDGM
$X_1^{(0)}$	0.01	0.000849	0.000027	0.000027	0.000027
$X_2^{(0)}$	0.02	0.003468	0.000013	0.000013	0.000013
$X_3^{(0)}$	0.03	0.007951	0.000018	0.000018	0.000018
$X_4^{(0)}$	0.04	0.014403	0.000004	0.000004	0.000003
$X_5^{(0)}$	0.05	0.022922	0.000016	0.000016	0.000016
$X_6^{(0)}$	0.10	0.100058	0.000008	0.000008	0.000008
$X_7^{(0)}$	0.15	0.244034	0.000009	0.000009	0.000009
$X_8^{(0)}$	0.20	0.467588	0.000003	0.000003	0.000007
$X_9^{(0)}$	0.25	0.783590	0.000005	0.000005	0.000006
$X_{10}^{(0)}$	0.30	1.205144	0.000004	0.000004	0.000010
$X_{11}^{(0)}$	0.35	1.745610	0.000006	0.000006	0.000010
$X_{12}^{(0)}$	0.40	2.418758	0.000004	0.000004	0.000010
$X_{13}^{(0)}$	0.45	3.238864	0.000007	0.000007	0.000008
$X_{14}^{(0)}$	0.50	4.220851	0.000011	0.000011	0.000008
$X_{15}^{(0)}$	0.55	5.380507	0.000003	0.000003	0.000003
$X_{16}^{(0)}$	0.60	6.734574	0.000016	0.000016	0.000011
$X_{17}^{(0)}$	0.65	8.301040	0.000009	0.000009	0.000006
$X_{18}^{(0)}$	0.70	10.099355	0.000021	0.000021	0.000021
$X_{19}^{(0)}$	0.80	14.478513	0.000015	0.000015	0.000015
$X_{20}^{(0)}$	0.90	20.068449	0.000016	0.000016	0.000022
$X_{21}^{(0)}$	1.00	27.110835	0.000047	0.000047	0.000047
$X_{22}^{(0)}$	1.10	35.908115	0.000040	0.000040	0.000035
$X_{23}^{(0)}$	1.20	46.844843	0.000105	0.000105	0.000105
$X_{24}^{(0)}$	1.50	98.188500	0.000129	0.000129	0.000129
$X_{25}^{(0)}$	1.80	–	0.000433	0.000433	0.000433

the discrete models will be smaller than that of Even Model GM(1,1)(EGM). From the simulation results of the three discrete models GM(1,1), we can see that with the increasing of the development coefficient, the simulation accuracy of Original Difference Model GM(1,1)(ODGM), and Even Difference Model GM(1,1)(EDGM) is higher than that of Discrete Model GM(1,1)(DGM) in most cases. The statistics for sorting the simulation error of different models with the sequence $Y_1^{(0)}$, $Y_2^{(0)}$, …,

Table 7.2 The simulation errors of the non-exponential increasing sequence of four kinds of GM(1,1) (%)

Code	$-$, $-a$, $+$	EGM	DGM	ODGM	EDGM
$Y_1^{(0)}$	0.01	0.030994	0.030429	0.030432	0.030430
$Y_2^{(0)}$	0.02	0.658978	0.659039	0.660095	0.659572
$Y_3^{(0)}$	0.03	0.495833	0.495773	0.495768	0.495770
$Y_4^{(0)}$	0.04	1.010474	1.010308	1.010329	1.010319
$Y_5^{(0)}$	0.05	1.550886	1.550331	1.550468	1.550401
$Y_6^{(0)}$	0.10	1.626294	1.704980	1.690324	1.697211
$Y_7^{(0)}$	0.15	1.343565	1.457800	1.458993	1.458442
$Y_8^{(0)}$	0.20	5.155856	5.100486	5.229480	5.171925
$Y_9^{(0)}$	0.25	4.353253	4.893857	4.743792	4.808361
$Y_{10}^{(0)}$	0.30	4.736323	5.345755	5.168529	5.244168
$Y_{11}^{(0)}$	0.35	5.236438	5.377225	5.192273	5.269577
$Y_{12}^{(0)}$	0.40	3.603875	4.166958	4.044567	4.096904
$Y_{13}^{(0)}$	0.45	12.834336	15.364230	13.520184	14.246584
$Y_{14}^{(0)}$	0.50	7.396770	8.276073	7.878017	8.044898
$Y_{15}^{(0)}$	0.55	10.218727	10.084749	10.188912	10.143461
$Y_{16}^{(0)}$	0.60	21.073070	23.709858	21.610863	22.440905
$Y_{17}^{(0)}$	0.65	6.637022	7.906483	7.629068	7.731359
$Y_{18}^{(0)}$	0.70	9.088900	11.000565	10.479505	10.677398
$Y_{19}^{(0)}$	0.80	21.156265	30.606589	28.554915	29.245194
$Y_{20}^{(0)}$	0.90	14.441947	20.378328	17.104000	18.188008
$Y_{21}^{(0)}$	1.00	11.685913	18.463203	17.357496	17.734931
$Y_{22}^{(0)}$	1.10	13.011857	20.620317	19.396248	19.782271
$Y_{23}^{(0)}$	1.20	17.176472	27.929743	26.163490	26.624283
$Y_{24}^{(0)}$	1.50	26.327218	51.915584	50.006882	50.471089
$Y_{25}^{(0)}$	1.80	62.460946	75.503705	73.434001	74.070128

$Y_{25}^{(0)}$ in ascending order are presented in Table 7.2. Table 7.3 shows the statistical results.

As can be seen from Table 7.3, among the four kinds of models, Even Model GM(1,1)(EGM) is the most suitable for modeling with a non-exponential increasing sequence, followed by the Original Differential Model GM(1,1) (ODGM) and Even Difference Model GM(1,1) (EDGM). The error is slightly larger when using

Table 7.3 Statistics for sorting the simulation error of the non-exponential increasing sequence of four kinds of model GM(1,1)

Error sorting	EGM	DGM	ODGM	EDGM
1	18	5	2	0
2	2	2	15	6
3	0	1	5	19
4	5	17	3	0

Table 7.4 Simulation errors of the vibration sequence of four kinds of model GM(1,1)

Code	$-, -a, +$	EGM	DGM	ODGM	EDGM
$z_1^{(0)}$	0.01	0.298392	0.299400	0.299118	0.299258
$z_2^{(0)}$	0.02	0.501223	0.505800	0.504877	0.505331
$z_3^{(0)}$	0.03	0.369630	0.378773	0.379089	0.378935
$z_4^{(0)}$	0.04	2.583662	2.586760	2.572109	2.579300
$z_5^{(0)}$	0.05	2.928035	2.953655	2.899369	2.925619
$z_6^{(0)}$	0.10	4.759929	4.791858	4.825226	4.807851
$z_7^{(0)}$	0.15	3.802630	3.770562	3.776330	3.773545
$z_8^{(0)}$	0.20	11.723459	11.946525	11.393483	11.642630
$z_9^{(0)}$	0.25	14.895391	14.979357	15.229595	15.130729
$z_{10}^{(0)}$	0.30	17.953543	17.992976	18.397577	18.241183
$z_{11}^{(0)}$	0.35	7.299184	8.980062	8.537865	8.708603
$z_{12}^{(0)}$	0.40	11.474779	11.519781	11.693309	11.619287
$z_{13}^{(0)}$	0.45	11.988111	12.321804	12.261075	12.286039
$z_{14}^{(0)}$	0.50	12.728220	11.753460	12.270432	12.038094
$z_{15}^{(0)}$	0.55	10.636507	10.285910	10.897796	10.623904
$z_{16}^{(0)}$	0.60	13.393234	13.515007	13.006751	13.227910
$z_{17}^{(0)}$	0.65	15.420377	15.457643	14.690315	15.004381
$z_{18}^{(0)}$	0.70	16.304197	16.365096	15.735103	15.998031
$z_{19}^{(0)}$	0.80	14.542100	14.579829	14.110548	14.310293
$z_{20}^{(0)}$	0.90	33.798587	33.160101	34.928437	34.293058
$z_{21}^{(0)}$	1.00	22.586380	22.384127	22.016157	22.145609
$z_{22}^{(0)}$	1.10	34.305920	34.481612	36.023522	35.484180
$z_{23}^{(0)}$	1.20	23.591927	24.133298	23.323921	21.511839
$z_{24}^{(0)}$	1.50	40.373380	40.475348	42.698005	41.917026
$z_{25}^{(0)}$	1.80	30.380522	54.851229	45.724311	48.579850

Table 7.5 Statistics for sorting the simulation error of the vibration sequence of four kinds of model GM(1,1)

Error sorting	EGM	DGM	ODGM	EDGM
1	12	4	8	1
2	1	7	6	11
3	9	1	2	13
4	3	13	9	0

the Discrete Model GM(1,1)(DGM) to simulate the non-exponential increasing sequence.

In theory, any simple model which describes a monotonous trend struggles to describe a change in the vibration sequence. Therefore, we add the limiting condition of the random number, then the research range is the vibration sequence $Z_1^{(0)}, Z_2^{(0)}, \ldots,$ $Z_{25}^{(0)}$. With that, when $k = 2, 3, \ldots, 5$, $z_i^{(0)}(k) < z_i^{(0)}(k-1), i = 1, 2, \ldots, 25$ will arise in the sequence data, but there is a growth trend as a whole. We can see from Table 7.4 that, for this specific vibration sequence, the simulation error of the four kinds of models is significantlyc higher than the non-exponential increasing sequence. Similar to the situation of the non-exponential increasing sequence, in most cases the simulation error of Even Model GM(1,1)(EGM) to the vibration sequence is smaller than that of the three discrete forms of Original Difference Model GM(1,1)(ODGM), Even Difference Model GM(1,1)(EDGM) and Discrete Model GM(1,1)(DGM). For the vibration sequence being close to the homogeneous exponential sequence, the simulation error of the discrete model is smaller than one of the difference, differential hybrid form of Even Model GM(1,1)(EGM).

The statistics for sorting the simulation error of different models with the vibration sequence $Z_1^{(0)}, Z_2^{(0)}, \ldots, Z_{25}^{(0)}$ in ascending order are presented in Table 7.4. Table 7.5 shows the statistical results.

As can be seen in Table 7.5, of the four kinds of models the Even Model GM(1,1)(EGM) is more suitable for modeling with vibration sequence than the other three discrete form models. The error using Discrete Model GM(1,1) (DGM) to simulate the vibration sequence is slightly larger than other two discrete form models.

The authors once tried to use the original form (7.1) of Model GM(1,1) to estimate the parameter vector $\hat{a} = [a, b]^T$ and, in accordance with the solution of whitenization Eq. (7.6) along with the time response formula of Even Model GM(1,1)(EGM), modeled the original Model GM(1,1). After simulating the above data we found that, even in cases where the development index is very small, the simulation error was still comparatively large. Also, as the development index increases, the simulation error increases rapidly. Based on even transformation of the accumulation data to build the Even Model GM(1,1), the simulation accuracy improves greatly. Then a new method which can accurately simulate and predict the uncertain system involving small data and poor information comes into being.

Among the four basic forms of model GM(1,1) discussed in Sects. 7.3 and 7.4, three discrete models can all accurately simulate the homogeneous exponential sequence. In the real world, a mass of practical data are not the simple homogeneous exponential sequence or close to it. This is the fundamental reason that people prefer to choose Even Model GM(1,1)(EGM) in the modeling process of the uncertain system involving small data and poor information, and it can reflect a satisfactory result in most cases.

In Sects. 7.3 and 7.4, the definitions of four basic forms of model GM(1,1) are put forward, and the properties and characteristics of different models are studied in-depth. The suitable sequences of different models are studied by simulation and analysis with homogeneous exponential sequences, non-exponential increasing sequences, and vibration sequences. The main conclusions of the research are as follows:

(1) The four basic forms of model GM(1,1), namely Even Model GM(1,1) (EGM), Original Difference Model GM(1,1)(ODGM), Even Difference Model GM(1,1)(EDGM) and Discrete Model GM(1,1)(DGM) are pairwise equivalent.
(2) Original Difference Model GM(1,1)(ODGM), Even Difference Model GM(1,1)(EDGM) and Discrete Model GM(1,1)(DGM) can all simulate the homogeneous exponential sequence accurately.
(3) For the non-exponential increasing sequences and vibration sequences, we should first choose the difference, differential hybrid form of Even Model GM(1,1)(EGM).
(4) For the non-exponential increasing sequences and vibration sequences which are close to the homogeneous exponential sequences, we should first choose the discrete form of Original Difference Model GM(1,1)(ODGM), Even Difference Model GM(1,1)(EDGM) or Discrete Model GM(1,1)(DGM).

The conclusions above can be the reference and basis for choosing an appropriate model in the actual modeling process. There is a modeling software corresponding to the models. Interested readers can download it for free from the website of the Institute for Grey System Studies of Nanjing University of Aeronautics and Astronautics (http://igss.nuaa.edu.cn) or from the website of the Marie Curie International Incoming Fellowship project (FP7-People-IIF-GA-2013-629051) (http://preview. dmu.ac.uk/research/research-faculties-and-institutes/technology/cci/projects/).

Example 6.2.1 Let sequences of $X_1^{(0)}$, $X_2^{(0)}$ and $X_3^{(0)}$ be as follows,

$$X_1^{(0)} = (x_1^{(0)}(1), x_1^{(0)}(2), x_1^{(0)}(3), x_1^{(0)}(4), x_1^{(0)}(5))$$
$$= (1.5, 2.1, 3.0, 4.5, 5.48)$$

$$X_2^{(0)} = (x_2^{(0)}(1), x_2^{(0)}(2), x_2^{(0)}(3), x_2^{(0)}(4), x_2^{(0)}(5), x_2^{(0)}(6))$$
$$= (1.5, 1.3, 3.0, 3.9, 7.2, 9.5)$$

Table 7.6 Simulation errors of four different models with $X_1^{(0)}$

Models	EGM	DGM	ODGM	EDGM
Mean relative errors (%)	4.7363	5.3458	5.1685	5.2442

$$X_3^{(0)} = (x_3^{(0)}(1), x_3^{(0)}(2), x_3^{(0)}(3), x_3^{(0)}(4), x_3^{(0)}(5))$$
$$= (2, 9, 32, 27, 55)$$

Try to build the Even Model GM(1,1)(EGM), Discrete Model GM(1,1)(DGM), Original Difference Model GM(1,1)(ODGM), and Even Difference Model GM(1,1)(EDGM) using sequences $X_1^{(0)}$, $X_2^{(0)}$ and $X_3^{(0)}$. Compare the simulation errors.

Solution (1) For $X_1^{(0)}$, we build Even Model GM(1,1)(EGM), Discrete Model GM(1,1)(DGM), Original Difference Model GM(1,1)(ODGM), and Even Difference Model GM(1,1)(EDGM) using 1.5, 2.1, 3.0, 4.5, 5.48. We then obtained the simulation results as follows.

Simulation results by EGM: $\hat{X}_1^{(0)} = (1.5000, 2.2459, 3.0428, 4.1225, 5.5853)$

Simulation results by DGM: $\hat{X}_1^{(0)} = (1.5000, 2.2746, 3.0844, 4.1827, 5.6719)$

Simulation results by ODGM: $\hat{X}_1^{(0)} = (1.5000, 2.2600, 3.0726, 4.1772, 5.6789)$

Simulation results by EDGM: $\hat{X}_1^{(0)} = (1.5000, 2.2662, 3.0776, 4.1795, 5.6760)$ (Table 7.6).

(2) For $X_2^{(0)}$, we build EGM, DGM, ODGM, and EDGM using 1.5, 1.3, 3.0, 3.9, 7.2, 9.5. Then we obtained the simulation results as follows.

Simulation results by EGM: $\hat{X}_2^{(0)} = (1.5000, 1.8632, 2.8290, 4.29556.5220, 9.9028)$

Simulation results by DGM: $\hat{X}_2^{(0)} = (1.5000, 1.9247, 2.9317, 4.4654, 6.8016, 10.3599)$

Simulation results by ODGM: $\hat{X}_2^{(0)} = (1.5000, 1.8793, 2.8771, 4.4047, 6.7433, 10.3236)$

Simulation results by EDGM: $\hat{X}_2^{(0)} = (1.5000, 1.8973, 2.8988, 4.4290, 6.7669, 10.3388)$ (Table 7.7).

(3) For $X_3^{(0)}$, we build EGM, DGM, ODGM, and EDGM using 2, 9, 32, 27, 55. Then we obtained the simulation results as follows.

Table 7.7 Simulation errors of four different models with $X_2^{(0)}$

Models	EGM	DGM	ODGM	EDGM
Mean relative errors (%)	11.9881	12.3218	12.2611	12.2860

Table 7.8 Simulation errors of four different models with $X_3^{(0)}$

Models	EGM	DGM	ODGM	EDGM
Mean relative errors (%)	27.2510	25.9994	26.4180	26.1794

Simulation results by EGM: $\hat{X}_3^{(0)} = (2.0000, 13.9767, 21.6340, 33.4864, 51.8323)$

Simulation results by DGM: $\hat{X}_3^{(0)} = (2.0000, 15.4516, 23.4647, 35.6332, 54.1122)$

Simulation results by ODGM: $\hat{X}_3^{(0)} = (2.0000, 13.4756, 21.3666, 33.8782, 53.7164)$

Simulation results by EDGM: $\hat{X}_3^{(0)} = (2.0000, 14.2602, 22.2313, 34.6581, 54.0311)$
(Table 7.8)

The simulation results with $X_1^{(0)}$, $X_2^{(0)}$ and $X_3^{(0)}$ confirmed the above conclusion once again.

7.3.2 Applicable Ranges of EGM

Proposition 7.3.1 When $(n-1)\sum_{k=2}^{n}\left[z^{(1)}(k)\right]^2 \to \left[\sum_{k=2}^{n} z^{(1)}(k)\right]^2$, the EGM(1,1) becomes invalid.

Proof By using the model parameters obtained by the least squared estimate, we have.

$$\hat{a} = \frac{\sum_{k=2}^{n} z^{(1)}(k) \sum_{k=2}^{n} x^{(0)}(k) - (n-1)\sum_{k=2}^{n} z^{(1)}(k)x^{(0)}(k)}{(n-1)\sum_{k=2}^{n}\left[z^{(1)}(k)\right]^2 - \left[\sum_{k=2}^{n} z^{(1)}(k)\right]^2}$$

$$\hat{b} = \frac{\sum_{k=2}^{n} x^{(0)}(k) \sum_{k=2}^{n}\left[z^{(1)}(k)\right]^2 - \sum_{k=2}^{n} z^{(1)}(k) \sum_{k=2}^{n} z^{(1)}(k)x^{(0)}(k)}{(n-1)\sum_{k=2}^{n}\left[z^{(1)}(k)\right]^2 - \left[\sum_{k=2}^{n} z^{(1)}(k)\right]^2}$$

When $(n-1)\sum_{k=2}^{n}\left[z^{(1)}(k)\right]^2 \to \left[\sum_{k=2}^{n} z^{(1)}(k)\right]^2$, $\hat{a} \to \infty$, $\hat{b} \to \infty$, so that the model parameters cannot be determined. Hence, the EGM(1,1) becomes invalid.

Proposition 7.3.2 When the development coefficient a of the EGM(1,1) model satisfies $|a| \geq 2$, the GM(1,1) model becomes invalid.

Proof From the following expression of the GM(1,1) model

$$x^{(0)}(k) = \left(\frac{1-0.5a}{1+0.5a}\right)^{k-2}\left(\frac{b-ax^{(0)}(1)}{1+0.5a}\right); k = 2, 3, \ldots, n$$

it can be seen that when $a = -2$, $x^{(0)}(k) \to \infty$; when $a = 2$, $x^{(0)}(k) = 0$; and when $|a| > 2$, $\frac{b-ax^{(0)}(1)}{1+0.5a}$ becomes a constant, while the sign of $\left(\frac{1-0.5a}{1+0.5a}\right)^{k-2}$ changes with k being even or odd. Thus, the sign of $x^{(0)}(k)$ flips with k being even or odd.

The discussion above indicates that $(-\infty, -2] \cup [2, \infty)$ is the forbidden area for the development coefficient $(-a)$ of the GM(1,1) model. When $a \in (-\infty, -2] \cup [2, \infty)$, the GM(1,1) model loses its validity. In general, when $|a| < 2$, the GM(1,1) model is meaningful. However, for different values of a, the prediction effect of the model is different. For the case of $-2 < a < 0$, let us respectively take $-a = 0.1, 0.2, 0.3, 0.4, 0.5, 0.6, 0.8, 1.5, 1.8$ to conduct a simulation analysis. By taking $k = 0, 1, 2, 3, 4, 5$, from $x_i^{(0)}(k+1) = e^{-ak}$, we obtain the following sequences:

If $-a = 0.1$, $X_1^{(0)} = (x_1^{(0)}(1), x_1^{(0)}(2), x_1^{(0)}(3), x_1^{(0)}(4), x_1^{(0)}(5), x_1^{(0)}(6))$
$= (1, 1.1051, 1.2214, 1.3499, 1.4918, 1.6487)$

If $-a = 0.2$, $X_2^{(0)} = (1, 1.2214, 1.4918, 1.8221, 2.2255, 2.7183)$.

If $-a = 0.3$, $X_3^{(0)} = (1, 1.3499, 1.8221, 2.4596, 3.3201, 4.4817)$.

If $-a = 0.4$, $X_4^{(0)} = (1, 1.4918, 2.225, 3.3201, 4.9530, 7.3890)$.

If $-a = 0.5$, $X_5^{(0)} = (1, 1.6487, 2.7183, 4.4817, 7.3890, 12.1825)$.

If $-a = 0.6$, $X_6^{(0)} = (1, 1.8821, 3.3201, 6.0496, 11.0232, 20.0855)$.

If $-a = 0.8$, $X_7^{(0)} = (1, 2.2255, 4.9530, 11.0232, 24.5325, 54.5982)$.

If $-a = 1$, $X_8^{(0)} = (1, 2.7183, 7.3890, 20.0855, 54.5982, 148.4132)$.

If $-a = 1.5$, $X_9^{(0)} = (1, 4.4817, 20.0855, 90.0171, 403.4288, 1808.0424)$.

If $-a = 1.8$, $X_{10}^{(0)} = (1, 6.0496, 36.5982, 221.4064, 1339.4308, 8103.0839)$.

Let us respectively apply $X_1^{(0)}$, $X_2^{(0)}$, ..., and $X_9^{(0)}$ to establish a GM(1,1) model and obtain the following time response sequences:

$$\hat{x}_1^{(1)}(k+1) = 10.50754e^{0.09992182k} - 9.507541,$$

$$\hat{x}_2^{(1)}(k+1) = 5.516431e^{0.1993401k} - 4.516431,$$

$$\hat{x}_3^{(1)}(k+1) = 3.85832e^{0.297769k} - 2.858321,$$

$$\hat{x}_4^{(1)}(k+1) = 3.033199e^{0.394752k} - 2.033199$$

$$\hat{x}_5^{(1)}(k+1) = 2.541474e^{0.4898382k} - 1.541474,$$

$$\hat{x}_6^{(1)}(k+1) = 2.216363e^{0.5826263k} - 1.216362,$$

$$\hat{x}_7^{(1)}(k+1) = 1.815972e^{0.7598991k} - 0.8159718,$$

$$\hat{x}_8^{(1)}(k+1) = 1.581973e^{0.9242348k} - 0.5819733,$$

$$\hat{x}_9^{(1)}(k+1) = 1.287182e^{1.270298k} - 0.2871823,$$

$$\hat{x}_{10}^{(1)}(k+1) = 0.198197e^{1.432596k} - 0.1981966.$$

From $\hat{x}_i^{(0)}(k+1) = \hat{x}_i^{(1)}(k+1) - \hat{x}_i^{(1)}(k)$, $i = 1, 2, \ldots, 10$, we obtain

$$\hat{x}_1^{(0)}(k+1) = 0.99918e^{0.09992182k}, \hat{x}_2^{(0)}(k+1) = 0.99698e^{0.1993401k},$$

$$\hat{x}_3^{(0)}(k+1) = 0.99362e^{0.297769k}, \hat{x}_4^{(0)}(k+1) = 0.989287e^{0.394752k},$$

$$\hat{x}_5^{(0)}(k+1) = 0.984248e^{0.4898382k}, \hat{x}_6^{(0)}(k+1) = 0.97868e^{0.5826263k},$$

$$\hat{x}_7^{(0)}(k+1) = 0.966617e^{0.7598991k}, \hat{x}_8^{(0)}(k+1) = 0.95419e^{0.9242348k},$$

$$\hat{x}_9^{(0)}(k+1) = 0.925808e^{1.270298k}, \hat{x}_{10}^{(0)}(k+1) = 0.91220e^{1.432596k}.$$

From the mean generation of $z^{(1)}(k) = \frac{1}{2}(x^{(1)}(k) + x^{(1)}(k-1))$ of GM(1,1) model $x^{(0)}(k) + az^{(1)}(k) = b$, it has the effect of weakening the growth for increasing sequences. For an exponential sequence, the established GM(1,1) has a small development coefficient.

Let us compare the errors between the original sequence $X_i^{(0)}$ and the simulation sequence $\hat{X}_i^{(0)}$, as seen in Table 7.9.

It can be seen that as the development coefficient increases, the simulation error grows drastically. When the development coefficient is smaller than or equal to 0.3, the simulation accuracy can reach above 98%. When the coefficient is smaller than or equal 0.5, the simulation accuracy can reach above 95%. When the coefficient is greater than 1, the simulation accuracy is lower than 70%. When the coefficient is

Table 7.9 The simulation errors of different development coefficients $(-a)$

Development coefficient $(-a)$	$\frac{1}{5}\sum_{i=2}^{6}[\hat{x}^{(0)}(k) - x^{(0)}(k)]$	Mean relative error $\frac{1}{5}\sum_{k=2}^{6}\Delta_k$ (%)
0.1	0.004	0.104
0.2	0.010	0.499
0.3	0.038	1.300
0.4	0.116	2.613
0.5	0.307	4.520
0.6	0.741	7.074
0.8	3.603	14.156
1	14.807	23.544
1.5	317.867	51.033
1.8	1632.240	65.454

Table 7.10 Prediction errors

$-a$	0.1	0.2	0.3	0.4	0.5	0.6	0.8	1	1.5	1.8
Step 1 error (%)	0.129	0.701	1.998	4.317	7.988	13.405	31.595	65.117	–	–
Step 2 error (%)	0.137	0.768	2.226	4.865	9.091	15.392	36.979	78.113	–	–
Step 5 error (%)	0.160	0.967	2.912	6.529	12.468	21.566	54.491	–	–	–
Step 5 error (%)	0.855	1.301	4.067	9.362	18.330	32.599	88.790	–	–	–

greater than 1.5, the simulation accuracy is lower than 50% (Liu & Deng, 2000; Liu, 2021).

Let us now further focus on the first step, second step, fifth step, and 10th step prediction errors. See Table 7.10.

It can be seen that when the development coefficient is smaller than 0.3, the step 1 prediction accuracy is above 98%, with both steps 2 and 5 accuracies above 97%. When $0.3 < -a \leq 0.5$, the steps 1 and 2 prediction accuracies are all above 90%; and the step 10 prediction accuracy also above 80%. When the development coefficient is greater than 0.8, the step 1 prediction accuracy is below 70%. The horizontal bars in Table 4.5 represent that the relevant errors are greater than 100%.

From this analysis, we can draw the following conclusions: When $-a \leq 0.3$, GM(1,1) can be applied to make mid- to long-term predictions; when $0.3 < -a \leq 0.5$, GM(1,1) can be applied to make short- and mid-term predictions with caution; when $0.5 < -a \leq 0.8$ and GM(1,1) is used to make short-term predictions, one needs to be very cautious about the prediction results; when $0.8 < -a \leq 1$, one should employ the remnant GM(1,1) model; and when $-a > 1$, GM(1,1) should not be applied (Liu & Deng, 2000; Liu et al., 2017).

7.4 Remnant GM(1,1) Model

When the accuracy of a GM(1,1) model does not meet the predetermined requirement, one can establish another GM(1,1) model using the error sequence to remedy the original model to improve the accuracy. We will use the remnant GM(1,1) of EGM(1) as an example (Deng, 1985; Liu, 2021).

Definition 7.4.1 Assume that $X^{(0)}$ is a sequence of raw data, $X^{(1)}$ the accumulation generated sequence based on $X^{(0)}$, and the time response formula of the GM(1,1) model is.

$$\hat{x}^{(1)}(k+1) = \left(x^{(0)}(1) - \frac{b}{a}\right) e^{-ak} + \frac{b}{a}$$

then

$$d\,\hat{x}^{(1)}(k+1) = (-a)\left(x^{(0)}(1) - \frac{b}{a}\right)e^{-ak} \tag{7.26}$$

is referred to as the restored value through derivatives.

Generally, $d\hat{x}^{(1)}(k+1) \neq \hat{x}^{(0)}(k+1)$, where $\hat{x}^{(0)}(k+1) = \hat{x}^{(1)}(k+1) - \hat{x}^{(1)}(k)$ stands for the restored value through inverse accumulation. This very fact implies that the GM(1,1) is neither a differential equation nor a difference equation. However, when $|a|$ is sufficiently small, from $1 - e^a \approx -a$, it follows that $d\,\hat{x}^{(1)}(k+1) \approx \hat{x}^{(0)}(k+1)$, meaning that the results of differentiation and difference are quite close. Therefore, the GM(1,1) model in this case can be seen as both a differential equation and a difference equation.

Because the restored values through derivatives and through inverse accumulation are different, to reduce possible errors caused by reciprocating operators, the errors of $X^{(1)}$ are often used to improve the simulated values $\hat{x}^{(1)}(k+1)$ of $X^{(1)}$.

Definition 7.4.2 Assume that $\varepsilon^{(0)} = (\varepsilon^{(0)}(1), \varepsilon^{(0)}(2), \ldots, \varepsilon^{(0)}(n))$, where $\varepsilon^{(0)}(k) = x^{(1)}(k) - \hat{x}^{(1)}(k)$, is the error sequence of $X^{(1)}$. If there is a k_0 satisfying that $n - k_0 \geq 4$ and $\forall k \geq k_0$, the signs of $\varepsilon^{(0)}(k)$ stay the same, and $(|\varepsilon^{(0)}(k_0)|, |\varepsilon^{(0)}(k_0 + 1)|, , \ldots, |\varepsilon^{(0)}(n)|)$ is referred to as the error sequence of modelability, which is and still denoted $\varepsilon^{(0)} = (\varepsilon^{(0)}(k_0), \varepsilon^{(0)}(k_0 + 1), \ldots, \varepsilon^{(0)}(n))$.

In this case, let the sequence $\varepsilon^{(1)} = (\varepsilon^{(1)}(k_0), \varepsilon^{(1)}(k_0 + 1), \ldots, \varepsilon^{(1)}(n))$ be accumulation generated on $\varepsilon^{(0)}$ with the following GM(1,1) time response formula:

$$\hat{\varepsilon}^{(1)}(k+1) = \left(\varepsilon^{(0)}(k_0) - \frac{b_\varepsilon}{a_\varepsilon}\right)\exp[-a_\varepsilon(k - k_0)] + \frac{b_\varepsilon}{a_\varepsilon}, k \geq k_0$$

Then the simulation sequence of $\varepsilon^{(0)}$ is given by $\hat{\varepsilon}^{(0)} = (\hat{\varepsilon}^{(0)}(k_0), \hat{\varepsilon}^{(0)}(k_0 + 1), \ldots, \hat{\varepsilon}^{(0)}(n))$, where

$$\hat{\varepsilon}^{(0)}(k+1) = (-a_\varepsilon)\left(\varepsilon^{(0)}(k_0) - \frac{b_\varepsilon}{a_\varepsilon}\right)\exp[-a_\varepsilon(k - k_0)], k \geq k_0.$$

Definition 7.4.3 If $\hat{\varepsilon}^{(0)}$ is used to improve $\hat{X}^{(1)}$, the modified time response formula

$$\hat{x}^{(1)}(k+1) = \begin{cases} \left(x^{(0)}(1) - \frac{b}{a}\right)e^{-ak} + \frac{b}{a}, & k < k_0 \\ \left(x^{(0)}(1) - \frac{b}{a}\right)e^{-ak} + \frac{b}{a} \pm a_\varepsilon\left(\varepsilon^{(0)}(k_0) - \frac{b_\varepsilon}{a_\varepsilon}\right)e^{-a_\varepsilon(k-k_0)}, & k \geq k_0 \end{cases} \tag{7.27}$$

is referred to as the GM(1,1) model with error modification, or simply remnant GM(1,1) for short, where the sign of the error modification value

$$\hat{\varepsilon}^{(0)}(k+1) = a_\varepsilon \times \left(\varepsilon^{(0)}(k_0) - \frac{b_\varepsilon}{a_\varepsilon}\right)\exp[-a_\varepsilon(k - k_0)]$$

needs to stay the same as those in $\varepsilon^{(0)}$.

If a modeling of the error sequence $\varepsilon^{(0)} = (\varepsilon^{(0)}(k_0), \varepsilon^{(0)}(k_0 + 1), \ldots, \varepsilon^{(0)}(n))$ of $X^{(0)}$ and $\hat{X}^{(0)}$ is used to modify the simulation value $\hat{X}^{(0)}$, then different methods of restoration from $\hat{X}^{(1)}$ to $\hat{X}^{(0)}$ can produce different time response sequences of error modification.

Definition 7.4.4 Let

$$\hat{x}^{(0)}(k) = \hat{x}^{(1)}(k) - \hat{x}^{(1)}(k - 1) = \left(1 - e^a\right)\left(x^{(0)}(1) - \frac{b}{a}\right)e^{-a(k-1)}$$

Then the corresponding time response sequence of error modification

$$\hat{x}^{(0)}(k + 1) = \begin{cases} \left(1 - e^a\right)\left(x^{(0)}(1) - \frac{b}{a}\right)e^{-ak}, & k < k_0 \\ \left(1 - e^a\right)\left(x^{(0)}(1) - \frac{b}{a}\right)e^{-ak} \pm a_\varepsilon\left(\varepsilon^{(0)}(k_0) - \frac{b_\varepsilon}{a_\varepsilon}\right)e^{-a_\varepsilon(k-k_0)}, & k \geq k_0 \end{cases}$$

$$(7.28)$$

is called the error modification model of inverse accumulation restoration.

Definition 7.4.5 Let

$$\hat{x}^{(0)}(k + 1) = (-a)\left(x^{(0)}(1) - \frac{b}{a}\right)e^{-ak},$$

then the corresponding time response sequence of error modification

$$\hat{x}^{(0)}(k + 1) = \begin{cases} (-a)\left(x^{(0)}(1) - \frac{b}{a}\right)e^{-ak}, & k < k_0 \\ (-a)\left(x^{(0)}(1) - \frac{b}{a}\right)e^{-ak} \pm a_\varepsilon\left(\varepsilon^{(0)}(k_0) - \frac{b_\varepsilon}{a_\varepsilon}\right)e^{-a_\varepsilon(k-k_0)}, & k \geq k_0 \end{cases}$$

$$(7.29)$$

is referred to as the error modification model of derivative restoration.

In the previous discussion, all the error simulation terms in remnant GM(1,1) have been taken as the derivative restoration. Of course, they can be taken as inverse accumulation restoration. That is, one can take

$$\hat{\varepsilon}^{(0)}(k + 1) = \left(1 - e^{a_\varepsilon}\right)\left(\varepsilon^{(0)}(k_0) - \frac{b_\varepsilon}{a_\varepsilon}\right)e^{-a_\varepsilon(k-k_0)}, \quad k \geq k_0$$

As long as $|a_\varepsilon|$ is sufficiently small, the effects of different error restoration methods on the modified $\hat{x}^{(0)}(k + 1)$ are almost the same.

Example 7.4.1 Let

$$X^{(0)} = (x^{(0)}(1), x^{(0)}(2), \ldots, x^{(0)}(13))$$

$$= (6, 20, 40, 25, 40, 45, 35, 21, 14, 18, 15.5, 17, 15)$$

be a sequence of raw data, and the creation of a EGM(1, 1) model produce the following time response sequence:

$$\hat{x}^{(1)}(k + 1) = -567.999e^{-0.06486k} + 573.999$$

The application of inverse accumulating restoration gives:

$$\hat{X}^{(0)} = \{\hat{x}^{(0)}(k)\}_2^{13} = (35.6704, 33.4303, 31.3308, 29.3682, 27.5192, 25.7900,$$
$$24.1719, 22.6534, 21.2307, 19.8974, 18.6478, 17.4768)$$

The errors and relative errors of the results can be seen in Table 7.11.

From Table 7.11, it can be seen that the simulation error is relatively large. Thus, it is necessary to apply a remnant model to remedy some of the errors.

Let $k_0 = 9$, we get the error sequence as follows

$$\varepsilon^{(0)} = (\varepsilon^{(0)}(9), \varepsilon^{(0)}(10), \varepsilon^{(0)}(11), \varepsilon^{(0)}(12), \varepsilon^{(0)}(13))$$
$$= (-8.6534, -3.2307, -4.3974, -1.6478, -2.4768)$$

which is an error sequence of modelability. Taking absolute value gives

$$\varepsilon^{(0)} = (8.6534, 3.2307, 4.3974, 1.6478, 2.4768)$$

In establishing a EGM(1,1) for $\varepsilon^{(0)}$, we have the time response sequence of $\varepsilon^{(1)}$

Table 7.11 The errors and relative errors of EGM(1,1)

No.	Real data $x^{(0)}(k)$	Simulated values $\hat{x}^{(0)}(k)$	Errors $\varepsilon(k) = x^{(0)}(k) - \hat{x}^{(0)}(k)$	Relative errors $\Delta k = \frac{|\varepsilon(k)|}{x^{(0)}(k)}$ (%)
2	20	35.6704	−15.6704	78.3540
3	40	33.4303	6.5697	16.4242
4	25	31.3308	−6.3308	25.3232
5	40	29.3682	10.6318	26.5795
6	45	27.5192	17.4808	38.8642
7	35	25.6901	9.2099	26.3140
8	21	24.1719	−3.1719	15.1043
9	14	22.6534	−8.6534	61.8100
10	18	21.2307	−3.2307	17.9483
11	15.5	19.8974	−4.3974	28.3703
12	17	18.6478	−1.6478	9.6926
13	15	17.4768	−2.4768	16.5120

Table 7.12 Improved results

No.	Real data $x^{(0)}(k)$	Simulated values $\hat{x}^{(0)}(k)$	Errors $\varepsilon(k) = x^{(0)}(k) - \hat{x}^{(0)}(k)$	Relative errors $\Delta_k = \frac{|\varepsilon(k)|}{x^{(0)}(k)}$
10	18	17.1858	0.8142	4.52%
11	15.5	16.4799	-0.9799	6.32%
12	17	15.6604	1.2396	7.29%
13	15	15.0372	-0.0372	0.25%

$$\hat{\varepsilon}^{(1)}(k+1) = -24e^{-0.16855(k-9)} + 32.7$$

whose restored value of derivatives is

$$\hat{\varepsilon}^{(0)}(k+1) = (-0.16855)(-24)e^{-0.16855(k-9)} = 4.0452e^{-0.16855(k-9)}$$

From

$$\hat{x}^{(0)}(k+1) = \hat{x}^{(1)}(k+1) - \hat{x}^{(1)}(k) = (1-e^a)\left(x^{(0)}(1) - \frac{b}{a}\right)e^{-ak} = 38.0614e^{-0.06486k}$$

We can obtain the remnant model of inverse accumulating restoration

$$\hat{x}^{(0)}(k+1) = \begin{cases} 38.0614e^{-0.06486k}, & k < 9 \\ 38.0614e^{-0.06486k} - 4.0452e^{-0.16855(k-9)}, & k \geq 9 \end{cases}$$

where the sign of $\hat{\varepsilon}^{(0)}(k+1)$ is the same as the original error sequence.

Based on this model, we can modify the four simulation values with k = 10, 11, 12, 13, with improved accuracy listed in Table 7.12.

From this table, we can compute the sum of squares of errors as follows,

$$s = \varepsilon^T \varepsilon = 3.1611$$

and the average relative error

$$\Delta = \frac{1}{12}\sum_{k=10}^{13} \Delta_k = 4.595\%$$

Here, the simulation accuracy of the remnant EGM(1, 1) has obviously increased. However, the current error sequence no longer satisfies the modeling requirement. Therefore, if the improved accuracy is still unsatisfactory, we will have to consider other models or some appropriate choice of data to the original sequence.

7.5 Group of GM(1,1) Models

In practice, one does not have to use all the available data in their modeling. Each subsequence of the original data can be employed to establish a model. Generally speaking, different subsequences lead to different models. Even though the same kind of GM(1,1) is applied, different subsequences lead to different a, b values. These changes reflect the fact that varied circumstances and conditions have different effects on the system under consideration.

For example, for the grain production in China, if we use the data values collected since 1949 to establish a model GM(1,1), the development coefficient $(-a)$ will be on the small side. However, if only the values collected after 1978 are used, the corresponding development coefficient $(-a)$ will obviously increase (Deng, 1985; Liu, 2021).

Definition 7.5.1 For a given sequence $X^{(0)} = (x^{(0)}(1), x^{(0)}(2), \ldots, x^{(0)}(n))$, if we take $x^{(0)}(n)$ as the origin of the time axis, then $t < n$ is seen as the past, $t = n$ the present, and $t > n$ the future.

Definition 7.5.2 Assume that $X^{(0)} = (x^{(0)}(1), x^{(0)}(2), \cdots, x^{(0)}(n))$ is a sequence of raw data, let

$$\hat{x}^{(0)}(k + 1) = \left(1 - e^a\right)\left(x^{(0)}(1) - \frac{b}{a}\right)e^{-ak}$$

be the restored values of inverse accumulation of the GM(1,1) time responses of $X^{(0)}$. Then:

(1) For $t \le n$, $\hat{x}^{(0)}(t)$ is referred to as the simulated value out of the model; and
(2) When $t > n$, $\hat{x}^{(0)}(t)$ is known as the prediction of the model.

The main purpose of modeling is to make predictions. To improve the prediction accuracy, one first needs to guarantee sufficiently high accuracy in his simulation, especially for the simulation of the time moment $t = n$. Therefore, in general, the data, including $x^{(0)}(n)$, used for modeling should be an equal-time-interval sequence.

Definition 7.5.3 Assume that $X^{(0)} = (x^{(0)}(1), x^{(0)}(2), \ldots, x^{(0)}(n))$ is a sequence of raw data, then:

(1) The GM(1,1) model established using the entire sequence $X^{(0)}$ is known as the all-data GM(1,1);
(2) $\forall k_0 > 1$, the GM(1,1) model established on the tail sequence $X^{(0)} = (x^{(0)}(k_0), x^{(0)}(k_0 + 1), \ldots, x^{(0)}(n))$ is known as a partial-data GM(1,1);
(3) If $x^{(0)}(n + 1)$ stands for a piece of new information, then the GM(1,1) model established on the prolonged sequence $X^{(0)} = (x^{(0)}(1), x^{(0)}(2), \ldots, x^{(0)}(n), x^{(0)}(n + 1))$ is known as a new-information GM(1,1);

(4) The GM(1,1) model established on $X^{(0)} = ((x^{(0)}(2), \ldots, x^{(0)}(n), x^{(0)}(n+1))$ with the new information added and the oldest piece $x^{(0)}(1)$ of information removed is known as a metabolic GM(1,1).

Example 7.5.1 Let

$$X^{(0)} = (60.7, 73.8, 86.2, 100.4, 123.3)$$

and $x^{(0)}(6) = 149.5$ is a piece of new information. Try to establish a model with $X^{(0)}$, a model of new information, and a metabolic EGM(1,1).

Solution (1) The model with $X^{(0)}$. From

$$X^{(0)} = (60.7, 73.8, 86.2, 100.4, 123.3)$$

We have

$$\hat{a} = (B^T B)^{-1} B^T Y = \begin{bmatrix} a \\ b \end{bmatrix} = \begin{bmatrix} -0.17241 \\ 55.889264 \end{bmatrix}$$

The time response sequence is as follows

$$\hat{x}^{(1)}(k) = \left(x^{(0)}(1) - \frac{b}{a} \right) e^{-a(k-1)} + \frac{b}{a} = 384.865028 e^{0.17241k} - 324.165028$$

Then we obtained the simulation sequence of $X^{(0)}$ as follows

$$\hat{X}^{(0)} = (60.7, 72.41804, 86.04456, 102.2351, 121.4721)$$

The corresponding error sequence is

$$\varepsilon = (0, 1.38196, 0.155434, -1.8351, 1.827829)$$

where $\varepsilon(k) = x^{(0)}(k) - \hat{x}^{(0)}(k)$.
    Therefore, we got the average relative error

$$\Delta = \frac{1}{4} \sum_{k=2}^{5} \Delta_k = 1.34\%$$

where $\Delta_k = \frac{|\varepsilon(k)|}{x^{(0)}(k)}$.

(2) The model of new information. In inserting a piece of new information $x^{(0)}(6) = 149.5$, the data sequence became

$$X^{(0)} = (60.7, 73.8, 86.2, 100.4, 123.3, 149.5)$$

We have

$$\hat{a} = (B^T B)^{-1} B^T Y = \begin{bmatrix} a \\ b \end{bmatrix} = \begin{bmatrix} -0.180888 \\ 54.254961 \end{bmatrix}$$

Its time response sequence is as follows:

$$\hat{x}^{(1)}(k) = \left(x^{(0)}(1) - \frac{b}{a}\right)e^{-a(k-1)} + \frac{b}{a} = 360.63748e^{0.180888k} - 299.93748$$

The simulation sequence of the new information sequence $X^{(0}$, the corresponding error sequence ε, and the average relative error $\triangle$ are as follows:

$$\hat{X}^{(0)} = (60.7, 71.50736, 85.68587, 102.6757, 123.0342, 147.429)$$

$$\varepsilon = (0, 2.29264, 0.514129, -2.2757, 0.265712, 2.07041)$$

$$\triangle = \frac{1}{5} \sum_{k=2}^{6} \triangle_k = 1.51\%$$

(3) The metabolic EGM(1,1). In adding a piece of new information $x^{(0)}(6) = 149.5$, and deleting a piece of old information $x^{(0)}(1) = 60.7$, we have

$$X^{(0)} = (73.8, 86.2, 100.4, 123.3, 149.5)$$

and

$$\hat{a} = (B^T B)^{-1} B^T Y = \begin{bmatrix} a \\ b \end{bmatrix} = \begin{bmatrix} -0.187862 \\ 62.830896 \end{bmatrix}$$

The corresponding time response sequence is:

$$\hat{x}^{(1)}(k) = \left(x^{(0)}(1) - \frac{b}{a}\right)e^{-a(k-1)} + \frac{b}{a} = 408.251645e^{0.187862k} - 334.451645$$

And the simulation sequence of the metabolic sequence $X^{(0}$, the corresponding error sequence ε, and the average relative error $\triangle$ are as follows:

$$\hat{X}^{(0)} = (73.8, 84.37234, 101.8093, 122.85, 148.2391)$$

$$\varepsilon = (0,\ 1.827657,\ -1.4093,\ 0.45,\ 1.2609)$$

$$\Delta = \frac{1}{4}\sum_{k=3}^{6}\Delta_k = 1.18\%$$

Compared with these different results, it implies that the simulation accuracy can be improved by appropriately choosing the data to be used in the process of modeling. From the three different error sequences, we can see that for the simulation accuracy of value $x^{(0)}(5)$, both the new information model and the metabolic model are better than the model in (1). This implies that the new information EGM(1,1) and the metabolic EGM(1,1) have better prediction abilities than the old model. As a matter of fact, in the development process of a grey system, there always exists some stochastic interferences or driving forces entering the system as time goes on, so that the consequent development of the system is accordingly affected.

Therefore, when using the EGM(1,1) model to do predictions, high accuracy can be achieved only for the first or the second data values after the last origin value $x^{(0)}(n)$. In general, the farther away into the future, and the farther away from the last origin value, the weaker the prediction ability of EGM(1,1) becomes. In practical applications, one needs to constantly consider those interferences and driving factors entering the system as time goes on and promptly add new pieces of information to the original sequence $X^{(0)}$ and establish consequent new information EGM(1,1) models.

From the simulation accuracy of value $x^{(0)}(6)$, it can be seen that the metabolic model is better than the new information model. From the angle of prediction, it can be seen that the metabolic model is the best prediction model. As the system develops further, the significance of the older data reduces so that, when new data are added, the older data are deleted promptly, and the constantly renewing modeling sequence can better reflect the current characteristics of the system. Specifically, as the accumulation of quantitative changes increases, a jump or sudden change in the system will occur. At this very moment, compared with the older system, the current system is completely different. Hence, the practice of deleting old data is very reasonable. Indeed, the ongoing replacement of old data can avoid computation difficulties in modeling due to the fact that increased information can increase computer storage space requirements tremendously.

7.6   The Fractional Grey Model

Definition 7.6.1 Assume that $X_1^{(0)} = (x_1^{(0)}(1),\ x_1^{(0)}(2),\ \cdots,\ x_1^{(0)}(n))$ is a non-negative sequence, then.

$$x^{\left(\frac{p}{q}\right)}(k) = \sum_{i=1}^{k} C_{k-i+\frac{p}{q}-1}^{k-i} x^{(0)}(i)$$

is called a $\frac{p}{q}$ order accumulation operator. Let $C_{\frac{p}{q}-1}^{0} = 1, C_{k}^{k+1} = 0, k = 0, 1, \ldots, n-1$,

$$C_{k-i+\frac{p}{q}-1}^{k-i} = \frac{\left(k-i+\frac{p}{q}-1\right)\left(k-i+\frac{p}{q}-2\right)\cdots\left(\frac{p}{q}+1\right)\frac{p}{q}}{(k-i)!}.$$

Then $X^{\left(\frac{p}{q}\right)} = \left(x^{\left(\frac{p}{q}\right)}(1), x^{\left(\frac{p}{q}\right)}(2), \cdots, x^{\left(\frac{p}{q}\right)}(n)\right)$ is called a $\frac{p}{q}$ order accumulation sequence (Wu et al., 2013).

Definition 7.6.2 Assume that $X^{(0)} = \left(x^{(0)}(1), x^{(0)}(2), \cdots, x^{(0)}(n)\right)$ is a non-negative sequence, then

$$\alpha^{(1)}x^{\left(1-\frac{p}{q}\right)}(k) = x^{\left(1-\frac{p}{q}\right)}(k) - x^{\left(1-\frac{p}{q}\right)}(k-1)$$

is called a $\frac{p}{q}\left(0 < \frac{p}{q} < 1\right)$ order inverse accumulation operator. And

$$\alpha^{\left(\frac{p}{q}\right)}X^{(0)} = \alpha^{(1)}X^{\left(1-\frac{p}{q}\right)} = \left(\alpha^{(1)}x^{\left(1-\frac{p}{q}\right)}(1), \alpha^{(1)}x^{\left(1-\frac{p}{q}\right)}(2), \cdots, \alpha^{(1)}x^{\left(1-\frac{p}{q}\right)}(n)\right)$$

is called a $\frac{p}{q}\left(0 < \frac{p}{q} < 1\right)$ order inverse accumulation sequence.

Definition 7.6.3 Assume that $X^{(0)} = \left(x^{(0)}(1), x^{(0)}(2), \cdots, x^{(0)}(n)\right)$ is a non-negative sequence, and $X^{\left(\frac{p}{q}\right)} = \left(x^{\left(\frac{p}{q}\right)}(1), x^{\left(\frac{p}{q}\right)}(2), \cdots, x^{\left(\frac{p}{q}\right)}(n)\right)$ is the $\frac{p}{q}$ order accumulation sequence of $X^{(0)}$, then the following

$$x^{\left(\frac{pq}{q}\right)}(k+1) = \beta_1 x^{\left(\frac{p}{q}\right)}(k) + \beta_2(k = 1, 2, \ldots n - 1) \tag{7.30}$$

is called a $\frac{p}{q}$ order accumulation discrete grey model (Wu et al., 2013).

Theorem 7.6.1 *Assume that* $x^{\left(\frac{p}{q}\right)}(k+1) = \beta_1 x^{\left(\frac{p}{q}\right)}(k) + \beta_2$ *is called a* $\frac{p}{q}$ *order accumulation discrete grey model, then*

$$\begin{bmatrix} \beta_2 \\ \beta_1 \end{bmatrix} = \left(B^T B\right)^{-1} B^T Y$$

where

$$
B = \begin{bmatrix} 1 & x^{\left(\frac{p}{q}\right)}(1) \\ 1 & x^{\left(\frac{p}{q}\right)}(2) \\ \vdots & \vdots \\ 1 & x^{\left(\frac{p}{q}\right)}(n-1) \end{bmatrix}, \quad Y = \begin{bmatrix} x^{\left(\frac{p}{q}\right)}(2) \\ x^{\left(\frac{p}{q}\right)}(3) \\ \vdots \\ x^{\left(\frac{p}{q}\right)}(n) \end{bmatrix}
$$

Definition 7.6.4 Assume that $X^{(0)} = (x^{(0)}(1), \ x^{(0)}(2), \ \ldots, x^{(0)}(n))$ is a non-negative sequence, $p(0 < p < 1)$, then.

$$
\alpha^{(1)}x^{(1-p)}(k) + az^{(0)}(k) = b \tag{7.31}
$$

is called a grey model of GM(p,1).

where $\alpha^{(1)}x^{(1-p)}(k)$ is the p order difference of $x^{(0)}(k)$. We can calculated the $1 - p$ order accumulation of $x^{(0)}(k)$ at first, then acted by the first order inverse accumulation operator on $x^{(1-p)}(k)$ $\alpha^{(1)}x^{(1-p)}(k) = x^{(1-p)}(k) - x^{(1-p)}(k-1)$, let

$$
\begin{bmatrix} a \\ b \end{bmatrix} = (B^T B)^{-1} B^T Y
$$

where

$$
B = \begin{bmatrix} -z^{(0)}(2) & 1 \\ -z^{(0)}(3) & 1 \\ \vdots & \vdots \\ -z^{(0)}(n) & 1 \end{bmatrix}, \quad Y = \begin{bmatrix} \alpha^{(1)}x^{(1-p)}(2) \\ \alpha^{(1)}x^{(1-p)}(3) \\ \vdots \\ \alpha^{(1)}x^{(1-p)}(n) \end{bmatrix}
$$

and $z^{(0)}(k) = \frac{x^{(0)}(k)+x^{(0)}(k+1)}{2}$.

The whitening equation of the model GM(p,1) as follows

$$
\frac{d^p x^{(0)}(t)}{dt^p} + ax^{(0)}(t) = b \tag{7.32}
$$

Let $\hat{x}^{(0)}(1) = x^{(0)}(1)$, we obtained the time response sequence of (7.32) by fractional Laplace transform

$$
x^{(0)}(k) = \left(x^{(0)}(1) - \frac{b}{a} \right) \sum_{i=0}^{\infty} \frac{(-at^p)^i}{\Gamma(pi+1)} + \frac{b}{a} \tag{7.33}
$$

where $\Gamma(pi+1)$ is Gamma function.

Example 7.6.1 Let

$$X^{(0)} = (247.839, \ 273.021, \ 289.014, \ 285.208, \ 288.818, \ 297.078)$$

Please try to build a 0.1 order accumulation discrete grey model.
Solution: The 0.1 order accumulation sequence of $X^{(0)}$ as follows

$$X^{(0.1)} = (247.839, \ 297.805, \ 329.947, \ 338.667, \ 351.141, \ 366.983),$$

From $\begin{bmatrix} \beta_2 \\ \beta_1 \end{bmatrix} = (B^T B)^{-1} B^T Y$, we have

$$\hat{x}^{(0.1)}(k+1) = -126.356 \times 0.6101^{k-1} + 374.195$$

The simulated sequence is $\hat{x}^{(0.1)}(k) = (247.839, \ 297.105, \ 327.163, \ 345.501,$ $356.689, 363.515)$, its 0.9 order accumulation sequence is

$$\hat{x}^{(1)}(k) = (247.839, \ 520.160, \ 806.460, \ 1098.811, \ 1392.639, \ 1685.479),$$

Acted by a first order inverse accumulation operator, we have

$$\hat{x}^{(0)}(k) = (247.839, \ 272.321, \ 286.299, \ 292.351, \ 293.828, \ 292.841).$$

7.7 The Models of GM(*r,h*)

7.7.1 The Model of GM(0,N)

Definition 7.7.1 Assume that $X_1^{(0)} = (x_1^{(0)}(1), x_1^{(0)}(2), \ldots, x_1^{(0)}(n))$ is a data sequence of a system's characteristic variable,

$$\begin{aligned} X_2^{(0)} &= (x_2^{(0)}(1), x_2^{(0)}(2), \ldots, x_2^{(0)}(n)) \\ X_3^{(0)} &= (x_3^{(0)}(1), x_3^{(0)}(2), \ldots, x_3^{(0)}(n)) \\ &\cdots \cdots \cdots \cdots \cdots \cdots \\ X_N^{(0)} &= (x_N^{(0)}(1), x_N^{(0)}(2), \ldots, x_N^{(0)}(n)) \end{aligned}$$

the data sequences of relevant factors, and $X_i^{(1)}$ the accumulation generated sequence of $X_i^{(0)}$, $i = 2, 3, \ldots, N$. Then

$$x_1^{(1)}(k) = a + b_2 x_2^{(1)}(k) + b_3 x_3^{(1)}(k) + \cdots + b_N x_N^{(1)}(k) \qquad (7.34)$$

is called the model of GM(0,N). Because this model does not contain any derivative, it is a static model. Although its form looks like a multivariate linear regression model, it is essentially different from any of the statistical models. In particular, the general multivariate linear regression model is established on the basis of the original data sequences, while the model of GM(0,N) is constructed on the accumulation generation of the original data (Deng, 1985; Liu, 2021).

Theorem 7.7.1 *Assume* $X_i^{(0)}$ *and* $X_i^{(1)} (i = 1, 2, \ldots, N)$ *as given in Definition 7.6.1, let*

$$
B = \begin{bmatrix} 1 & x_2^{(1)}(2) & x_3^{(1)}(2) & \cdots & x_N^{(1)}(2) \\ 1 & x_2^{(1)}(3) & x_3^{(1)}(3) & \cdots & x_N^{(1)}(3) \\ \cdots & \cdots & \cdots & & \cdots \cdots \\ 1 & x_2^{(1)}(n) & x_3^{(1)}(n) & \cdots & x_N^{(1)}(n) \end{bmatrix}, Y = \begin{bmatrix} x_1^{(1)}(2) \\ x_1^{(1)}(3) \\ \vdots \\ x_1^{(1)}(n) \end{bmatrix}
$$

then the least squares estimate of the parametric sequence $\hat{a} = [a, b_1, b_2, \ldots, b_N]^T$ *is given by*

$$
\hat{a} = (B^T B)^{-1} B^T Y.
$$

Example 7.7.1 Let

$$
X_1^{(0)} = (2.874, 3.278, 3.307, 3.39, 3.679) = \{x_1^{(0)}(k)\}_1^5
$$

be a data sequence of a system's characteristic variable, and

$$
X_2^{(0)} = (7.04, 7.645, 8.075, 8.53, 8.774) = \{x_2^{(0)}(k)\}_1^5
$$

the data sequences of a relevant factor. Try to establish the model of GM(0,2).

Solution Assume the model of GM(0,2) as follows:

$$
X_1^{(1)} = bX_2^{(1)} + a
$$

From

$$
B = \begin{bmatrix} x_2^{(1)}(2) & 1 \\ x_2^{(1)}(3) & 1 \\ x_2^{(1)}(4) & 1 \\ x_2^{(1)}(5) & 1 \end{bmatrix} = \begin{bmatrix} 14.685 & 1 \\ 22.76 & 1 \\ 31.29 & 1 \\ 40.064 & 1 \end{bmatrix}, Y = \begin{bmatrix} x_1^{(1)}(2) \\ x_1^{(1)}(3) \\ x_1^{(1)}(4) \\ x_1^{(1)}(5) \end{bmatrix} = \begin{bmatrix} 6.152 \\ 9.459 \\ 12.849 \\ 16.528 \end{bmatrix}
$$

We have

$$
\hat{b} = \begin{bmatrix} b \\ a \end{bmatrix} = (B^T B)^{-1} B^T Y = \begin{bmatrix} 0.412435 \\ -0.482515 \end{bmatrix}
$$

Table 7.13 Simulation results with errors

Ordinality	Real data $x^{(0)}(k)$	Simulated values $\hat{x}^{(0)}(k)$	Errors $\varepsilon(k) = x^{(0)}(k) - \hat{x}^{(0)}(k)$	Relative errors $\Delta_k = \frac{|\varepsilon(k)|}{x^{(0)}(k)}$ (%)
2	3.278	3.153	0.125	3.8
3	3.307	3.331	−0.024	0.7
4	3.390	3.518	−0.128	3.8
5	3.679	3.619	0.06	1.6

It follows that

$$\hat{x}_1^{(1)}(k) = 0.412435x_2^{(1)}(k) - 0.482515$$

Therefore, the simulation results are as shown in Table 7.13.
The average relative error is

$$\overline{\Delta} = \frac{1}{4}\sum_{k=2}^{5}\Delta_k = \frac{1}{4}\sum_{k=2}^{5}\frac{|\varepsilon(k)|}{x^{(0)}(k)} == 2.475\%$$

7.7.2 The Model of GM(1, N)

Definition 7.7.2 Assume that $X_i^{(0)}$ and $X_i^{(1)}(i = 1, 2, \ldots, N)$ as given in Definition 7.6.1. Let $X_i^{(1)}$ be the accumulated sequences of $X_i^{(0)}$, $i = 1, 2, ..., N$, and $Z_1^{(1)}$ the adjacent neighbor average sequence of $X_1^{(1)}$. Then,

$$x_1^{(0)}(k) + az_1^{(1)}(k) = \sum_{i=2}^{N}b_i x_i^{(1)}(k) \tag{7.35}$$

is called the model of GM(1,N) (Deng, 1985; Liu, 2021).

The constant $(-a)$ is known as the system's development coefficient, $b_i x_i^{(1)}(k)$ the driving term, b_i the driving coefficient, and $\hat{a} = [a, b_1, b_2, \ldots, b_N]^T$ the sequence of parameters.

Theorem 7.7.2 *For the previously defined terms $X_i^{(0)}$, $X_i^{(1)}$, and $Z_1^{(1)}$, $i = 1, 2, ..., N$, let*

$$B = \begin{bmatrix} -z_1^{(1)}(2) \ x_2^{(1)}(2) \ \cdots \ x_N^{(1)}(2) \\ -z_1^{(1)}(3) \ x_2^{(1)}(3) \ \cdots \ x_N^{(1)}(3) \\ \cdots \qquad \cdots \qquad \cdots \cdots \\ -z_1^{(1)}(n) \ x_2^{(1)}(n) \ \cdots \ x_N^{(1)}(n) \end{bmatrix}, \ Y = \begin{bmatrix} x_1^{(0)}(2) \\ x_1^{(0)}(3) \\ \vdots \\ x_1^{(0)}(n) \end{bmatrix}$$

Then the least squares estimate of the sequence $\hat{a} = [a, b_1, b_2, \ldots, b_N]^T$ of parameters satisfies

$$\hat{a} = (B^T B)^{-1} B^T Y.$$

Example 7.7.2 Let

$$X_1^{(0)} = (2.874, 3.278, 3.307, 3.39, 3.679) = \{x_1^{(0)}(k)\}_1^5$$

is a data sequence of a system's characteristic variable, and

$$X_2^{(0)} = (7.04, 7.645, 8.075, 8.53, 8.774) = \{x_2^{(0)}(k)\}_1^5$$

the data sequences of a relevant factor. Try to establish the model of GM(1,2).

Solution Assume that the model of GM(1,2) is as follows:

$$x_1^{(0)}(k) + az_1^{(1)}(k) = bx_2^{(1)}(k)$$

From

$$X_1^{(1)} = [x_1^{(1)}(1), x_1^{(1)}(2), x_1^{(1)}(3), x_1^{(1)}(4), x_1^{(1)}(5)]$$
$$= (2.874, 6.152, 9.459, 12.849, 16.528)$$

$$X_2^{(1)} = [x_2^{(1)}(1), x_2^{(1)}(2), x_2^{(1)}(3), x_2^{(1)}(4), x_2^{(1)}(5)]$$
$$= (7.04, 14.685, 22.76, 31.29, 40.064)$$

We have

$$Z_1^{(1)} = [z_1^{(1)}(2), z_1^{(1)}(3), z_1^{(1)}(4), z_1^{(1)}(5)]$$
$$= (4.513, 7.8055, 11.154, 14.6885)$$

It follows that

$$B = \begin{bmatrix} -z_1^{(1)}(2) \ x_2^{(1)}(2) \\ -z_1^{(1)}(3) \ x_2^{(1)}(3) \\ -z_1^{(1)}(4) \ x_2^{(1)}(4) \\ -z_1^{(1)}(5) \ x_2^{(1)}(5) \end{bmatrix} = \begin{bmatrix} -4.513 & 14.685 \\ -7.8055 & 22.76 \\ -11.154 & 31.29 \\ -14.6885 & 40.064 \end{bmatrix}, \ Y = \begin{bmatrix} x_1^{(0)}(2) \\ x_1^{(0)}(3) \\ x_1^{(0)}(4) \\ x_1^{(0)}(5) \end{bmatrix} = \begin{bmatrix} 3.278 \\ 3.307 \\ 3.390 \\ 3.679 \end{bmatrix}$$

Table 7.14 Simulation results with errors

Ordinality	Real data $x^{(0)}(k)$	Simulated values $\hat{x}^{(0)}(k)$	Errors $\varepsilon(k) = x^{(0)}(k) - \hat{x}^{(0)}(k)$	Relative errors $\Delta_k = \frac{|\varepsilon(k)|}{x^{(0)}(k)}$ (%)
2	3.278	3.265	0.013	0.4
3	3.307	3.254	0.053	1.6
4	3.390	3.530	−0.140	4.1
5	3.679	3.614	0.065	1.8

Therefore, we have

$$\hat{a} = \begin{bmatrix} a \\ b \end{bmatrix} = (B^T B)^{-1} B^T Y = \begin{bmatrix} 2.2273 \\ 0.9068 \end{bmatrix}$$

and

$$x_1^{(0)}(k) + 2.2273 z_1^{(1)}(k) = 0.9068 x_2^{(1)}$$

That is,

$$\hat{x}_1^{(0)}(k) = -2.2273 z_1^{(1)}(k) + 0.9068 x_2^{(1)}$$

The simulation results are as shown in Table 7.14.
The average relative error is

$$\overline{\Delta} = \frac{1}{4} \sum_{k=2}^{5} \Delta_k = \frac{1}{4} \sum_{k=2}^{5} \frac{|\varepsilon(k)|}{x^{(0)}(k)} = 1.975\%$$

7.7.3 The Grey Verhulst Model

The GM(1,1) model is suitable for sequences that show an obvious exponential pattern and can be used to describe monotonic changes. As for non-monotonic wavelike development sequences, or saturated sigmoid sequences, one can consider establishing a grey Verhulst model.

Definition 7.7.3 Assume that $X^{(0)}$ is a sequence of raw data, $X^{(1)}$ the accumulation sequence of $X^{(0)}$, and $Z^{(1)}$ the adjacent neighbor average sequence of $X^{(1)}$. Then,

$$x^{(0)}(k) + az^{(1)}(k) = b[z^{(1)}(k)]^{\alpha} \tag{7.36}$$

is known as the power model of GM(1,1). Also,

$$dx^{(1)}/dt + ax^{(1)} = b(x^{(1)})^\alpha \tag{7.37}$$

is known as the shadow equation of the power model of GM(1,1) (Deng, 1985).

Theorem 7.7.3 *The solution of the whitenization equation of the power model of GM(1,1) is*

$$x^{(1)}(t) = \left\{ e^{-(1-a)at}[(1-a)\int be^{(1-a)at}dt + c] \right\}^{\frac{1}{1-a}} \tag{7.38}$$

Theorem 7.7.4 *Let* $X^{(0)}$, $X^{(1)}$, *and* $Z^{(1)}$ *be defined as above. Let*

$$B = \begin{bmatrix} -z^{(1)}(2) & [z^{(1)}(2)]^\alpha \\ -z^{(1)}(3) & [z^{(1)}(3)]^\alpha \\ \vdots & \vdots \\ -z^{(1)}(n) & [z^{(1)}(n)]^\alpha \end{bmatrix}, Y = \begin{bmatrix} x^{(0)}(2) \\ x^{(0)}(3) \\ \vdots \\ x^{(0)}(n) \end{bmatrix}$$

Then the least squares estimate of the parametric sequence $\hat{a} = [a, b]^T$ of the power model of GM(1,1) is

$$\hat{a} = (B^T B)^{-1} B^T Y.$$

Definition 7.7.4 When the power $\alpha = 2$ in the power model of GM(1,1), the resultant model

$$x^{(0)}(k) + az^{(1)}(k) = b(z^{(1)}(k))^2 \tag{7.39}$$

is known as the grey Verhulst model; and

$$dx^{(1)}/dt + ax^{(1)} = b(x^{(1)})^2 \tag{7.40}$$

is known as the whitenization equation of the grey Verhulst model (Deng, 1985).

Theorem 7.7.5

(1) *The solution of the Verhulst whitenization equation is*

$$x^{(1)}(t) = \frac{1}{e^{at}[\frac{1}{x^{(1)}(0)} - \frac{b}{a}(1 - e^{-at})} = \frac{ax^{(1)}(0)}{e^{at}[a - bx^{(1)}(0)(1 - e^{-at})}$$

That is

$$x^{(1)}(t) = \frac{ax^{(1)}(0)}{bx^{(1)}(0) + [a - bx^{(1)}(0)]e^{-at}} \tag{7.41}$$

(2) *The time response sequence* of the *grey Verhulst model is*

$$\hat{x}^{(1)}(k+1) = \frac{ax^{(1)}(0)}{bx^{(1)}(0) + [a - bx^{(1)}(0)]e^{-ak}} \tag{7.42}$$

The Verhulst model is mainly used to describe and study processes with saturated states (or sigmoid processes). For instance, this model is often used in the prediction of human populations, biological growth, reproduction, and economic life span of consumable products. From the solution of the Verhulst equation, it can be seen that when $t \to \infty$, if $a > 0$, then $x^{(1)}(t) \to 0$; if $a < 0$, then $x^{(1)}(t) \to \frac{a}{b}$. That is, there is a sufficiently large t such that for any $k > t$, both $x^{(1)}(k+1)$ and $x^{(1)}(k)$ will be sufficiently close to each other. In this case, $x^{(0)}(k+1) = x^{(1)}(k+1) - x^{(1)}(k) \approx 0$, which means that the system approaches distinction.

In practice, one often faces sigmoid processes in the original data sequences. When such an instance appears, we can simply take the original sequence as $X^{(1)}$ with its accumulation generation as $X^{(0)}$ to establish a grey Verhulst model to directly simulate $X^{(1)}$.

Example 7.7.3 Assume that the expenditures on the research of a certain kind of torpedo are given in Table 7.15. Try to employ the grey Verhulst model to simulate the data and make predictions (Liang et al., 2005).

The accumulated expenditures are given in Table 7.16.
From Theorem 7.7.5, we compute the parameters as follows:

$$\hat{a} = [a, b]^T = \begin{bmatrix} -0.98079 \\ -0.00021576 \end{bmatrix}$$

so that the whitenization equation is

$$dx^{(1)}/dt - 0.98079x^{(1)} = -0.00021576(x^{(1)})^2.$$

By taking $x^{(1)}(0) = x^{(0)}(1) = 496$, we obtain the time response sequence

Table 7.15 Expenditures on the research of a certain kind of torpedo (in million Yuan)

Year	1995	1996	1997	1998	1999	2000	2001	2002	2003	2004
Expenditure	496	779	1187	1025	488	255	157	110	87	79

Table 7.16 Accumulated expenditures (in ten thousand Yuan)

Year	1995	1996	1997	1998	1999	2000	2001	2002	2003	2004
Expenditure	496	1275	2462	3487	3975	4230	4387	4497	4584	4663

Table 7.17 The simulation results with errors

Ordinality	Actual data $x^{(0)}(k)$	Simulated data $\hat{x}^{(0)}(k)$	Error $\varepsilon(k) = x^{(0)}(k) - \hat{x}^{(0)}(k)$	Relative error $\Delta_k = \frac{|\varepsilon(k)|}{x^{(0)}(k)}$
2	1275	1119.1	155.9	0.12226
3	2462	2116	346	0.14053
4	3487	3177.5	309.5	0.08876
5	3975	3913.7	61.3	0.01541
6	4230	4286.2	−56.2	0.01328
7	4387	4444.8	−57.8	0.01318
8	4497	4507.4	−10.4	0.0023
9	4584	4531.3	52.7	0.0115
10	4663	4540.3	122.7	0.02631

$$\hat{x}^{(1)}(k+1) = \frac{ax^{(1)}(0)}{bx^{(1)}(0) + [a - bx^{(1)}(0)]e^{-ak}} = \frac{-486.47}{-0.10702 - 0.87378e^{-0.98079k}}.$$

On the basis of this formula, we produce the simulated values $\hat{x}^{(1)}(k)$ as shown in Table 7.17.

From Table 7.17, we can obtain the average relative error

$$\Delta = \frac{1}{9} \sum_{k=2}^{10} \Delta_k = 4.3354\%$$

and predict the research expenditure for the year of 2005 on the special kind of torpedo as

$$\hat{x}_1^{(0)}(11) = \hat{x}_1^{(1)}(11) - \hat{x}_1^{(1)}(10) = 9.0342.$$

This value indicates that the research work on the torpedo is nearing its conclusion.

7.7.4 The Self-memory Grey Model

For unimodal series or nonlinear saturated growth series, self memory GM(1,1) power model can also be established to describe its evolution law (Guo et al., 2015; Liu et al., 2017; Liu, 2021).

Definition 7.7.5 Assume that

$$F(x,t) = -ax^{(1)} + b(x^{(1)})^\gamma \tag{7.43}$$

where x is variable, t is time, then formula (7.43) is called a self-memory dynamic equation.

Definition 7.7.6 Assume that $\beta(t)$, $|\beta(t)| \le 1$ is a memory function, the variable x, memory function $\beta(t)$ and self-memory dynamic equation $F(x, t)$ all meet the conditions of continuity, differentiability, integrability, then following

$$\beta_t x_t - \beta_{-p} x_{-p} - \sum_{i=-p}^{0} x_i^m (\beta_{i+1} - \beta_i) - \int_{t_{-p}}^{t} \beta(\tau) F(x, \tau) d\tau = 0 \qquad (7.44)$$

is called a self-memory prediction model.

Where $T = \{t_{-p}, t_{-p+1}, \ldots, t_{-1}, t_0, t\}$ is the time set.
Let $x_{-p-1}^m \equiv x_{-p}$, $\beta_{-p-1} \equiv 0$, we can obtained the following

$$x_t = \frac{1}{\beta_t} \sum_{i=-p-1}^{0} x_i^m (\beta_{i+1} - \beta_i) + \frac{1}{\beta_t} \int_{t_{-p}}^{t} \beta(\tau) F(x, \tau) d\tau = S_1 + S_2 \qquad (7.45)$$

where S_1 is the self-memory item which represents the influence of historical statistical data on the predicted value x_t, S_2 is other effective item which represents the influence of the dynamic equation $F(x, t) = -ax^{(1)} + b(x^{(1)})^\gamma$ on the predicted value x_t within the backtracking period $[t_{-p}, t_0]$.

In (7.45), we use addition to approximately replace integration and difference to approximately replace differentiation and let $x_i^m = \frac{1}{2}(x_{i+1} + x_i) \equiv y_i$, $\Delta t_i = t_{i+1} - t_i = 1$ further, then we obtained the discrete self-memory prediction model as follows

$$x_t = \sum_{i=-p-1}^{-1} \alpha_i y_i + \sum_{i=-p}^{0} \theta_i F(x, i) \qquad (7.46)$$

where $\alpha_i = (\beta_{i+1} - \beta_i)/\beta_t$, $\theta_i = \beta_i/\beta_t$, and $F(x, t) = -ax^{(1)} + b(x^{(1)})^\gamma$.

Theorem 7.7.6 *Assume that*

$$\underset{L \times 1}{X_t} = \begin{bmatrix} x_{t1} \\ x_{t2} \\ \vdots \\ x_{tL} \end{bmatrix}, \quad \underset{L \times (p+1)}{Y} = \begin{bmatrix} y_{-p-1,1} & y_{-p,1} & \cdots & y_{-1,1} \\ y_{-p-1,2} & y_{-p,2} & \cdots & y_{-1,2} \\ \vdots & \vdots & \ddots & \vdots \\ y_{-p-1,L} & y_{-p,L} & \cdots & y_{-1,L} \end{bmatrix}, \quad \underset{(p+1) \times 1}{A} = \begin{bmatrix} \alpha_{-p-1} \\ \alpha_{-p} \\ \vdots \\ \alpha_{-1} \end{bmatrix}$$

$$\underset{L \times (p+1)}{\Gamma} = \begin{bmatrix} F(x, -p)_1 & F(x, -p+1)_1 & \cdots & F(x, 0)_1 \\ F(x, -p)_2 & F(x, -p+1)_2 & \cdots & F(x, 0)_2 \\ \vdots & \vdots & \ddots & \vdots \\ F(x, -p)_L & F(x, -p+1)_L & \cdots & F(x, 0)_L \end{bmatrix}, \quad \underset{(p+1) \times 1}{\Theta} = \begin{bmatrix} \theta_{-p} \\ \theta_{-p+1} \\ \vdots \\ \theta_0 \end{bmatrix}$$

Let $Z = [Y, \Gamma]$, $W = \begin{bmatrix} A \\ \Theta \end{bmatrix}$.

then the least squares estimate of the parametric vector $W = \begin{bmatrix} A \\ \Theta \end{bmatrix}$ satisfies

$$W = (Z^T Z)^{-1} Z^T X_t \qquad (7.47)$$

7.7.5 The Models of GM(r,h)

In this subsection, we focus on the investigation of the structure of the models of GM(r,h), and its relationships with models GM(1,1), GM(1,N), GM(0,N), and the grey Verhulst model.

Definition 7.7.7 Assume that $X_i^{(0)} = (x_i^{(0)}(1), x_i^{(0)}(2), \cdots, x_i^{(0)}(n))$, $i = 1, 2, ..., h$, where $X_1^{(0)}$ stands for a data sequence of a system's characteristic, and $X_i^{(0)}$, $i = 2, 3, \cdots, h$ data sequences of relevant factors. Let.

$$\alpha^{(1)} \hat{x}_1^{(1)}(k) = \hat{x}_1^{(1)}(k) - \hat{x}_1^{(1)}(k-1) = \hat{x}_1^{(0)}(k)$$

$$\alpha^{(2)} \hat{x}_1^{(1)}(k) = \alpha^{(1)} \hat{x}_1^{(1)}(k) - \alpha^{(1)} \hat{x}_1^{(1)}(k-1) = \hat{x}_1^{(0)}(k) - \hat{x}_1^{(0)}(k-1)$$

$$\cdots \quad \cdots \quad \cdots \quad \cdots \quad \cdots$$

$$\alpha^{(r)} \hat{x}_1^{(1)}(k) = \alpha^{(r-1)} \hat{x}_1^{(1)}(k) - \alpha^{(r-1)} \hat{x}_1^{(1)}(k-1) = \alpha^{(r-2)} \hat{x}_1^{(0)}(k) - \alpha^{(r-2)} \hat{x}_1^{(0)}(k-1)$$

and $z^{(1)}(k) = \frac{1}{2}(x^{(1)}(k) + x^{(1)}(k-1))$, then

$$\alpha^{(r)} \hat{x}_1^{(1)}(k) + \sum_{i=1}^{r-1} a_i \alpha^{(r-i)} x_1^{(1)}(k) + a_r z_1^{(1)}(k) = \sum_{j=1}^{h-1} b_j x_{j+1}^{(1)}(k) + b_h \qquad (7.48)$$

is referred to as the model of GM(r,h). The GM(r,h) model is a rth order grey model in h variables.

Definition 7.7.8 In the model of GM(r,h), $-\hat{a} = [-a_1, -a_2, \ldots, -a_r]^T$ is referred to as the development coefficient vector, $\sum_{j=1}^{h-1} b_j x_{j+1}^{(1)}(k)$ the driving term, and $\hat{b} = [b_1, b_2, \ldots, b_h]^T$ the vector of driving coefficients.

Theorem 7.7.7 *Let $X_1^{(0)}$ be a data sequence of a system's characteristic, $X_i^{(0)}$, $i = 2, 3, \ldots, h$, the data sequences of relevant factors, $X_i^{(1)}$ the accumulation sequence of $X_i^{(0)}$, $Z_1^{(1)}$ the adjacent neighbor average sequence from $X_1^{(1)}$, and $\alpha^{(r-i)} X_1^{(1)}$ the $(r-i)$th order inverse accumulation sequence of $X_1^{(1)}$. Define*

$$B = \begin{bmatrix} -\alpha^{(r-1)}x_1^{(1)}(2) & -\alpha^{(r-2)}x_1^{(1)}(2) & \cdots & -\alpha^{(1)}x_1^{(1)}(2) & -z_1^{(1)}(2) & x_2^{(1)}(2) & \cdots & x_h^{(1)}(2) & 1 \\ -\alpha^{(r-1)}x_1^{(1)}(3) & -\alpha^{(r-2)}x_1^{(1)}(3) & \cdots & -\alpha^{(1)}x_1^{(1)}(3) & -z_1^{(1)}(3) & x_2^{(1)}(3) & \cdots & x_h^{(1)}(3) & 1 \\ \cdots & \cdots & \cdots & \cdots & \cdots & \cdots & \cdots & \cdots & \cdots \\ -\alpha^{(r-1)}x_1^{(1)}(n) & -\alpha^{(r-2)}x_1^{(1)}(n) & \cdots & -\alpha^{(1)}x_1^{(1)}(n) & -z_1^{(1)}(n) & x_2^{(1)}(n) & \cdots & x_h^{(1)}(n) & 1 \end{bmatrix},$$

$$Y = \begin{bmatrix} \alpha^{(r)}x_1^{(1)}(2) \\ \alpha^{(r)}x_1^{(1)}(3) \\ \vdots \\ \alpha^{(r)}x_1^{(1)}(n) \end{bmatrix}$$

then the parametric sequence $\hat{c} = [-\hat{a}, \hat{b}]^T = [-a_1, -a_2, \ldots, -a_r; b_1, b_2, \ldots, b_h]^T$ *of the least squares estimate satisfies*

$$\hat{a} = (B^T B)^{-1} B^T Y.$$

The model of GM(r,h) is the general form of grey systems models. In particular,

(1) When $r = 1$ and $h = 1$, the previous (7.48) reduces to:

$$dx_1^{(1)}/dt + a_1 x_1^{(1)} = b_1 \quad \text{and} \, \alpha^{(1)}x_1^{(1)}(k) + a_1 z_1^{(1)}(k) = b_1$$

which is the model of GM(1,1).

(2) When $r = 1$ and $h = N$, the previous (7.48) takes the form of

$$x_1^{(0)}(k) + a_1 z_1^{(1)}(k) = \sum_{i=2}^{N} b_i x_i^{(1)}(k)$$

which is the GM(1,N) model.

(3) When $r = 0$ and $h = $ N, the previous model (7.48) is

$$x_1^{(1)}(k) = b_1 x_2^{(1)}(k) + b_2 x_3^{(1)}(k) + \cdots + b_{N-1} x_N^{(1)}(k) + b_N$$

which is the GM(0,N) model.

(4) When $r = 1$ and $h = 1$, and b_1 in the model of GM(1,1) is changed to $b(z^{(1)}(k))^2$, then we have the following grey Verhulst model:

$$x^{(0)}(k) + az^{(1)}(k) = b(z^{(1)}(k))^2.$$

Based on this discussion, it can be seen that models GM(1,1), GM(1,N), GM(0,N), etc., are all special cases of model GM(r,h). So, it is very important to further the study of model GM(r,h).

7.8   Practical Applications

Example 7.8.1 (Liu, 1991) Let us look at the revenue predictions of private enterprises at Changge County, Henan Province, The People's Republic of China, which we mentioned in Example 4.3.1. For the years from 1983 to 1986, the overall business revenue of private enterprises in Changge county was recorded as.

$$X = (10155, 12588, 23480, 35388)$$

We obtained the following second-order buffered sequence

$$XD^2 = (27260, 29547, 32411, 35388)$$

in Example 4.3.1 by a second-order average weakening buffer operator (AWBO) as follows:

$$x(k)d = \frac{1}{n-k+1}[x(k) + x(k+1) + \cdots + x(n)], k = 1, 2, \ldots, n$$

We denote the XD^2 as $X^{(0)}$, that is, let

$$X^{(0)} = (27260, 29547, 32411, 35388).$$

Then the 1-AGO sequence $X^{(1)}$ of $X^{(0)}$ is as follows

$$X^{(1)} = (x^{(1)}(1), x^{(1)}(2), x^{(1)}(3), x^{(1)}(4)) = (27260, 56807, 89218, 124606).$$

Assume that

$$x^0(k) + az^1(k) = b$$

Based on the least squares method, we obtain the estimated values for a and b as follows:

$$\hat{a} = -0.089995, \hat{b} = 25790.28$$

Thus, the resultant whitenization equation of EGM(1, 1) is given by

$$\frac{dx^{(1)}}{dt} - 0.089995x^{(1)} = 25790.28$$

and its time response sequence is

$$\begin{cases} \hat{x}^{(1)}(k+1) = 313834\, e^{0.089995k} - 286574 \\ \hat{x}^{(0)}(k+1) = \hat{x}^{(1)}(k+1) - \hat{x}^{(1)}(k) \end{cases}$$

From these results, we obtain the simulated sequence

$$\hat{X} = (\hat{x}(1), \hat{x}(2), \hat{x}(3), \hat{x}(4)) = (27260, 29553, 32337, 35381)$$

with the sequence of errors

$$\varepsilon^{(0)} = (\varepsilon^{(0)}(1), \varepsilon^{(0)}(2), \varepsilon^{(0)}(3), \varepsilon^{(0)}(4)) = (0 - 6, 74, 7)$$

The sequence of relative errors

$$\Delta = \left[\left|\frac{\varepsilon^{(0)}(1)}{x^{(0)}(1)}\right|, \left|\frac{\varepsilon^{(0)}(2)}{x^{(0)}(2)}\right|, \left|\frac{\varepsilon^{(0)}(3)}{x^{(0)}(3)}\right|, \left|\frac{\varepsilon^{(0)}(4)}{x^{(0)}(4)}\right|\right]$$
$$= (0, 0.0002, 0.00228, 0.0002)$$

And the average relative error

$$\overline{\Delta} = \frac{1}{4}\sum_{k=1}^{n4} \Delta_k = 0.00067 = 0.067\% < 0.01$$

$$\Delta_4 = 0.0002 = 0.02\% < 0.01$$

Therefore, the accuracy of our simulation is in level one.
Now, we can compute the absolute degree ε of grey incidences of X and $\hat{X}$.

$$|s| = \left|\sum_{k=2}^{3}[x(k) - x(1)] + \frac{1}{2}[x(4) - x(1)]\right| = 11502$$

$$|\hat{s}| = \left|\sum_{k=2}^{3}[\hat{x}(k) - \hat{x}(1)] + \frac{1}{2}[\hat{x}(4) - \hat{x}(1)]\right| = 11430.5$$

$$|\hat{s} - s| = \left|\sum_{k=2}^{3}[x(k) - x(1) - (\hat{x}(k) - \hat{x}(1))] + \frac{1}{2}[x(4) - x(1) - (\hat{x}(4) - \hat{x}(1))]\right|$$
$$= 71.5$$

Thus,

$$\varepsilon = \frac{1 + |s| + |\hat{s}|}{1 + |s| + |\hat{s}| + |\hat{s} - s|} = \frac{1 + 11502 + 11430.5}{1 + 11502 + 11430.5 + 71.5} = 0.997 > 0.90$$

That is, the degree of incidence is in level one.
Compute the ratio of mean square deviations C:

$$\bar{x} = \frac{1}{4}\sum_{k=1}^{4} x(k) = 31151.5, \ S_1^2 = \frac{1}{4}\sum_{k=1}^{4}(x(k) - \bar{x})^2 = 37252465, \ S_1 = 6103.48$$

$$\bar{\varepsilon} = \frac{1}{4}\sum_{k=1}^{4} \varepsilon(k) = 18.75, \ S_2^2 = \frac{1}{4}\sum_{k=1}^{4}(\varepsilon(k) - \bar{\varepsilon})^2 = 4154.75, \ S_2 = 64.46$$

It follows that

$$C = \frac{S_2}{S_1} = \frac{64.46}{6103.48} = 0.01 < 0.35$$

which is in level one.

Compute the small error probability. From

$$0.6745 S_1 = 4116.80$$

$$|\varepsilon(1) - \bar{\varepsilon}| = 18.75, \ |\varepsilon(2) - \bar{\varepsilon}| = 24.75, \ |\varepsilon(3) - \bar{\varepsilon}| = 55.25, \ |\varepsilon(4) - \bar{\varepsilon}| = 11.75$$

Therefore

$$p = P(|\varepsilon(k) - \bar{\varepsilon}| < 0.6745 S_1) = 1 > 0.95$$

With our accuracy checks in place, we can apply the grey model

$$\begin{cases} \hat{x}^{(1)}(k+1) = 313834 e^{0.089995k} - 286574 \\ \hat{x}^{(0)}(k+1) = \hat{x}^{(1)}(k+1) - \hat{x}^{(1)}(k) \end{cases}$$

to make predictions. Here, we list five predicted values as follows:

$$\hat{X}^{(0)} = [\hat{x}^{(0)}(5), \hat{x}^{(0)}(6), \hat{x}^{(0)}(7), \hat{x}^{(0)}(8), \hat{x}^{(0)}(9)]$$
$$= (38714, 42359, 46348, 50712, 55488)$$

These predictions indicated an average 9.4% annual growth. When we look back today, this predicted rate of growth agreed very well with the recorded values over the time span of our predictions.

Example 7.8.2 Subgrade settlement prediction (Guo et al., 2015).

Subgrade settlement is one important indicator affecting road safety because the major hidden danger could result in road traffic accidents. So subgrade settlement prediction is one of the major research topics in the field of geotechnical engineering. Three monitoring points (Points A, B and C) at certain roadbed sections of Beijing-Harbin freeway (G102 line) were arranged, the method of single point extensometer was employed to monitor its subgrade settlement.

Table 7.18 The accumulated subgrade settlement data of monitoring points A, B and C (unit: mm)

Period	Number of days	Accumulated subgrade settlement value		
		Point A	Point B	Point C
1	35	13.42	9.89	12.03
2	50	15.38	12.20	15.60
3	65	22.18	16.27	19.57
4	80	23.30	17.66	20.80
5	95	24.55	19.07	22.03
6	110	25.41	20.85	23.38
7	125	26.91	21.91	24.60
8	140	28.02	23.40	25.79
9	155	28.44	23.77	26.36
10	170	28.64	24.12	27.16

The three groups of accumulated subgrade settlement data at different monitoring points are listed in Table 7.18 (Liu et al., 2013).

Step 1: Analyze the coupling relationship between the data of different monitoring points.

Analyze the coupling relationship between the data of different monitoring points to determine whether the data of monitoring points A, B and C are relevant. Let

$X_1 = (13.42, 15.38, 22.18, 23.30, 24.55, 25.41, 26.91, 28.02, 28.44, 28.64)$

$X_2 = (9.89, 12 \cdot 20, 16 \cdot 27, 17.66, 19.07, 20.85, 21 \cdot 91, 23.40, 23.77, 24.12)$

$X_3 = (12.03, 15.60, 19.57, 20.80, 22.03, 23.38, 24.60, 25.79, 26.36, 27.16)$

Calculate the grey absolute relational degree between X_1, X_2, and X_3 respectively, we have

$$\varepsilon_{12} = 0.9923, \quad \varepsilon_{13} = 0.9721, \quad \varepsilon_{25} = 0.9648$$

The results shows that there is coupling relationship and certain relationship exists of the data at monitoring point A, B and C.

Step 2: Determining the self-memory dynamic equation.

$$\begin{cases} \dfrac{dx_1^{(1)}}{dt} + 4.0920x_1^{(1)} + 2.8789x_2^{(1)} - 7.0870x_3^{(1)} = 7.2671 \\[2mm] \dfrac{dx_2^{(1)}}{dt} + 1.7787x_1^{(1)} + 1.5361x_2^{(1)} - 3.3707x_3^{(1)} = 7.4767 \\[2mm] \dfrac{dx_3^{(1)}}{dt} + 1.9224x_1^{(1)} + 1.5285x_2^{(1)} - 3.5145x_3^{(1)} = 10.9312 \end{cases} \qquad (7.49)$$

The matrix form $dX^{(1)}/dt = -AX^{(1)} + B$ of Eq. (7.49) was taken as the dynamic kernel $F(X, t)$ of the self-memory equation of the SMGM(1,3) model.

Step 3: Deducing the self-memory prediction equation system.

The value of retrospective order is determined as p = 1 by trial calculation method under the principle of minimum error of fitting root-mean-square. Then the self-memorization equation system of the SMGM(1,3) model can be established for subgrade settlement forecasting as follows.

$$\begin{cases} x_{1t} = \displaystyle\sum_{i=-2}^{-1} \alpha_{1i} y_{1i} + \sum_{i=-1}^{0} \theta_{1i} F_1(x, i) \\[4mm] x_{2t} = \displaystyle\sum_{i=-2}^{-1} \alpha_{2i} y_{2i} + \sum_{i=-1}^{0} \theta_{2i} F_2(x, i) \\[4mm] x_{3t} = \displaystyle\sum_{i=-2}^{-1} \alpha_{3i} y_{3i} + \sum_{i=-1}^{0} \theta_{3i} F_3(x, i) \end{cases} \qquad (7.50)$$

Step 4: Estimate the memory coefficients matrix by the least square method.

$$W = [W_1, W_2, W_3] = \begin{bmatrix} \alpha_{1,-2} & \alpha_{2,-2} & \alpha_{3,-2} \\ \alpha_{1,-1} & \alpha_{2,-1} & \alpha_{3,-1} \\ \theta_{1,-1} & \theta_{2,-1} & \theta_{3,-1} \\ \theta_{1,0} & \theta_{2,0} & \theta_{3,0} \end{bmatrix} = \begin{bmatrix} -0.0520 & 0.0890 & 0.0260 \\ 1.0468 & 0.9067 & 0.9712 \\ 0.2141 & 0.3616 & 0.2931 \\ 1.2761 & 1.2769 & 1.2702 \end{bmatrix}$$

Step 5: Simulation

Substituting the memory coefficient matrix into Eq. (7.50), The simulation values of original subgrade settlement data matrix $X^{(0)}$ can be obtained.

The simulated values and their corresponding APE of three compared models, SMGM(1,3), GM(1,1) and MGM(1,3) are presented in Tables 7.19, 7.20, and 7.21 respectively.

Step 6: Simulation accuracy comparison

Table 7.19 The simulated values and APE of the SMGM(1,3), GM(1,1) and MGM(1,3) at point A (unit: mm)

No.	Actual value	GM(1,1)		MGM(1,3)		SMGM(1,3)	
		Simulated value	APE/%	Simulated value	APE/%	Simulated value	APE/%
1	13.42	–	–	–	–	–	–
2	15.38	18.931	23.088	16.283	5.871	–	–
3	22.18	20.333	8.327	21.182	4.500	22.168	0.055
4	23.30	21.839	6.270	23.276	0.103	23.331	0.132
5	24.55	23.456	4.456	24.601	0.208	24.625	0.307
6	25.41	25.193	0.854	25.715	1.200	25.105	1.201
7	26.91	27.059	0.554	26.775	0.502	27.229	1.184
8	28.02	29.063	3.722	27.824	0.700	27.913	0.381
9	28.64	31.216	8.994	29.057	1.456	28.594	0.161
10	28.44	30.298	6.533	30.340	6.681	29.746	4.591

Table 7.20 The simulated values and APE of the SMGM(1,3), GM(1,1) and MGM(1,3) at point B (unit: mm)

No.	Actual value	GM(1,1)		MGM(1,3)		SMGM(1,3)	
		Simulated value	APE / %	Simulated value	APE / %	Simulated value	APE / %
1	9.89	—	—	—	—	—	—
2	12.20	14.171	16.156%	12.625	3.484%	—	—
3	16.27	15.484	4.831%	15.789	2.956%	16.258	0.075%
4	17.66	16.920	4.190%	17.713	0.300%	17.737	0.434%
5	19.07	18.488	3.052%	19.256	0.975%	19.165	0.496%
6	20.85	20.202	3.108%	20.662	0.902%	20.373	2.288%
7	21.91	22.075	0.753%	22.001	0.415%	22.222	1.423%
8	23.40	24.121	3.081%	23.300	0.427%	23.520	0.512%
9	23.77	26.357	10.883%	24.654	3.719%	24.527	3.186%
10	24.12	26.624	10.381%	25.131	4.192%	24.842	2.993%

The values of accuracy criteria (MSE, AME and MAPE) of different subgrade settlement prediction models are shown in Table 7.22.

From the viewpoint of error analysis, the multi-variable models of MGM(1,3) and SMGM(1,3) always show lower error values than the uni-variable model GM(1,1). It is shown that the multi-point prediction models can take the relationship among variables into account, and are able to adequately reflect the integral evolution laws of subgrade settlement system. The self-memory technique helped model SMGM(1,3) to further reduce the modeling errors compared with the traditional model MGM(1,3).

Table 7.21 The simulated values and APE of the SMGM(1,3), GM(1,1) and MGM(1,3) at point C (unit: mm)

No	Actual value	GM(1,1)		MGM(1,3)		SMGM(1,3)	
		Simulated value	APE/%	Simulated value	APE/%	Simulated value	APE/%
1	12.03	–	–	–	–	–	–
2	15.60	17.443	11.814	16.044	2.846	–	–
3	19.57	18.696	4.466	19.085	2.478	19.564	0.032
4	20.80	20.039	3.659	20.810	0.048	20.836	0.171
5	22.03	21.479	2.501	22.153	0.558	22.107	0.351
6	23.38	23.022	1.531	23.378	0.009	23.084	1.266
7	24.60	24.676	0.309	24.558	0.171	24.771	0.694
8	25.79	26.449	2.555	25.719	0.275	25.886	0.374
9	26.36	28.349	7.546	26.954	2.253	26.813	1.720
10	27.16	29.087	7.095	27.445	1.049	27.412	0.927

Table 7.22 Simulation error of different subgrade settlement prediction models

Monitoring point	Model	MSE	AME	MAPE (%)
Point A	MGM(1,3)	3.924	1.595	10.174
	OMGM(1,3)	3.088	1.511	9.277
	SMGM(1,3)	0.170	0.316	1.572
Point B	MGM(1,3)	4.538	1.665	10.096
	OMGM(1,3)	3.336	1.542	9.017
	SMGM(1,3)	0.232	0.370	1.741
Point C	MGM(1,3)	5.200	1.735	10.051
	OMGM(1,3)	3.700	1.607	8.961
	SMGM(1,3)	0.343	0.451	2.026
Subgrade system	MGM(1,3)	4.554	1.665	10.107
	OMGM(1,3)	3.375	1.553	9.085
	SMGM(1,3)	0.248	0.379	1.780

Meanwhile, the model SMGM(1,3) has passed the modeling simulation and prediction accuracy test, and the single-step and two-step rolling prediction precisions are also generally superior than that of the other two grey models. In summary, the model SMGM (1,3) markedly promoted the predictive performance compared with other grey prediction models.

References

Dang, Y. G., Liu, B., & Guang, Y. Q. (2005). On strengthening buffer operators. *Control and Decision, 20*(12), 1332–1336.

Deng, J. L. (1982). Control problems of grey systems. *Systems & Control Letters, 1*(5), 288–294.

Deng, J. L. (1985). *Grey control systems.* Press of Huazhong University of Science and Technology.

Guo, X. J., Liu, S. F., & Wu, L. F., et al. (2015). A multi-variable grey model with a self-memory component and its application on engineering prediction. *Engineering Applications of Artificial Intelligence, 42*, 82–93.

Ji, P. R., Huang, W. S., & Hu, X. Y. (2001). On characteristics of grey prediction models. *Systems Engineering: Theory and Practice, 21*(9), 105–108.

Li, J. F., & Dai, W. Z. (2004). Multi-period grey modeling method based on stepwise ratios and its application in Chinese GDP modeling. *Systems Engineering: Theory and Practice, 24*(9), 98–102.

Liang, Q. W., Song, B. W. & Jia, Y. (2005). The grey Verhulst model of the costs on the research of torpedoes. *Journal of System Simulation, 17*(2), 257–258.

Liu, S. F. (1991). The three axioms of buffer operator and their application. *Journal of Grey System, 3*(1), 39–48.

Liu, S. F. (2021). *Grey system theory and its application.* (9th ed.). Beijing: Science Press

Liu, S. F., & Deng, J. L. (2000). The range suitable for GM(1,1). *Systems Engineering Theory & Practice, 20*(5), 121–124.

Liu, S. Q., Sheng, K. Q., & Forrest, J. (2012). On uncertain systems and uncertain models. *Kybernetes, 41*(5-6), 548–558.

Liu, H. B., Xiang, Y. M., & Ruan, Y. X. (2013). A multivariable grey model based on background value optimization and its application to subgrade settlement prediction. *Rock Soil Mechanics, 34*(1), 173–181.

Liu, S. F., Zeng, B., & Liu, J. F., et al. (2015a). Four basic models of GM(1, 1) and their suitable sequences. *Grey Systems: Theory and Application, 5*(2), 141–156.

Liu, J. F., Liu, S. F., & Fang, Z. G. (2015b). Fractional-order reverse accumulation generation GM(1,1) model and its applications. *The Journal of Grey System, 27*(4), 52–62.

Liu, S. F., Yang Y. J., & Forrest, J. (2017). Grey data analysis. Singapore: Springer-Verlag.

Luo, Y. X. (2010). Non-equidistant step by step optimum new information GM(1, 1) and its application. *Systems Engineering Theory & Practice, 30*(12), 2254–2258.

Mao, S. H., Gao, M. Y., Xiao, X. P. et al. (2016). A novel fractional grey system model and its application. *Applied Mathematical Modelling, 40*(7–8), 5063–5076.

Salmeron, J. L. (2010). Modelling grey uncertainty with fuzzy grey cognitive maps. *Expert Systems with Applications., 37*, 7581–7588.

Song, Z. M., Xiao, X. P., & Deng, J. L. (2002). The character of opposite direction AGO and class ratio. *The Journal of Grey System, 4*(1), 9–14.

Tan, G. J. (2005). The structure method and application of background value in grey system GM(1, 1) model. *Systems Engineering Theory & Practice, 25*(1), 98–103.

Wang, Y. N. (2003). An extended step by step optimum direct modeling method of GM(1, 1). *Systems Engineering Theory & Practice, 23*(2), 120–124.

Wang, Y. M., Dang, Y. G., & Wang, Z. X. (2008). Optimization of the background values of non-equal-distant GM(1,1) model. *Management Science of China, 16*(4), 159–162.

Wu, L. F., Liu, S. F, & Yao, L. G., et al. (2013). Grey system model with the fractional order accumulation. *Communications in Nonlinear Science and Numerical Simulation, 18*(7), 1775–1785.

Xiao, X. P. (2000). On parameters in grey models. *The Journal of Grey System, 11*(4), 73–78.

Xie, N. M., & Liu, S. F. (2005). Discrete GM(1,1) model and modeling mechanism of grey prediction models. *Systems Engineering: Theory and Practice, 28*(4), 93–99.

Xie, N. M., & Liu, S. F. (2009). Discrete grey forecasting model and its optimization. *Applied Mathematical Modelling, 33*(1), 1173–1186.

Zadeh, L. A. (1994). Soft computing and fuzzy logic. *IEEE Software, 11*(6), 48–56.
Zhang, Q. S. (2007). Improving the precision of GM(1, 1) model using particle swarm optimization. *Chinese Journal of Management Science, 15*(5), 126–129.

Chapter 8
Combined Grey Models

Along with the disciplinary development of systems science and systems engineering, methods and modeling techniques established for systems evaluation, prediction, decision-making, and optimization are enriched constantly. Generally, each method and every model have their strengths and weaknesses, so in practical applications several different methods and modeling techniques are combined to form hybrid methods or techniques in order to successfully deal with the problems at hand. Such combinations and mixtures are used to capitalize upon the strengths and advantages of different methods so that they complement each other and at the same time improve the weaknesses of individual methods and modeling techniques. This explains why combined or mixed systems are superior to individual component methods. Additionally, the availability of many different methods and modeling techniques also provides us with different ways to deal with information and systems. Therefore, how to combine and mix different methods and techniques has become a research direction with wide-ranging applicability in areas of data mining and knowledge discovery.

8.1 Grey Econometrics Models

8.1.1 Determination of Variables Using the Grey Relational Principles

In analyzing systems, due to the complications of mutually crossing influences of the endogenous variables, at the very start of modeling, the first problem that needs to be addressed is how to select the variables that will be part of the eventual model. To revolve this problem, the researcher needs not only rely on his qualitative analysis of the system, but also have sufficiently adequate tools for conducting quantitative

© The Author(s), under exclusive license to Springer Nature Singapore Pte Ltd. 2022
S. Liu, *Grey Systems Analysis*, Series on Grey System,
https://doi.org/10.1007/978-981-19-6160-1_8

analysis. Grey relational analysis model provide an effective method for this class of problems.

Let y be an endogenous variable of the system of our concern (for systems with many endogenous variables, these variables can be studied individually), and $x_1, x_2, \ldots, x_n$ be pre-images of influencing factors that are correlated either positively or negatively to y. Calculate the grey relational degree ε_i between y and $x_i, i = 1, 2, \ldots, n$, at first. For a chosen lower threshold value ε_0, when $\varepsilon_i < \varepsilon_0$, remove the variable x_i out of consideration. By doing so, some of the system's endogenous variables with weak grey relational degrees with y can be removed from further consideration. Assume that the remaining illustrative variables of y are $x_{i_1}, x_{i_2}, \ldots, x_{i_m}$. Next, consider the grey relational degrees $\varepsilon_{i_j i_k}(i_j, i_k = i_1, i_2, \ldots,)i_m$ between these remaining variables. For a chosen threshold value ε_0', when $\varepsilon_{i_j i_k} \geq \varepsilon_0'$, the variables x_{i_j} and x_{i_k} are seen as the same kind so that the remaining variables are divided into several subsets. Now, choose one representative from each of these subsets to enter into the eventual model. By going through this possess, the resultant econometrics model can be greatly simplified without losing the needed power of explanation. At the same time, to a certain degree the difficult problem of collinearity of the variables can be avoided.

8.1.2 Grey Econometrics Models

In econometrics, there are many different kinds of models, such as linear regression models in one or multiple variables, nonlinear models, systems of equations, among others. When estimating the parameters of these models, one often faces phenomena that are difficult to explain. For instance, the coefficients of the major illustrative variables are nearly zero; the signs of some estimated values of the parameters do not agree with reality or contradict theoretical economic analysis; small vibrations in a few individual observations cause drastic changes in many other estimated parametric values. Among the main reasons underlying these difficulties are:

(1) During the time period the observations are done, the internal structure of the system goes through major changes;
(2) There is a problem of collinearity between the illustrative variables; and
(3) There are randomness and noise in the observed data.

For the first two scenarios, there is a need to repeat the investigation of the model structure or a need to recheck the illustrative variables. For the third scenario, one can consider establishing models using the GM(1,1) simulated values of the original observations to eliminate the effect of the randomness or noise existing in the available data. The combined grey econometrics model, obtained this way, can more accurately reflect the relationship between the system's variables. At the same time, the prediction results made on the endogenous variables of the grey econometrics model system, which is based on the GM(1,1) predicted values of the illustrative variables, possess more solid scientific foundation than qualitative estimate values of the illustrative variables. Besides, by comparing the results of grey predictions of the endogenous variables with those obtained out of econometrics models, one can further improve the reliability of the predictions.

The steps for establishing and applying grey econometrics models are as follows:

Step 1: Design the theoretical model. Study the economic activity of interest closely. According to the purpose of the investigation, select the variables that will potentially enter the model. Discover the relationships between these variables based on theories of economic behavior and experience and/or analyze the sampled data. Develop the mathematical expressions, which are the theoretical model, that describe the relationships between these variables. This stage is the most important and difficult phase of the entire modeling process, and the following work need to be done:

(1) Study relevant theories of economics

Theoretical models summarize the fundamental characteristics and laws of development of the objective matters. They are abstract pictures of reality. Therefore, in the stage of model design, one first needs to conduct a qualitative analysis using economic theories. With different theories, various models can be established. For instance, according to the theory of equilibrium of labor markets, the rate y of wage increase is related to the unemployment rate x_1 and inflation rate x_2, that is, $y = f(x_1, x_2)$. The greater the unemployment rate increases, the smaller the rate of wage increase due to the fact that the supply of labor is clearly greater than the demand. This is the well-established Alban W. Phillips curve, which has been widely accepted and applied in the economic models of Western countries. However, this model may not necessarily hold true in the socialist market economy of China. As a second example, according to Keynes's theory of consumption, it is believed that, on average, when income grows, people tend to increase their consumption. However, the degree of increase in consumption is not as high as that of income. Assume that y stands for consumption, and x for income. Then, a mathematical expression for the relationship between these variables is

$$y = f(x) = b_0 + b_1 x + \varepsilon$$

where the parameter $b_1 = dy/dx$ stands for the marginal consumption tendency, and ε a random noise, representing the inherent randomness of consumption. According to Keynes, $0 < b_1 < 1$. However, Simon Kuznets does not agree with Keynes's opinion of a declining marginal consumption tendency. His work indicates that there is a stable proportion of increase between consumption and income. That is, the previous model is only a product of Keynes's theory.

(2) Variables and the form of the eventual model

The established model should reflect the objective economic activity. However, it is impossible for such a reflection to include all details. This is why we need reasonable assumptions. Employing the method of this section to select the major variables to be included in the model using grey relational analysis will help to eliminate minor relationships and factors. It focuses on the dominant connections while simplifying the eventual model, making it convenient to handle and apply.

The specific works of this stage of model design include: (i) Determine which variables to include, which ones are dependent variables, and which ones are independent. Here, each independent variable is also known as illustrative variable. (ii) Determine the number of parameters to be included in the model and their (positive or negative) signs. (iii) Determine the mathematical form of the model expression. Is it linear or nonlinear?

(3)   Collection and organization of statistical data

After having decided on which variables to consider, one needs to collect all the relevant data. That is the foundation of establishing models. Generally speaking, all the collected raw data need to be statistically categorized and organized so that they become the empirical evidence of the characteristics of the problem of concern and are systematically usable for the purpose of modeling. The basic types of statistical data, as discussed in Chap. 3, include behavioral sequences, time series, index sequences, horizontal sequences, among others.

Step 2: Establish the GM(1,1) model and obtain its simulated values. In order to eliminate the random effect or error noise existing in the observational values of individual variables of the model, establish the GM(1,1) models for the individually observed sequences and then apply the simulated values of these GM(1,1) models as the base sequences on which to construct the eventual model.

Step 3: Estimate the parameters. After having designed the econometrics model, the next task is to estimate the parameters, which are the constant coefficients of the quantitative relationship between the chosen variables of the model. They connect the individual variables within the model. More specifically, these parameters explain how independent variables affect the dependent variable. Before using observed data to make estimations, these parameters are unknown. After the form of the model is established on the basis of the GM(1,1) model, simulated sequences solve the estimated values of the parameters using an appropriate method, such as that of least squares estimate. As soon as the parameters are clearly specified, the relationships between model variables become known and the model can be determined.

The estimated values of the parameters provide realistic and empirical contents and verification for the theories of economics. For instance, in the previously mentioned consumption model, if the estimated value of parameter b_1 is $\hat{b}_1 = 0.8$, it not only classifies the realistic content of the marginal consumption tendency, but also provides a piece of evidence for the assumption of Keynes's theory of consumption that this parameter is between 0 and 1.

Step 4: Test the model. After the parameters are estimated, the abstract model becomes specific and determined. However, to determine whether or not the model agrees with objective reality, and whether or not it can explain realistic economic processes, it still has to go through tests. The tests consist of two aspects, the test of economic meanings and statistical tests. The test of economic meanings checks whether or not the individual estimated values of the parameters agree with economic theories and relevant experiences. Statistical tests check the reliability of the estimate, the effectiveness of the data sequence simulation, the correctness of various econometrics assumptions, as well as the overall structure of the model and its prediction ability using the principles of statistical reasoning. It is only after the model passes through these tests that it can be applied in practice. If the model does not pass the tests, then the model needs to be modified and improved.

Step 5: Apply the established model. Grey econometrics models have been mainly employed to analyze economic structures, evaluate policies and decisions, simulate economic systems, and predict economic development. Each application process is also a process of verifying the model and its underlying theory. If the prediction contains small errors, it means that the model is of high accuracy and quality, with a strong ability to explain reality and an underlying theory that agrees with reality. Otherwise, the model and the economic theory on which the model was initially developed need to be modified.

Combined grey econometrics models can be employed not only to situations of known system structures, but also to situations of system structures that need further study and exploration. Combined grey econometrics models have produced satisfactory results in practical applications. To this end, please consult Liu and Lin (2006, pp. 247–254) and Liu and Zhu (1996) to see how applications are carried out.

8.2 Combined Grey Linear Regression Models

Combined grey linear regression models can improve the weakness of original linear regression models where no exponential growth is considered. They can also improve the weakness of GM(1,1) models that do not involve enough linear factors. Thus, such combined models are suited for studying sequences with both linear tendencies and exponential growth tendencies. For such a sequence, the modeling process can be described as follows.

Definition 8.2.1 Assume that $X^{(0)} = \left\{x^{(0)}(1), x^{(0)}(2), ..., x^{(0)}(n)\right\}$ is a sequence of raw data. Its first order accumulation sequence is $X^{(1)} = \left\{x^{(1)}(1), x^{(1)}(2), ..., x^{(1)}(n)\right\}$

$$\hat{x}^{(1)}(k) = C_1 e^{-vk} + C_2 k + C_3 \tag{8.1}$$

is called a combined grey linear regression model, where v, C_1, C_2, C_3, are parameters that need to be estimated.

In fact, combined grey linear regression model (8.1) is a simulation model of $X^{(1)}$, which can be seen as the sum of a linear regression model of $y = ak + b$ and an exponential model of $y = C_1 e^{-ak} + C_2$.

From the model GM(1,1), we can obtain

$$\hat{x}^{(1)}(k+1) = \left(x^{(0)}(1) - \frac{b}{a}\right)e^{-ak} + \frac{b}{a} \tag{8.2}$$

Let $C_1 = \left(x^{(0)}(1) - \frac{b}{a}\right)$, $C_3 = \frac{b}{a}$, which can be written as shown below:

$$\hat{x}^{(1)}(k+1) = C_1 e^{-ak} + C_3 \tag{8.3}$$

By adding a linear term $C_2 k$ to formula (8.3), we can obtain the same formula as (8.1).

Lemma 8.2.1 *Assume that $X^{(0)}$ and $X^{(1)}$ are the same as in Definition 8.2.1, then the parameter v in formula (8.1) can be estimated by the following formula (8.4):*

$$\hat{v} = \frac{\sum_{m=1}^{n-3} \sum_{k=1}^{n-2-m} \tilde{V}_m(k)}{(n-2)(n-3)/2} \tag{8.4}$$

where $\tilde{V}_m(k) = \ln[y_m(k+1)/y_m(k)]$, $y_m(k) = x^{(1)}(k+m+1) - x^{(1)}(k+m) - x^{(1)}(k+1) + x^{(1)}(k)$, $k, m = 1, 2, ..., n-3$.

Theorem 8.2.1 *Assume that $X^{(0)}$ and $X^{(1)}$ are the same as in Definition 8.2.1. Let*

$$X^{(1)} = \begin{bmatrix} x^{(1)}(1) \\ x^{(2)}(2) \\ \vdots \\ x^{(1)}(n) \end{bmatrix}, \quad C = \begin{bmatrix} C_1 \\ C_2 \\ C_3 \end{bmatrix}, \quad A = \begin{bmatrix} e^v & 1 & 1 \\ e^{2v} & 2 & 1 \\ \vdots & \vdots & \vdots \\ e^{in} & n & 1 \end{bmatrix},$$

then we have the matrix form (8.5) of (8.1):

$$X^{(1)} = AC \tag{8.5}$$

Therefore, we have

$$C = (A^T A)^{-1} A^T X^{(1)} \tag{8.6}$$

With the estimated values of parameters, v, C_1, C_2, C_3, the following formula (8.7) can be used as a simulating or forecasting model:

$$\hat{x}^{(1)}(k) = C_1 e^{-\hat{v}k} + C_2 k + C_3 \tag{8.7}$$

From Eq. (8.7), it can be seen that if $C_1 = 0$, then the first order accumulation sequence stands for a linear regression model. If $C_2 = 0$, then the accumulation sequence stands for a GM(1,1) model. This new model improves the weaknesses

Table 8.1 The original sequence of recorded subsides

Time	9502	9504	9506	9508	9510	9512	9602	9604
Amount of subside	12	22	31	43	51	57	75	83

of the original linear regression model with no exponential growth and that of the GM(1,1) model where no linear factors are considered.

By applying the inverse accumulation generation operator to Eq. (8.7), we can obtain the simulated and predicted values $\hat{X}^{(0)}$ of the original sequence.

Example 8.2.1 At a certain observation station of ore and rock movement, the sequence of recorded subsides of a specific location from February 1995 to April 1996 is given in Table 8.1. Try to make predictions for the sinking dynamics of this specific location (Han & He, 1997).

Solution Due to the small amount of available data, grey systems models are the most appropriate models for this prediction task. However, grey systems models employ exponential functions to simulate accumulation generated sequences. They are generally only suitable for modeling situations of exponential development, as it is difficult for such models to describe linear tendencies of change. Therefore, in this case study we will apply a grey linear exponential regression model to predict the subsides of the specified location.

The original sequence of data is.

$$X^{(0)} = (12, 22, 31, 43, 51, 57, 75, 83).$$

Its first order accumulation sequence is

$$X^{(1)} = (12, 34, 65, 108, 159, 216, 291, 374).$$

For different m values, from Eqs. (8.6) and (8.7) we obtain the estimated value $\hat{V} = 0.02058096$ for v. Also, from Eq. (6.10), we obtain the estimated value of C:

$$C = (A^T A)^{-1} A^T X^{(1)} = (21750.995, -439.9523, -21751.078)$$

Thus, the combined model of the first order accumulation generation sequence is

$$\hat{x}^{(1)}(k) = 21750.995 e^{0.020580966} - 439.9523k - 21751.078$$

Out of this model, we obtain the simulated and predicted values for each of the time moments as listed in Table 8.2.

Table 8.2 Simulated and predicted values and their errors

Time	9502	9504	9506	9508	9510	9512	9602	9604	9606	9608
$x^{(0)}(k)$	12	22	31	43	51	57	75	83		
$\hat{x}^{(0)}(k)$	12.34	21.75	31.35	41.15	51.15	61.36	71.79	82.43	93.29	104.38
Error (%)	−2.85	1.15	−1.12	4.31	−0.30	−7.66	4.28	0.69		

8.3 Grey Cobb–Douglas Model

In this section, we study the Cobb–Douglas or production function model. Let K be the capital input, L the labor input, and Y the production output. Then,

$$Y = A_0 e^{\gamma t} K^\alpha L^\beta$$

is known as the C-D production function model, where α stands for capital elasticity, β labor elasticity, and γ the parameter for the progress of technology. The log-linear form of this production function model is given below:

$$\ln Y = \ln A_0 + \gamma t + \alpha \ln K + \beta \ln L$$

For given time series data of the production output Y, capital input K, and labor input L,

$$Y = (y(1), y(2), \ldots, y(n)), \ K = (k(1), k(2), \ldots, k(n)), \text{ and } L = (l(1), l(2), \ldots, l(n))$$

one can employ the method of multivariate least squares estimate to approximate the parameters $\ln A_0$, γ, α, and β.

When Y, K, and L represent the time series of a specific department, district, or business, it is often the case that, due to severe fluctuations existing in the data, the estimated parameters contain errors leading to incorrect results. For instance, the estimated coefficient γ for progress of technology is too small or becomes a negative number; the estimated values α and β for elasticity go beyond their reasonable ranges. Under such circumstances, if one considers using the GM(1,1) simulated data of Y, K, and L as the original data for their least squares estimates, then to a certain degree they can eliminate some of the random fluctuations, produce more reasonable estimated parameter values, and obtain a model that can more accurately reflect the relationship between the production output and labor, and capital inputs and the progress of technology.

Definition 8.3.1 Assume that

$$\hat{Y} = (\hat{y}(1), \hat{y}(2), \ldots, \hat{y}(n)),$$
$$\hat{K} = (\hat{k}(1), \hat{k}(2), \ldots, \hat{k}(n)), \text{ and}$$
$$\hat{L} = (\hat{l}(1), \hat{l}(2), \ldots, \hat{l}(n))$$

are respectively the GM(1,1) simulated sequences of Y, K, and L. Then $\hat{Y} = A_0 e^{\gamma t} \hat{K}^\alpha \hat{L}^\beta$ is known as the grey model of production function.

In the grey production function model, although no grey parameters appear explicitly, it stands for an expression that combines the idea of grey systems modeling into the C-D production function model. That is, this model possesses a very deep intension of the greyness. It embodies the non-uniqueness principle of solutions and the absoluteness principle of greyness. This is why, in practical applications, this model has produced satisfactory results. To this end, please consult Liu et al. (2004) and Liu and Lin, 2006, pp. 256–258) to see how applications are carried out.

8.4 Grey Artificial Neural Network Models

8.4.1 BP Artificial Neural Model and Computational Schemes

Each artificial neural network is made up of a large amount of elementary information processors, known as neurons or nodes. The model with multi-layered nodes, or the scheme known as error back propagation, represents the currently well developed and widely employed artificial neural network system and computational method. It translates the input–output problem of an available sample into a nonlinear optimization problem. It is a powerful tool that can be employed to uncover the laws and patterns hidden in large amounts of data. The use of artificial neural networks to simulate data sequences has several latent advantages. First, it has the ability to model multiple kinds of functions, including nonlinear functions, piecewise defined functions, among others. Secondly, artificial neural networks are unlike the traditional methods of distinguishing data sequences, which, to work properly, must have presumed types of functional relationships between data sequences. This means that artificial neural networks can establish the needed functional relationship by using the attributes and intension naturally existing in the provided data variables, without presuming the kinds of distributions the parameters satisfy. Thirdly, this method possesses the advantage of making use of available information very efficiently, while avoiding the problem of losing the real meanings and pictures of the data due to various combinations, such as additions of positive and negative values of data mining methods. That is, the artificial neural networks method is especially useful for improving the GM(1,1) model.

Figure 8.1 shows a back propagation network with three layers. The network consists of an input layer, an implicit (or latent) layer, and an output layer. An entire process of learning consists of forward and backward propagation. The particular scheme of learning is given below:

(1) Apply random numbers to initialize W_{ij} (the connection weight between nodes i and j of different layers) and θ_j (the threshold value of node j);

Fig. 8.1 A back propagation
neural network

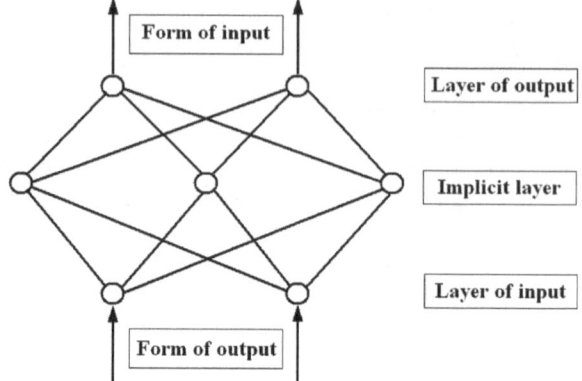

(2) Feed in the preprocessed training samples $\{X_{PL}\}$ and $\{Y_{PK}\}$;
(3) Compute the output of the nodes of each layer, $O_{pj} = f \sum_i (W_{ij} I_{pi} - \theta_j)$ for
    the pth sample point, where I_{pi} stands for the output of node i and the input of
    node j;
(4) Compute the information error of each layer. For the input layer, $\delta_{pk} =$
    $O_{pk}(y_{pk} - O_{pk})(1 - O_{pk})$; for the latent layer, $O_{pi} = O_{pi}(1 - O_{pi}) \sum_i \delta_{pi} W_{ij}$;
(5) For the backward propagation, the modifiers of the weights are $W_{ij}(t + 1) =$
    $\alpha \delta_{pi} O_{pi} + W_{ij}(t)$, and the modifiers of the thresholds $\theta_j(t + 1) = \theta_j(t) + \beta \delta_{pi}$,
    where α stands for the learning factor and β the momentum factor for accelerated
    convergence; and
(6) Calculate the error $E_p = (\sum_p \sum_k)(O_{pk} - Y_{pk})^2/2$.

8.4.2 Steps in Grey BP Neural Network Modeling

The steps to establish a grey BP neural network model are as follows:

Step 1: Assume that a time series $\{x^{(0)}(i)\}, i = 1, 2, \ldots, n$, is given. We then obtain
the restored values $\hat{x}^{(0)}(t), i = 1, 2, \ldots, n$, using the outputs of the GM(1,1) model.

Step 2: Establish the back propagation network model for the error sequence
$\{e^{(0)}(k) = x^{(0)}(k) - \hat{x}^{(0)}(k)\}, k = 1, 2, \ldots, n$.

  If the order of prediction is S, it means that we use the information of $e^{(0)}(i - 1)$,
$e^{(0)}(i - 2)$, ..., $e^{(0)}(i - S)$ to predict the value at the ith moment; we will treat
$e^{(0)}(i-1), e^{(0)}(i-2), \ldots, e^{(0)}(i-S)$ as the input sample points of the back propagation
network training, while using the value of $e^{(0)}(i)$ as the expected prediction of the
back propagation network training. By using the back propagation computational
scheme outlined earlier, train this network through enough amount of cases of error
sequences so that output values (along with empirical test values) are produced in
ways that correspond to different input vectors. The resultant weights and thresholds
represent the correct internal representations through the self-learning and adaptation

of the network. A well trained back propagation network model can be an effective tool for error sequence prediction.

Step 3: Determine the simulation values of $\{e^{(0)}(k) = x^{(0)}(k) - \hat{x}^{(0)}(k)\}$, $k = 1, 2, \ldots, n$. Assume that the simulation sequence is $\{\hat{e}^{(0)}(k)\}$, $k = 1, 2, \ldots, n$, which is obtained by the BP neural network.

Step 4: Based on $\{\hat{x}^{(0)}(i)\}$ and $\{\hat{e}^{(0)}(k)\}$, $i, k = 1, 2, \ldots, n$, we have the following result

$$\hat{x}^{(0)}(i, k) = \hat{x}^{(0)}(i) + \hat{e}^{(0)}(k) \tag{8.8}$$

which is the predicted sequence of the grey artificial neural network model.

Example 8.4.1 Given the actual yearly investments in environmental protection over a period of time of a certain location, and the GM(1,1) simulations and relevant errors in Table 8.3, establish an artificial neural network model for the error sequence (Dong & Yang, 1998).

Solution Based on and using the GM(1,1) error sequence data given in Table 8.3, we apply the previously outlined method to establish a back propagation network model. Our projected back propagation network will have three characteristic parameters, one latent layer, within which there are 6 nodes, and one input layer within which there is one node. Let the learning rate be 0.6, the convergence rate 0.001, and the variance limited within the range of 0.01. Let us conduct the training and testing of the network on a computer. Then, Table 8.4 lists the simulation results of the combined back propagation network model.

Table 8.3 The GM(1,1) simulations and errors

Year	Investment $x^{(0)}(i)$	GM(1,1) simulation $\hat{x}^{(0)}(i)$	Errors $e^{(0)}(k)$
1985	110.20	110.20	0
1986	146.34	164.39	−19.05
1987	185.36	187.65	−2.29
1988	221.14	214.22	6.92
1989	255.16	244.54	10.52
1990	289.18	279.17	9.01
1991	320.54	319.69	1.85
1992	352.79	363.81	−11.02

Table 8.4 Simulation results of the grey artificial neural network model

Year	Actual value $x^{(0)}(i)$	Simulated value $\hat{x}(i, k)$	Relative errors (%)
1988	221.14	221.12	0.01
1989	255.16	255.29	0.05
1990	289.18	289.11	0.02
1991	320.54	320.79	0.08
1992	352.79	352.70	0.03

8.5 Grey Markov Model

8.5.1 Grey Moving Probability Markov Model

Definition 8.5.1 Assume that $\{X_n, n \in T\}$ is a stochastic process. If for any whole number $n \in T$ and any states $i_0, i_1, \ldots, i_{n+1} \in I$, the following conditional probability satisfies

$$P(X_{n+1} = i_{n+1}|X_0 = i_0, X_1 = i_1, \ldots, X_n = i_n) = P(X_{n+1} = i_{n+1}|X_n = i_n) \tag{8.9}$$

then $\{X_n, n \in T\}$ is known as a Markov chain. Equation (8.9) is seen without any post-effect. It means that the future state of the system at $t = n + 1$ is only related to the current state at $t = n$, without any influence from any other earlier state $t \leq n - 1$.

For any $n \in T$ and states $i, j \in I$, the following

$$p_{ij}(n) = P(X_{n+1} = j|X_n = i) \tag{8.10}$$

is known as the transition probability of the Markov chain. If the transition probability $p_{ij}(n)$ in this equation does not have anything to do with the index n, then $\{X_n, n \in T\}$ is known as a homogeneous Markov chain. For such a Markov chain, the transition probability $p_{ij}(n)$ is often denoted as p_{ij}. Because our discussion will be mainly on homogeneous Markov chains, the word "homogeneous" will be omitted. When all the transition probabilities $p_{ij}(n)$ are placed in a matrix, such as $P = [p_{ij}]$, this matrix is referred to as the transition probability matrix of the system's state.

Proposition 6.1 The entries of the transition probability matrix P satisfy

(1) $p_{ij} \geq 0, i, j \in I$; and
(2) $\sum_{j \in I} p_{ij} = 1, i \in I$

The probability $p_{ij}^{(n)} = P(X_{m+n} = j|X_m = i), i, j \in I, n \geq 1$ is known as the nth step transition probability of the given Markov chain, and $P^{(n)} = [p_{ij}^{(n)}]$ the nth step transition probability matrix.

Proposition 6.2 *The nth step transition probability matrix* $P^{(n)}$ *satisfies*

(1) $p_{ij}^{(n)} \geq 0, i, j \in I$;

(2) $\sum_{j \in I} p_{ij}^{(n)} = 1, i \in I$; and

(3) $P^{(n)} = P^n$

Any Markov chain with grey transition probabilities is known as a grey Markov chain. When studying practical problems, due to a lack of sufficient information, it is often difficult to determine the exact values of the transition probabilities. In such cases, it might be possible to determine the grey ranges $p_{ij}(\otimes)$ of these uncertain probabilities based on available information. When the transition probability matrix is grey, the entries of its whitenization $\tilde{P}(\otimes) = [\tilde{P}_{ij}(\otimes)]$ are generally required to satisfy the following properties:

(1) $\tilde{P}_{ij}(\otimes) \geq 0, i, j \in I$; and

(2) $\sum_{j \in I} \tilde{P}_{ij}(\otimes) = 1, i \in I$.

Proposition 6.3 *Assume that the initial distribution of a finite-state grey Markov chain is* $P^T(0) = (p_1, p_2, \ldots, p_n)$ *and the transition probability matrix* $P(\otimes) = [P_{ij}(\otimes)]$. *Then, the system's distribution of the next sth state is*

$$P^T(s) = P^T(0)P^s(\otimes) \tag{8.11}$$

That is, when the system's initial distribution and the transition probability matrix are known, one can predict the system's distribution for any future state.

8.5.2 Grey State Markov Model

Assume that a stationary process $X^{(0)}$ satisfies the condition of Markov chains. If we divide it into n states and each of the states $\otimes_i$ is expressed by

$$\otimes_i = [a_i, b_i], (i = 1, 2, \ldots, s)$$

where a_i, b_i are constants and determined according to the states. The steps to establish a grey state Markov model are outlined next.

Step 1: Determine the states for a stationary process $X^{(0)}$ which satisfies the condition of Markov chains

$$\otimes_i = [a_i, b_i], (i = 1, 2, \ldots, s)$$

Step 2: Compute the initial probability distribution. Assume that there are s different states $\otimes_1, \otimes_2, \ldots, \otimes_s$. If state $\otimes_i (i = 1, 2, \ldots, s)$ occurs M_i times in total in M experimentations, then the frequency of M_i can be calculated by

$$f_i = \frac{M_i}{M} (i = 1, 2, \ldots, s)$$

We can use $f_i (i = 1, 2, \ldots, s)$ as an approximation of the initial probability $p_i (i = 1, 2, \ldots, s)$, that is, let $f_i \approx p_i (i = 1, 2, \ldots, s)$

Step 3: Compute the transition probability. Just like computing the initial probability, we take the frequency as an approximation of the transition probability.

Firstly, we calculate the one step transition frequency of $\otimes_i \rightarrow \otimes_j$ (from state $\otimes_i$ transfer to state $\otimes_j$ through one step) by

$$f_{ij} = f(\otimes_j | \otimes_i)$$

If state $\otimes_i (i = 1, 2, \ldots, s)$ occurs M_i times in total in M experimentations, let M_{ij} be the number of transfers to the state $\otimes_j$ from M_i state $\otimes_i$. Then we have

$$f_{ij} = \frac{M_{ij}}{M_i}$$

Then, if $f_{ij} \approx p_{ij}$, we have the transition probability matrix $P = (p_{ij})_{s \times s}$. Similarly, we can calculate the approximation of m steps transition probability as follows (8.12):

$$p_{ij}(m) = \frac{M_{ij}(m)}{M_i}, (i = 1, 2, \ldots, s) \tag{8.12}$$

where $M_{ij}(m)$ is the number of transfers to the state $\otimes_j$ from M_i state $\otimes_i$ through m steps.

Step 4: Prediction using the transition probability. Assume that the object of prediction is located at state $\otimes_k$, then consider the kth row of P. If

$$\max_j p_{kj} = p_{kl}$$

then it can be inferred that, at the next time moment, the system will most likely transform from state $\otimes_k$ to state $\otimes_l$. If there are two or more entries in the kth row of P that are equal or roughly equal, then the direction of change in the system's state is difficult to determine. In this case, one needs to look at the two-step or n-step transition probability matrix $P^{(2)}$ or $P^{(n)}$, where $n \geq 3$.

8.6   Combined Grey-Rough Model

Grey systems theory and rough set theory are two mathematical tools developed to address uncertain and incomplete information. To a certain degree they complement each other. They both apply the idea of lowering the preciseness of expression of the available data to gain the extra generality of the expression. In particular, grey systems theory employs the method of grey sequence generations to reduce the accuracy of data expressions, while rough set theory makes use of the idea of data scattering to uncover patterns hidden in the data by ignoring unnecessary details. Neither grey systems theory nor rough set theory requires any prior knowledge, such as probability distribution or degree of membership. On one hand, rough set theory investigates rough, non-intersecting classes and concepts of roughness, with emphasis placed on the indistinguishability of objects. On the other hand, grey systems theory focuses on grey sets with clear extension and unclear intension, with emphasis placed on uncertainties caused by insufficient information. Thus, if rough set theory and grey systems methodology are mixed, their individual weaknesses both in theory and application can be improved so that greater theoretical strength and practical applicability can be achieved (Jian & Liu, 2005).

8.6.1   Rough Membership, Grey Membership and Grey Numbers

Rough set theory can be seen as an expansion of the classic set theory. It makes use of rough membership functions to define rough sets, where each membership function is explained and understood as those of conditional probabilities.

The concepts of rough approximation sets and rough membership functions of the rough set theory are closely related to those of greyness of grey numbers. When either $\mu_X(x) = 0$ or $\mu_X(x) = 1$, the object is assured either to belong or not to belong to set X. In such cases, the classification is definite and clear; the involved greyness is the smallest. If $0 < \mu_X(x) < 1$, then object x belongs to set X with the degree of confidence $\mu_X(x)$. In this case, object x projects a kind of grey state of transition between definitely being in set X and definitely not being in X. When $\mu_X(x) = 0.5$, the probability of object x to either belong to set X or not to belong to X is 50%. For this situation, the degree of uncertainty is the highest. That is, the degree of greyness is the highest. When the rough membership function $\mu_X(x)$ is near 1 or 0, the uncertainty for object x to belong or not to belong to set X is decreased, and the corresponding degree of greyness should also decrease. The closer to 0.5 the rough membership is, the greater the uncertainty for object x to belong or not to belong to set X; the corresponding degree of greyness is also greater in such cases. We categorize all rough membership functions into two groups: upper and lower rough membership functions, where a rough membership function is upper if its values come from the interval [0.5,1], denoted $\overline{\mu}_X(x)$; the corresponding grey membership function is also referred to as upper and denoted by $\overline{g}_X(x)$. A lower rough membership function is one that takes values from the interval [0,0.5], denoted $\underline{\mu}_X(x)$. The corresponding grey membership function is referred to as a lower grey membership function, denoted $\underline{g}_X(x)$.

Evidently, upper, lower and general rough membership functions satisfy the following properties:

(1) $\overline{\mu}_X(x) = 1 - \underline{\mu}_X(x)$;
(2) $\mu_{X \cup Y}(x) = \mu_X(x) + \mu_Y(x) - \mu_{X \cap Y}(x)$; and
(3) $\max(0, \mu_X(x) + \mu_Y(x) - 1) \le \mu_{X \cap Y}(x) \le \min(1, \mu_X(x) + \mu_Y(x))$.

Based on the discussion above, we introduce the following definition of grey membership functions using the concept of rough membership functions.

Definition 8.6.1 Assume that x is an object with its field of discourse U. That is, $x \in U$. Let X be a subset of U. Then mappings from U to the closed interval [0, 1]:

$$\overline{\mu}_X : U \to [0.5, 1], \mu| \to \overline{g}_X(x) \in [0, 1], \text{ and}$$

$$\underline{\mu}_X : U \to [0, 0.5], \mu \Big| \to \underline{g}_X(x) \in [0, 1]$$

are respectively referred to as upper and lower grey membership functions of X, where $\overline{\mu}_X \ge \underline{\mu}_X$; $\overline{g}_X(x)$ and $\underline{g}_X(x)$ are respectively referred to as upper and lower grey membership functions of object x with respect to X.

The defined concept of grey membership functions based on rough membership functions is depicted in Fig. 8.2.

Definition 8.6.2 Assume that $x \in U$, $X \subseteq U$, the grey number scale of the uncertainty for x to belong to X is g_c, the grey number scale of the upper grey membership function $\overline{g}_X(x)$ is $\overline{g}_c$, and the grey number scale of the lower grey membership function $\underline{g}_X(x)$ is $\underline{g}_c$. Then the greyness scales $\overline{g}_c$ and $\underline{g}_c$ of the upper grey number

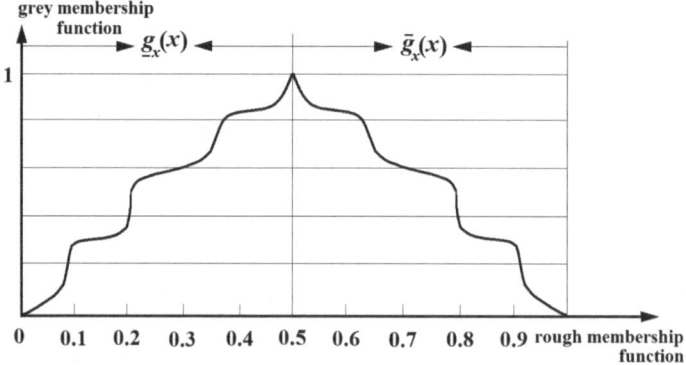

Fig. 8.2 A conceptual depiction of grey membership functions

and the lower grey number of the greyness scales g_c of different grey numbers are respectively given as outlined below.

The greyness of white numbers ($g_c = 0$): if $\mu_X(x) = 0$, then $\underline{g}_c = 0$; if $\mu_X(x) = 1$, then $\overline{g}_c = 0$.

For first class grey numbers ($g_c = 1$): if $\mu_X(x) \in (0,0.1]$, then $\underline{g}_c = 1$; if $\mu_X(x) \in [0.9,1)$, then $\overline{g}_c = 1$.

For second class grey numbers ($g_c = 2$): if $\mu_X(x) \in (0.1,0.2]$, then $\underline{g}_c = 2$; if $\mu_X(x) \in [0.8,0.9)$, then $\overline{g}_c = 2$.

For third class grey numbers ($g_c = 3$): if $\mu_X(x) \in (0.2,0.3]$, then $\underline{g}_c = 3$; if $\mu_X(x) \in [0.7,0.8)$, then $\overline{g}_c = 3$.

For fourth class grey numbers ($g_c = 4$): if $\mu_X(x) \in (0.3,0.4]$, then $\underline{g}_c = 4$; if $\mu_X(x) \in [0.6, 0.7)$, then $\overline{g}_c = 4$.

For fifth class grey numbers ($g_c = 5$): if $\mu_X(x) \in (0.4,0.5)$, then $\underline{g}_c = 5$; if $\mu_X(x) \in (0.5, 0.6)$, then $\overline{g}_c = 5$.

The greyness of black numbers ($g_c > 5$): if $\mu_X(x) = 0.5$, then $\underline{g}_c = \overline{g}_c > 5$.

When $\mu_X(x) \in [0,1]$, $\underline{g}_X(x) = 0$ and $\overline{g}_X(x) = 1$. In this case, there is no uncertain information, so it is referred to as the greyness of white numbers. That is, $g_c = \underline{g}_c = \overline{g}_c = 0$. When $\mu_X(x) = 0.5$, $\underline{g}_X(x) = \overline{g}_X(x) = 1$, the degree of uncertainty for object x to belong or not to belong to set X reaches its maximum, which is referred to as the greyness of black numbers $g_c > 5$. From Definition 8.6.1, it follows that the higher the greyness of a grey number, the less clear the information is; the lower the greyness of a grey number, the clearer the information is.

From Definition 8.6.1, it can be readily obtained that $\overline{\mu}_X(x) = 1 - \underline{\mu}_X(x)$. If we use the greyness of the upper grey number to represent the degree of uncertainty for object x to belong to set X, and the greyness of the lower grey number to illustrate the degree of uncertainty for object x not to belong to set X, then these two degrees of uncertainty are supplementary.

According to Definition 8.6.2, the scale of the greyness of a grey number is determined by the grey interval to which the maximum rough membership value of the information granularity could belong. Thus, the whitenizations of grey numbers of different degrees of greyness are defined as the maximum possible rough membership value of the grey numbers of corresponding scales. For example, if the possible maximum rough membership value of a certain conditional subset computed out of the available decision-making table is $\mu_X(x) = 0.75$, because $0.75 \in [0.7,0.8)$, then $\mu_X(x) = 0.75$ stands for the white value of such a grey number whose upper greyness is $\overline{g}_c = 3$.

8.6.2 Grey Rough Approximation

Definition 8.6.3 Assume that $S = (U, A, V, f)$, $A = C \cup D$, $X \subseteq U$, $P \subseteq C$, and the greyness scale $g_c \leq 5$ of a grey number. Then

$$apr_P^{g_c}(X) = \bigcup \left\{ \frac{|I_P(x) \cap X|}{|I_P(x)|} \leq \overline{g}_c \right\} \tag{8.13}$$

and

$$\overline{apr}_P^{g_c}(X) = \cup \left\{ \frac{|I_P(x) \cap X|}{|I_P(x)|} > \underline{g}_c \right\} \tag{8.14}$$

are respectively referred to as the g_c-lower approximation and g_c-upper approximation of X with respect to I_P, where the upper rough membership function corresponding to the upper scale $\overline{g}_c$ of grey-number greyness satisfies $\overline{\mu}_X(x) \in (0.5, 1]$, and the lower rough membership function corresponding to the lower scale $\underline{g}_c$ of grey-number greyness satisfies $\underline{\mu}_X(x) \in [0, 0.5)$.

The g_c-lower approximation of the set $X \subseteq U$ under the grey-number greyness scale g_c equals the union of all the equivalence classes of U that belong to X, with grey-number greyness scales less than or equal to the upper grey-number greyness scale $\overline{g}_c$. The g_c-upper approximation is equal to the intersection of all the equivalence classes of U that belong to X, with grey-number greyness scales greater than the lower grey-number greyness scale $\underline{g}_c$.

Definition 8.6.4 The quality of g_c-classification is

$$\gamma_P^{g_c}(P, D) = \frac{|\cup \left\{ \frac{|X \cap I_P(x)|}{|I_P(x)|} \leq \overline{g}_c \right\}|}{|U|} \tag{8.15}$$

The classification quality $\gamma_P^{g_c}(P, D)$ measures the percentage of the knowledge in the field of discourse that can be clearly classified for a given grey-number greyness scale $g_c \leq 5$, in the totality of current knowledge.

For a given grey-number greyness scale $g_c \leq 5$, let approximate reduction $red_P^{g_c}(C, D)$ stand for the set of attributes with the minimum condition that still produces clear classification without containing any extra attributes.

In rough set theory, the classification of the elements located along the boundary regions is not clear. Whether or not an element in such a region can be clearly classified is determined most commonly by the pre-fixed greyness scale. The concept of grey rough approximation so defined is analogous to that of variable precision rough approximation. When the interval grey numbers in which the upper greyness scale $\overline{g}_c$ and the lower greyness scale $\underline{g}_c$ of the grey-number greyness g_c of the grey rough approximation respectively belong to their corresponding white values, the grey rough approximation is consequently transformed into rough approximation under the meaning of variable precision rough sets. Evidently, variable precision rough approximation can be seen as a special case of grey rough approximation. When compared to models of variable precision rough sets of the sets of variable precision, whether or not elements in a relatively rough set X can be correctly classified is mostly determined by the pre-fixed maximum critical confidence threshold parameter

β. This is where classification can be done if smaller than or equal to the upper bound of β, and indistinguishability appears when this upper bound is surpassed. However, the parameter of the maximum critical confidence threshold β in general is difficult to determine beforehand, especially for large data sets. In other words, the parameter of maximum critical confidence threshold β generally stands for a grey number. Thus, the concept of interval grey numbers provides a practical quantitative tool which appoints upper and lower endpoints. For cases where we cannot obtain much information about the degree of accuracy of the actual data, this method of representation becomes extremely useful.

Proposition 8.6.1 *Given the greyness scale $g_c \leq 5$, the following hold true:*

(1) $\overline{apr}_P^{g_c}(X \cup Y) \supseteq \cdot \overline{apr}_P^{g_c}(X) \cup \overline{apr}_P^{g_c}(Y)$;

(2) $\underline{apr}_P^{g_c}(X \cap Y) \subseteq \underline{apr}_P^{g_c}(X) \cap \underline{apr}_P^{g_c}(Y)$;;

(3) $\underline{apr}_P^{g_c}(X \cup Y) \supseteq \underline{apr}_P^{g_c}(X) \cup \underline{apr}_P^{g_c}(Y)$; and

(4) $\overline{apr}_P^{g_c}(X \cap Y) \subseteq \overline{apr}_P^{g_c}(X) \cap \overline{apr}_P^{g_c}(Y)$

Proof

(1) For any $X \subseteq U$ and $Y \subseteq U$, and given the greyness scale g_c, we have

$$\frac{|I_P(x) \cap (X \cup Y)|}{|I_P(x)|} \geq \frac{|I_P(x) \cap X|}{|I_P(x)|}$$

and

$$\frac{|I_P(x) \cap (X \cup Y)|}{|I_P(x)|} \geq \frac{|I_P(x) \cap Y|}{|I_P(x)|}.$$

Therefore, $\overline{apr}_P^{g_c}(X \cup Y) \supseteq \overline{apr}_P^{g_c}(X) \cup \overline{apr}_P^{g_c}(Y)$.

(2) For any $X, Y \subseteq U$, and given the greyness scale $g_c \leq 5$, we have

$$\frac{|I_P(x) \cap (X \cap Y)|}{|I_P(x)|} \leq \frac{|I_P(x) \cap X|}{|I_P(x)|}$$

and

$$\frac{|I_P(x) \cap (X \cap Y)|}{|I_P(x)|} \leq \frac{|I_P(x) \cap Y|}{|I_P(x)|}.$$

Therefore, $\underline{apr}_P^{g_c}(X \cap Y) \subseteq \underline{apr}_P^{g_c}(X) \cap \underline{apr}_P^{g_c}(Y)$. Similarly, we can prove (3) and (4). QED.

Proposition 8.6.2 $$\underline{apr}_P^{g_c}(X) \subseteq \overline{apr}_P^{g_c}(X)$$

Proof Let $x \in \underline{apr}_P^{g_c}(X)$. Because the equivalence relation I_P is reflective, we have $x \in I_P(x)$. From Definition 8.6.2, it follows that $g_c \leq 5$ and that the interval grey number to which the rough membership value corresponding to the upper grey-number greyness scale belongs is greater than the interval grey number to which the rough membership value corresponding to the lower grey-number greyness scale belongs. Hence, we have $x \in \overline{apr}_P^{g_c}(X)$ and consequently $\underline{apr}_P^{g_c}(X) \subseteq \overline{apr}_P^{g_c}(X)$. QED

8.6.3 Combined Grey Clustering and Rough Set Model

When employing the expansion dominant rough set model to probabilistic decision-making, one needs to have a multi-criteria decision-making table. However, in many practical applications involving uncertain multi-criteria decision-making, the researcher has to rely on existing data sets to generate his multi-criteria information table instead of being able to obtain their own multi-criteria decision making table. For instance, we can easily collect the financial data of a publically-traded company, such as income per share, net asset per share, net profit, reliability, operating profit, and so on. Based on the collected financial data, we can establish a multi-criteria information table. Given that such a company's style of decision-making is unknown ahead of time, it is difficult to classify it according to whether it presents a high risk, moderate risk, or low risk decision-making style. Thus, it is also difficult, if not impossible, to generate a relevant multi-criteria decision-making table. Therefore, dominant rough set models and expanded dominant rough set models cannot be directly employed to conduct decision-making analysis of these problems. However, the method of grey clustering of grey systems theory generally groups objects into different preference categories by considering attribute preference information and decision-makers' preference behaviors. In particular, the method of grey fixed weight clustering provides an effective way to transform a multi-criteria information table, which is made of preferred attributes of various dimensions, into a multi-criteria decision-making table. For instance, based on the collected financial data of companies, the distributions of the preferred attributes' values of the criteria, and the preferred behaviors of the decision-makers, we can establish possibility functions. On this basis, we can group the companies into different risk classes, such as high risk, moderate risk, and low risk class.

When considering the strengths of the methods of dominant rough sets and grey fixed weight clustering, we can construct a hybrid method combining grey fixed weight clustering and dominant rough sets, where grey fixed weight clustering can be seen as a processing tool used before the method of dominant rough sets is employed. The purpose of doing so is to generalize the dominant rough sets to a method that can be employed to conduct decision-making analysis based on multi-criteria information tables, and to extract the most precise expression of knowledge from the multi-criteria information table.

By following the steps below, one can establish the needed model combining grey fixed weight clustering and dominant rough sets:

(1) Develop a system of knowledge expressions using the values of preferred conditional attributes (criteria);
(2) Determine the ordered decision-making evaluation grey classes g according to the specific circumstances;
(3) Establish the possibility function for the field of each criterion. Let the possibility function of the kth subclass of the jth criteria be $f_j^k(\cdot)$ ($j = 1, 2, \ldots, m; k = 1, 2, \ldots, g$);
(4) Determine the clustering weight $\eta_j, j = 1, 2, \ldots, m$, for each criterion;
(5) Based on the observed value x_{ij}, $i = 1, 2, \ldots, n, j = 1, 2, \ldots, m$, of object i with respect to criterion j, compute the coefficients $\sigma_i^k = \sum_{j=1}^m f_j^k(x_{ij})\eta_j$ of the grey fixed weight clustering $i = 1, 2, \ldots, n, k = 1, 2, \ldots, g$;
(6) Obtain the clustering coefficient vector

$$\sigma_i = (\sigma_i^1, \sigma_i^2, \ldots, \sigma_i^g) = \left(\sum_{j=1}^m f_j^1(x_{ij})\eta_j, \sum_{j=1}^m f_j^2(x_{ij})\eta_j, \ldots \sum_{j=1}^m f_j^g(x_{ij})\eta_j \right);$$

(7) Generate the clustering coefficient matrix

$$\sum = (\sigma_i^k) = \begin{bmatrix} \sigma_1^1 & \sigma_1^2 & \cdots & \sigma_1^g \\ \sigma_2^1 & \sigma_2^2 & \cdots & \sigma_2^g \\ \vdots & \vdots & \vdots & \vdots \\ \sigma_n^1 & \sigma_n^2 & \cdots & \sigma_n^g \end{bmatrix};$$

(8) Based on the clustering coefficient matrix Σ, determine the classes to which individual objects belong. If $\max_{1 \le k \le g} \{\sigma_i^k\} = \sigma_i^{k^*}$, then object i belongs to grey class k^*;
(9) Establish the decision-making table using preferred conditional attributes and preferred decision-making grey classes; and
(10) Employ the method of dominant rough sets to conduct decision-making analysis.

8.7 Practical Applications

Example 8.7.1 Let us look at how to choose regional key technologies using a hybrid model combining the methods of grey fixed weight clustering and dominant rough sets. For a specific geographic area, the evaluation criteria system and relevant evaluation values for its key technologies are given in Table 8.5 (Liu & Jian, 2009).

Table 8.5 The criteria system for evaluating key regional technologies

Code	Meaning of criterion	Criterion weight	Evaluation values
a_1	Time lag of technology	0.1	A: >10 years; B: 5–10 years; C: 3–5 years; D: <3 years
a_2	Time length technological bottleneck existed	0.09	A: >10 years; B: 5–10 years; C: 3–5 years; D: <3 years
a_3	Ability to create own knowledge right	0.14	A: complete own right; B: partial right; C: no right at all
a_4	Coverage of technology	0.09	A: widely applicable; B: applied in profession; C: special technique
a_5	Promotion and lead of technological fields	0.11	A: strong; B: relatively strong; C: general; D: weak
a_6	Time needed for technology transfer	0.07	A: within 1 year; B: 1–3 years; C: 4–5 years; D: >5 years
a_7	Input/output ratio	0.13	A: high; B: relatively high; C: normal; D: low
a_8	Effect on environmental protection	0.12	A: strong; B: relatively strong; C: normal; D: weak

Based on the evaluations of relevant experts on 11 key technologies candidates, we generate the knowledge expression system as shown in Table 8.6.

In the following graph we present a decision-making analysis for this region's key technologies candidates.

For the evaluation criteria of the region's key technologies, the preference orders are the same as A > B > C > D. Quantify the set of criteria evaluations by letting the set be $V = (A, B, C, D) = (7, 5, 3, 1)$. According to practical needs, we divide each criterion into three grey classes of decision-making: the class of weak need

Table 8.6 The knowledge system on key regional technologies

U	a_1	a_2	a_3	a_4	a_5	a_6	a_7	a_8
a_1	B	B	C	B	D	C	B	A
a_2	D	D	B	B	C	B	C	D
a_3	D	D	B	B	C	A	D	A
a_4	B	C	C	B	B	B	A	C
a_5	B	B	B	B	C	B	B	C
a_6	D	D	B	B	B	B	B	C
a_7	D	D	B	C	D	A	C	C
a_8	C	B	B	C	C	C	B	C
a_9	B	B	B	B	A	B	A	B
a_{10}	C	B	B	B	B	B	B	B
a_{11}	B	B	B	B	C	B	C	B

Fig. 8.3 Possibility
functions of the three grey
classes

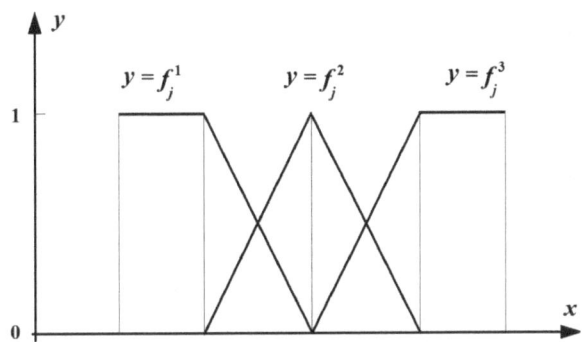

for key technologies (coded with 1), the class of general need for key technologies
(coded with 2), and the class of strong need for key technologies (coded with 3).
Let us take the possibility function of the class of weak need as the measurement of
the low bound, that of the class of general need as the moderate measurement, and
that of the class of strong need as the measurement of the upper bound. For details,
see Fig. 8.3. Based on decision-making goals and specific distributions of experts'
evaluation values, we introduce the possibility functions for each grey class as shown
in Table 8.7.

From formula $\sigma_i^k = \sum_{j=1}^m f_j^k(x_{ij}) \cdot \eta_j$, we can compute the clustering coefficient
for each grey class of each key technology. Based on such coefficients we can estab-
lish the evaluation decision-making (Table 8.8) for choosing key technologies for
the region.

Because the values of all the conditional attributes have the preference order A >
B > C > D, these attributes contain preference information. Based on the decision-
making attributes, the comprehensive evaluation can be divided into three preference
ordered classes: $Cl_1 = \{1\}, Cl_2 = \{2\}, Cl_3 = \{3\}$. Based on this result, we divide the
field of discourse and obtain the following unions of the decision-making classes:

$Cl_1^\le = Cl_1$, with the comprehensive evaluation 1 (the need for key technologies
is weak);

$Cl_2^\le = Cl_1 \cup Cl_2$, with the comprehensive evaluation ≤ 2 (the need for key
technologies is at most moderate);

$Cl_2^\ge = Cl_2 \cup Cl_3$, with the comprehensive evaluation ≥ 2 (the need for key
technologies is at least moderate);

$Cl_3^\le = Cl_1 \cup Cl_2 \cup Cl_3$, with the comprehensive evaluation ≤ 3 (the need for key
technologies is at most strong); and.

$Cl_3^\ge = Cl_3$, with the comprehensive evaluation 3 (the need for key technologies
is strong).

A reduction found by using the method of dominant rough sets is $\{a_2, a_7\}$. The
sets $D_\ge$ and $D_\le$ of the least amounts of preference rules generated from this reduction
are respectively given in Tables 8.9 and 8.10.

Based on the set $D_\ge$ of preference decision-making rules generated by employing
our hybrid model that combines grey fixed weight clustering and dominant rough

Table 8.7 Possibility functions for key regional technologies

Criterion name	Weak need class (1)	Moderate need class (2)	Strong need class (3)
Time lag of technology	$f(x) =$ $\begin{cases} 1 & 1 \le x < 2 \\ 3 - x & 2 \le x < 3 \\ 0 & \text{otherwise} \end{cases}$	$f(x) =$ $\begin{cases} x - 2 & 2 \le x < 3 \\ 4 - x & 3 \le x < 4 \\ 0 & \text{otherwise} \end{cases}$	$f(x) =$ $\begin{cases} 0.5x - 1.5 & 3 \le x < 5 \\ 1 & 5 \le x \le 7 \\ 0 & \text{otherwise} \end{cases}$
Time length technological bottleneck existed	$f(x) =$ $\begin{cases} 1 & 1 \le x < 2 \\ 3 - x & 2 \le x < 3 \\ 0 & \text{otherwise} \end{cases}$	$f(x) =$ $\begin{cases} x - 2 & 2 \le x < 3 \\ 4 - x & 3 \le x < 4 \\ 0 & \text{otherwise} \end{cases}$	$f(x) =$ $\begin{cases} 0.5x - 1.5 & 3 \le x < 5 \\ 1 & 5 \le x \le 7 \\ 0 & \text{otherwise} \end{cases}$
Ability to create own knowledge right	$f(x) =$ $\begin{cases} 1 & 3 \le x < 4 \\ 5 - x & 4 \le x < 5 \\ 0 & \text{otherwise} \end{cases}$	$f(x) =$ $\begin{cases} x - 4 & 4 \le x < 5 \\ 6 - x & 5 \le x < 6 \\ 0 & \text{otherwise} \end{cases}$	$f(x) =$ $\begin{cases} x - 5 & 5 \le x < 6 \\ 1 & 6 \le x \le 7 \\ 0 & \text{otherwise} \end{cases}$
Coverage of technology	$f(x) =$ $\begin{cases} 1 & 3 \le x < 4 \\ 5 - x & 4 \le x < 5 \\ 0 & \text{otherwise} \end{cases}$	$f(x) =$ $\begin{cases} x - 4 & 4 \le x < 5 \\ 6 - x & 5 \le x < 6 \\ 0 & \text{otherwise} \end{cases}$	$f(x) =$ $\begin{cases} x - 5 & 5 \le x < 6 \\ 1 & 6 \le x \le 7 \\ 0 & \text{otherwise} \end{cases}$
Promotion and lead of technological fields	$f(x) =$ $\begin{cases} 1 & 1 \le x < 2 \\ 2 - 0.5x & 2 \le x < 4 \\ 0 & \text{otherwise} \end{cases}$	$f(x) =$ $\begin{cases} 0.5x - 1 & 2 \le x < 4 \\ 3 - 0.5x & 4 \le x < 6 \\ 0 & \text{otherwise} \end{cases}$	$f(x) =$ $\begin{cases} 0.5x - 2 & 4 \le x < 6 \\ 1 & 6 \le x \le 7 \\ 0 & \text{otherwise} \end{cases}$

(continued)

Table 8.7   (continued)

Criterion name	Weak need class (1)	Moderate need class (2)	Strong need class (3)
Time needed for technology transfer	$f(x) =$ $\begin{cases} 1 & 1 \le x < 2 \\ 2 - 0.5x & 2 \le x < 4 \\ 0 & \text{otherwise} \end{cases}$	$f(x) =$ $\begin{cases} 0.5x - 1 & 2 \le x < 4 \\ 3 - 0.5x & 4 \le x < 6 \\ 0 & \text{otherwise} \end{cases}$	$f(x) =$ $\begin{cases} 0.5x - 2 & 4 \le x < 6 \\ 1 & 6 \le x \le 7 \\ 0 & \text{otherwise} \end{cases}$
Input/output ratio	$f(x) =$ $\begin{cases} 1 & 1 \le x < 3 \\ 4 - x & 3 \le x < 4 \\ 0 & \text{otherwise} \end{cases}$	$f(x) =$ $\begin{cases} x - 3 & 3 \le x < 4 \\ 3 - 0.5x & 4 \le x < 6 \\ 0 & \text{otherwise} \end{cases}$	$f(x) =$ $\begin{cases} 0.5x - 2 & 4 \le x < 6 \\ 1 & 6 \le x \le 7 \\ 0 & \text{otherwise} \end{cases}$
Effect on environmental protection	$f(x) =$ $\begin{cases} 1 & 1 \le x < 3 \\ 2.5 - 0.5x & 3 \le x < 5 \\ 0 & \text{otherwise} \end{cases}$	$f(x) =$ $\begin{cases} 0.5x - 1.5 & 3 \le x < 5 \\ 3.5 - 0.5x & 5 \le x < 7 \\ 0 & \text{otherwise} \end{cases}$	$f(x) =$ $\begin{cases} 0.5x - 25 & 5 \le x \le 7 \\ 0 & otherwise \end{cases}$

Table 8.8   Evaluation decision-making table for key regional technologies

a	a_1	a_2	a_3	a_4	a_5	a_6	a_7	a_8	a
a_1	B	B	C	B	D	C	B	A	3
a_2	D	D	B	B	C	B	C	D	1
a_3	D	D	B	B	C	A	D	A	1
a_4	B	C	C	B	B	B	A	C	3
a_5	B	B	B	B	C	B	B	C	2
a_6	D	D	B	B	B	B	B	C	1
a_7	D	D	B	C	D	A	C	C	1
a_8	C	B	B	C	C	C	B	C	2
a_9	B	B	B	B	A	B	A	B	3
a_{10}	C	B	B	B	B	B	B	B	2
a_{11}	B	B	B	B	C	B	C	B	2

Table 8.9 Set $D_{\geq}$ of preference rules

Rule	Confidence (%)	Support number
If the length of time for technology bottleneck to exist $\geq$ C and input/output ratio $=$ A, then the urgency for needing key technologies $=$ 3 (strong)	100	2
If the length of time for technology bottleneck to exist $\geq$ B and input/output ratio $\geq$ C, then the urgency for needing key technologies $\geq$ 2 (moderate)	100	5
If the length of time for technology bottleneck to exist $=$ D, then the urgency for needing key technologies $=$ 1 (weak)	100	4

Table 8.10 Set $D_{\leq}$ of preference rules

Rule	Confidence (%)	Support number
If the length of time for technology bottleneck to exist $\leq$ C and input/output ratio $=$ A, then the need for key technologies $=$ 3 (strong)	100	2
If the length of time for technology bottleneck to exist $\leq$ B and input/output ratio $\leq$ C, then the need for key technologies $\leq$ 2 (moderate)	100	1
If the length of time for technology bottleneck to exist $=$ D, then the need for key technologies $=$ 1 (weak)	100	4

sets, all the 11 key technologies considered are correctly classified. That is, the quality of classification is 100%. Based on the set $D_{\leq}$ of preference decision-making rules, a total of 7 key technologies are classified correctly so that the quality of classification is 63.6%.

References

Dong, J. R., & Yang, X.T. (1998). Prediction of environmental protection investment trend using grey combination model. *Chongqing Environmental Science, 20*(5), 30–32.

Han, X. D., & He, Z. L. (1997). Combination model of GM(1,1) and linear regression and its application to deformation prediction. *Journal of Huainan Mining Institute, 17*(4), 51–54.

Jian, L. R., & Liu, S.F. (2005). Extended rough set method for preference multi-attribute decision analysis. *Journal of Nanjing University of Aeronautics and Astronautics, 37*(4), 270–275.

Liu, S. F., & Jian, L.R. (2009). Evaluation and selection of regional key technologies for international cooperation. *Science & Technology Progress and Policy. 26*(10), 126–129.

Liu, S. F., & Zhu, Y. D. (1996). Grey-econometrics combined model. *Journal of Grey System, 8*(1), 103–110.

Liu, S. F., Li, B. J., & Dang, Y. G. (2004). The G-C.D model and technical advance. *Kybernetes, 33*(2), 303–309.

Liu, S. F., Lin, Y., et al. (2006). On measures of information content of grey numbers. *Kybernetes, 35*(5), 899–904.

Chapter 9
Techniques for Grey Systems Forecasting

9.1 Introduction

No matter what needs to be done, one should always get familiar with the situation, think through the details, make educated predictions, and lay out a detailed plan before he could potentially arrive at his desired successful conclusions. For matters as great as international affairs, national events and citizens' lives, the development of regional or business entities, and for matters as small as daily work or living arrangements, scientifically sound predictions are needed everywhere.

Prediction is about foretelling the possible course of development of societal events, political matters, economic ups and downs, and so on, using scientific methods and techniques based on attainable historical and present data so that appropriate actions can be planned and carried out. In short, prediction is about making scientific inferences regarding the evolution of materials and events ahead of time. General prediction includes not only static inference about unknown matters based on what is known within a specific time frame, but also dynamic inference about the future based on history and the present state of affairs of a certain matter. A specific prediction is a dynamic forecast within which a scientific inference about the future evolution of a certain event is given.

Grey prediction makes scientific, quantitative forecasts about the future states of systems based on understandings of unascertained characteristics of such systems. It makes use of sequence operators on the original data sequences in order to generate, treat, and excavate the hidden laws of systems evolution, so that grey systems models can be established to predict future outcomes. All the methods of the grey systems theory studied so far can be employed to make predictions. For a given problem, the appropriate prediction model is chosen by making use of the conclusions of a sufficiently and carefully done qualitative analysis. Also, the choice of models should vary along with changing conditions. Each model chosen has to be tested through many different methods in order to decide its appropriateness and effectiveness. Only the models that pass various tests can be meaningfully employed to make predictions (Deng, 1990; Liu & Guo, 1991; Liu, 2021).

© The Author(s), under exclusive license to Springer Nature Singapore Pte Ltd. 2022   229
S. Liu, *Grey Systems Analysis*, Series on Grey System,
https://doi.org/10.1007/978-981-19-6160-1_9

Definition 9.1.1 Let $X^{(0)} = (x^{(0)}(1), x^{(0)}(2), \ldots, x^{(0)}(n))$ be a sequence of raw data, $\hat{X}^{(0)} = (\hat{x}^{(0)}(1), \hat{x}^{(0)}(2), \ldots, \hat{x}^{(0)}(n))$ the simulated data out of a chosen prediction model, $\varepsilon^{(0)} = (\varepsilon(1), \varepsilon(2), \ldots \varepsilon(n)) = \left(x^{(0)}(1) - \hat{x}^{(0)}(1), x^{(0)}(2) - \hat{x}^{(0)}(2), \ldots, x^{(0)}(n) - \hat{x}^{(0)}(n)\right)$ the error sequence, and

$$\Delta = \left(\left|\frac{\varepsilon(1)}{x^{(0)}(1)}\right|, \left|\frac{\varepsilon(2)}{x^{(0)}(2)}\right|, \ldots, \left|\frac{\varepsilon(n)}{x^{(0)}(n)}\right|\right) = \{\Delta_k\}_1^n$$

the relative error sequence. Then:

(1) For $k \leq n$, $\Delta_k = \left|\frac{\varepsilon(k)}{x^{(0)}(k)}\right|$ is known as relative error of the simulation at point k, and $\overline{\Delta} = \frac{1}{n}\sum_{k=1}^{n}\Delta_k$ the average relative error;

(2) $1 - \overline{\Delta}$ is known as the average relative accuracy, and $1 - \Delta_k$ the simulation accuracy at point k, $k = 1, 2, \ldots, n$; and

(3) For a given α, when $\overline{\Delta} < \alpha$ and $\Delta_n < \alpha$ hold true, the prediction model is said to be error-satisfactory.

Definition 9.1.2 Let ε stand for the absolute grey relational degree between the raw data $X^{(0)}$ and the simulated values $\hat{X}^{(0)}$. If for a given $\varepsilon_0 > 0$ the absolute grey relational degree ε satisfies $\varepsilon > \varepsilon_0$, then the simulation model is said to be grey relational satisfactory.

Definition 9.1.3 Assume that the sequences $X^{(0)}$, $\hat{X}^{(0)}$, and $\varepsilon^{(0)}$ are the same as above, and consider the relevant means and variances

$$\overline{x} = \frac{1}{n}\sum_{k=1}^{n}x^{(0)}(k), \quad S_1^2 = \frac{1}{n}\sum_{k=1}^{n}(x^{(0)}(k) - \overline{x})^2$$

and

$$\overline{\varepsilon} = \frac{1}{n}\sum_{k=1}^{n}\varepsilon(k), \quad S_2^2 = \frac{1}{n}\sum_{k=1}^{n}(\varepsilon(k) - \overline{\varepsilon})^2.$$

(1) If for a given $C_0 > 0$, the ratio of root-mean-square deviation (RMSD) is $C = \frac{S_2}{S_1} < C_0$, then the model is said to be RMSD ratio satisfactory.

(2) If $p = P(|\varepsilon(k) - \overline{\varepsilon}| < 0.6745S_1)$ is seen as a small error probability and for a given $p_0 > 0$, when $p > p_0$, then the model is said to be small-error probability satisfactory.

The discussion above shows three different ways to test a chosen model. Each of them is based on observations of the error to determine the accuracy of the model. For both the mean relative error $\overline{\Delta}$ and the simulation error, the smaller they are, the better. With regards to the grey relational degree ε, the greater it is the better. As

Table 9.1 Commonly used scales of accuracy for model testing

Accuracy scale	Threshold			
	Relative error α	Grey relational degree ε_0	RMSD C_0	Small error probability p_0
1st level	0.01	0.90	0.35	0.95
2nd level	0.05	0.80	0.50	0.80
3rd level	0.10	0.70	0.65	0.70
4th level	0.20	0.60	0.80	0.60

for the RMSD ratio C, the smaller the value is, the better. This is because a small C indicates that S_2 is relatively small, while S_1 is relatively large. This means that the error variance is small while the variance of the original data is large, so that the errors are relatively more concentrated with little fluctuation compared to the original data. Therefore, for better simulation results, the smaller S_2 is when compared to S_1, the better. With regards to small error probability p, as soon as a set of α, ε_0, C_0, and p_0 values are chosen, a scale of accuracy for testing models is determined. The most commonly used scales of accuracy for testing models are listed in Table 9.1.

In most applications published so far in the area of grey systems, the most commonly used is the criterion of relative errors.

9.2 Interval Forecasting

If a given sequence of raw data is chaotic and it is difficult for any model to pass the accuracy test, the researcher will then have trouble producing accurate quantitative predictions. In this case, one can consider providing a range for future values to fall within (Deng, 1985; Liu et al., 2017).

Definition 9.2.1 Let $X(t)$ be a zigzagged line. If there are smooth and continuous curves $f_u(t)$ and $f_s(t)$, satisfying that for any t, $f_u(t) < X(t) < f_s(t)$, then $f_u(t)$ is known as the lower bound function of $X(t)$, $f_s(t)$ the upper bound function, and $S = \{(t, X(t))|X(t) \in [f_u(t), f_s(t)]\}$ the value domain of $X(t)$. If the upper and lower bound of $X(t)$ are the same kind of function, then S is known as a uniform domain. When S is a uniform band with exponential functions as its upper and lower bounds $f_u(t)$ and $f_s(t)$, then S is known as a uniform exponential domain. If a uniform band S has linear upper and lower bound functions $f_u(t)$ and $f_s(t)$, then S is known as a uniform linear domain or a straight domain for short. If for $t_1 < t_2$, $f_s(t_1) - f_u(t_1) < f_s(t_2) - f_u(t_2)$ always holds true, then S is known as a trumpet-like domain.

Example 9.2.1 Let $X^{(0)} = (x^{(0)}(1), x^{(0)}(2), \ldots, x^{(0)}(n))$ be a sequence of raw data, and its accumulation generation be $X^{(1)} = (x^{(1)}(1), x^{(1)}(2), \ldots, x^{(1)}(n))$. Define

Fig. 9.1 A trumpet-like
domain

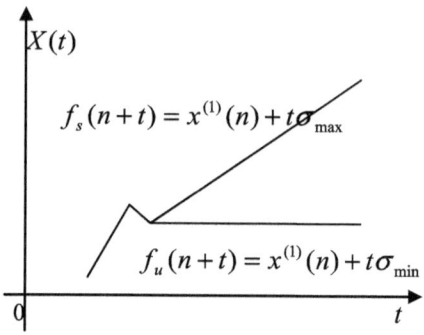

$$\sigma_{\max} = \max_{1 \le k \le n} \{x^{(0)}(k)\}, \quad \sigma_{\min} = \min_{1 \le k \le n} \{x^{(0)}(k)\}$$

and respectively take the upper and lower bound functions $f_u(n + t)$ and $f_s(n + t)$ of $X^{(1)}$ as follows:

$$f_u(n + t) = x^{(1)}(n) + t\sigma_{\min}, \quad f_s(n + t) = x^{(1)}(n) + t\sigma_{\max}.$$

That is, both the upper and lower bound functions of a proportional band are increasing straight lines of time with slopes $\sigma_{\min}$ and $\sigma_{\max}$, respectively.

Then $S = \{(t, X(t)) | t > n, X(t) \in [f_u(t), f_s(t)]\}$ is known as the proportional domain (see Fig. 9.1).

Example 9.2.2 For a sequence $X^{(0)}$ of raw data, let $X_u^{(0)}$ be the sequence corresponding to the curve that connects all the low points of $X^{(0)}$, and $X_s^{(0)}$ the sequence corresponding to the curve of all the upper points of $X^{(0)}$. Assume that

$$\hat{x}_u^{(1)}(k + 1) = \left(x_u^{(0)}(1) - \frac{b_u}{a_u} \right) \exp(-a_u k) + \frac{b_u}{a_u}$$

and

$$\hat{x}_s^{(1)}(k + 1) = \left(x_s^{(0)}(1) - \frac{b_s}{a_s} \right) \exp(-a_s k) + \frac{b_s}{a_s}$$

are respectively the GM(1,1) time response sequences of $X_u^{(0)}$ and $X_s^{(0)}$. Then

$$S = \{(t, X(t)) \big| X(t) \in [\hat{X}_u^{(1)}(t), \hat{X}_s^{(1)}(t)]\}$$

is known as a wrapping domain (see Fig. 9.2).

Example 9.2.3 For a given sequence $X^{(0)}$ of raw data, let us take m different subsequences to establish m GM(1,1) models with the corresponding parameters $\hat{a}_i = [a_i, b_i]^T$; $i = 1, 2, \ldots, m$. Let

Fig. 9.2 A wrapping
domain

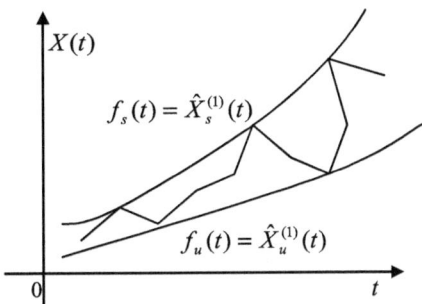

$$-a_{\max} = \max_{1 \le i \le m} \{-a_i\}, \quad -a_{\min} = \min_{1 \le i \le m} \{-a_i\}, \text{ and}$$

$$\hat{x}_u^{(1)}(k+1) = \left(x_u^{(0)}(1) - \frac{b_{\min}}{a_{\min}} \right) \exp(-a_{\min}k) + \frac{b_{\min}}{a_{\min}}$$

$$\hat{x}_s^{(1)}(k+1) = \left(x_s^{(0)}(1) - \frac{b_{\max}}{a_{\max}} \right) \exp(-a_{\max}k) + \frac{b_{\max}}{a_{\max}}$$

Then $S = \{(t, X(t)) | X(t) \in [\hat{X}_u^{(1)}(t), \hat{X}_s^{(1)}(t)]\}$ is known as a development domain. The wrapping domain and development domain are exponential domains.

Definition 9.2.2 For a sequence $X^{(0)} = (x^{(0)}(1), x^{(0)}(2), \ldots, x^{(0)}(n))$ of raw data, let $f_u(t)$ and $f_s(t)$ be a upper and a lower bound function of the accumulation sequence $X^{(1)}$ of $X^{(0)}$. For any $k > 0$,

$$\hat{x}^{(0)}(n+k) = \frac{1}{2}[f_u(n+k) + f_s(n+k)]$$

is known as basic prediction value, and $\hat{x}_u^{(0)}(n+k) = f_u(n+k)$ and $\hat{x}_s^{(0)}(n+k) = f_s(n+k)$, respectively, the lowest and highest predicted values.

Example 9.2.4 The data (in tens of thousands) for car sales in a certain city are given as follows:

$$X^{(0)} = (x^{(0)}(1), x^{(0)}(2), x^{(0)}(3), x^{(0)}(4), x^{(0)}(5), x^{(0)}(6))$$
$$= (5.0810, 4.6110, 5.1177, 9.3775, 11.0574, 11.3524)$$

where $x^{(0)}(1) = 5.0810$ is the annual sales for the year of 2010, ..., and $x^{(0)}(6) = 11.3524$ for the year of 2015. Try to make a prediction using development domain.

Solution: Take the following sub-sequences

$$X_1^{(0)} = (x^{(0)}(1), x^{(0)}(2), x^{(0)}(3), x^{(0)}(4), x^{(0)}(5), x^{(0)}(6))$$
$$= (5.0810, 4.6110, 5.1177, 9.3775, 11.0574, 11.3524)$$

$$X_2^{(0)} = (x^{(0)}(1), x^{(0)}(2), x^{(0)}(3), x^{(0)}(4), x^{(0)}(5))$$
$$= (5.0810, 4.6110, 5.1177, 9.3775, 11.0574)$$
$$X_3^{(0)} = (x^{(0)}(2), x^{(0)}(3), x^{(0)}(4), x^{(0)}(5), x^{(0)}(6))$$
$$= (4.6110, 5.1177, 9.3775, 11.0574, 11.3524)$$
$$X_4^{(0)} = (x^{(0)}(3), x^{(0)}(4), x^{(0)}(5), x^{(0)}(6))$$
$$= (5.1177, 9.3775, 11.0574, 11.3524)$$

Based on each of these sub-sequences, let us establish the corresponding the models of EGM(1,1):

$$\frac{dx^{(1)}}{dt} + a_i x^{(1)} = b_i, \quad i = 1, 2, 3, 4$$

Their individual parameters $\hat{a}_i = [a_i, b_i]^T, i = 1, 2, 3, 4$, are given below:

$$\hat{a}_1 = [a_1, b_1]^T = [-0.2202, 3.4689]^T, \hat{a}_2 = [a_2, b_2]^T = [-0.3147, 2.1237]^T$$
$$\hat{a}_3 = [a_3, b_3]^T = [-0.2013, 5.0961]^T, \hat{a}_4 = [a_4, b_4]^T = [-0.0911, 8.7410]^T$$

Because

$$-a_{\min} = \min_{1 \leq i \leq 4}\{-a_i\} = \min\{0.2202, 0.3147, 0.2013, 0.0911\} = 0.0911 = -a_4$$
$$-a_{\max} = \max_{1 \leq i \leq 4}\{-a_i\} = \max\{0.2202, 0.3147, 0.2013, 0.0911\} = 0.3147 = -a_2$$

the upper bound time response sequence of the development domain is

$$\begin{cases} \hat{x}_s^{(1)}(k+1) = \left(x^{(0)}(1) - \dfrac{b_2}{a_2}\right)e^{-a_2 k} + \dfrac{b_2}{a_2} = 11.8293e^{0.3147k} - 6.7483 \\ \hat{x}_s^{(0)}(k+1) = \hat{x}_s^{(1)}(k+1) - \hat{x}_s^{(1)}(k) \end{cases}$$

That is, $\hat{x}_s^{(0)}(k+1) = 11.8293e^{0.3147k} - 11.8293e^{0.3147k-0.3147} = 3.1938e^{0.3147k}$. Thus, the highest predicted values are $\hat{x}_s^{(0)}(7) = 21.1029$, $\hat{x}_s^{(0)}(8) = 28.9078$, and $\hat{x}_s^{(0)}(9) = 39.5993$. Because the starting value of $X_4^{(0)}$ is $x^{(0)}(3)$, the lower bound time response sequence of the development domain is

$$\begin{cases} \hat{x}_u^{(1)}(k+3) = \left(x^{(0)}(3) - \dfrac{b_4}{a_4}\right)e^{-a_4 k} + \dfrac{b_4}{a_4} = 101.0672e^{0.0911k} - 95.9495 \\ \hat{x}_u^{(0)}(k+3) = \hat{x}_u^{(1)}(k+3) - \hat{x}_u^{(0)}(k+2) \end{cases}$$

That is, $\hat{x}_u^{(0)}(k+3) = 101.0672e^{0.0911k} - 101.0672e^{0.0911k-0.0911} = 8.8003e^{0.0911k}$. Therefore, we obtain the lowest predicted values: $\hat{x}_u^{(0)}(7) = 12.6694$, $\hat{x}_u^{(0)}(8) =$

13.8777, and $\hat{x}_u^{(0)}(9) = 15.2014$. From the highest and lowest predicted values, we obtain the basic prediction values:

$$\hat{x}^{(0)}(7) = \frac{1}{2}[\hat{x}_s^{(0)}(7) + \hat{x}_u^{(0)}(7)] = 16.8862$$

$$\hat{x}^{(0)}(8) = \frac{1}{2}[\hat{x}_s^{(0)}(8) + \hat{x}_u^{(0)}(8)] = 21.3928$$

$$\hat{x}^{(0)}(9) = \frac{1}{2}[\hat{x}_s^{(0)}(9) + \hat{x}_u^{(0)}(9)] = 27.4004$$

Based on the qualitative analysis of the estimated amount of car ownership in the given city and the improvement in public transportation systems, we conclude that the lowest predicted values are the most reliable.

9.3 Grey Distortion Forecasting

The basic idea of grey distortion prediction is essentially the prediction of abnormal values. The kinds of values that are considered abnormal are commonly determined based on individuals' experiences. The objective of grey distortion predictions is to provide the time moments of the forthcoming abnormal values so that relevant parties can prepare for the worst ahead of time (Deng, 1985; Liu et al., 2017).

Definition 9.3.1 Let $X = (x(1), x(2), \ldots, x(n))$ be a sequence of raw data. Then

(1) For a given upper abnormal value ξ, the sub-sequence of X

$$X_\xi = (x[q(1)], x[q(2)], \ldots, x[q(m)])$$
$$= \{x[q(i)]|x[q(i)] \geq \xi; \quad i = 1, 2, \ldots, m\}$$

is known as the upper distortion sequence.

(2) For a given lower abnormal value ζ, the sub-sequence

$$X_\zeta = (x[q(1)], x[q(2)], \ldots, x[q(l)]) = \{x[q(i)]|x[q(i)] \geq \zeta; \quad i = 1, 2, \ldots, l\}$$

is known as the lower distortion sequence. Together, these upper and lower distortion sequences are referred to as distortion sequences. Because the idea behind the discussion of distortion sequences is the same, in the following discussion we will not distinguish between upper and lower distortion sequences.

Definition 9.3.2 Assume that $X = (x(1), x(2), \ldots, x(n))$ is a sequence of raw data. The following sub-sequence of X.

$$X_\xi = (x[q(1)], x[q(2)], \ldots, x[q(m)]) \subset X$$

is a distortion sequence. Then,

$$Q^{(0)} = (q(1), q(2), \ldots, q(m))$$

will be referred to as the distortion date sequence. Distortion prediction is about finding patterns, if any, through the study of distortion date sequences to predict future dates of occurrences of distortion. In grey system theory, each distortion prediction is realized through establishing GM(1,1) models for relevant distortion date sequences.

Definition 9.3.3 If $Q^{(0)} = (q(1), q(2), \ldots, q(m))$ is a distortion date sequence, the following

$$Q^{(1)} = (q(1)^{(1)}, q(2)^{(1)}, \ldots, q(m)^{(1)})$$

is the 1-AGO sequence of the distortion date sequence $Q^{(0)}, Z^{(1)}$ is the adjacent neighbor mean generated sequence of $Q^{(1)}$, and

$$q(k) + az^{(1)}(k) = b$$

is referred to as a distortion model of GM(1,1). For the available sequence $X = (x(1), x(2), \ldots, x(n))$ of raw data, if n stands for the present and the last entry $q(m)(\le n)$ in the corresponding distortion date sequence $Q^{(0)}$ represents when the last abnormal value occurred, then the predicted value $\hat{q}(m + 1)$ represents the next forthcoming abnormal value and for any $k > 0$, $\hat{q}(m + k)$ stands for the predicted date for the kth abnormal value to occur in the future.

Example 9.3.1 The following sequence gives the annual average precipitations (in mm) of a certain region for 17 years, where x(1), x(2), ...,x(17) are respectively the data for the years of 2005, 2006, ..., 2021:

$$X = (x(1), x(2), x(3), x(4), x(5), x(6), x(7), x(8), x(9), x(10)$$
$$x(11), x(12), x(13, x(14), x(15), x(16), x(17))$$
$$= (390.6, 412.0, 320.0, 559.2, 380.8, 542.4, 553.0, 310.0, 561.0, 300.0$$
$$632.0, 540.0, 406.2, 313.8, 576.0, 586.6, 318.5)$$

Take $\xi = 320$ mm as a lower abnormal (drought) value. Carry out a drought prediction for this specific region.

Solution: If $\xi = 320$, we obtain the following lower distortion sequence

$$X_\xi = (x(3), x(8), x(10), x(14), x(17)) = (320.0, 310.0, 300.0, 313.8, 318.5)$$

with the corresponding distortion date sequence

$$Q^{(0)} = (q(1), q(2), q(3), q(4), q(5)) = (3, 8, 10, 14, 17)$$

and it's 1-AGO sequence

$$Q^{(1)} = (3, 11, 21, 35, 52)$$

The mean sequence based on consecutive neighbors of $Q^{(1)}$ is given by

$$Z^{(1)} = (7, 16, 28, 43.5)$$

Let $q(k) + az^{(1)}(k) = b$. From

$$B = \begin{bmatrix} -7 & 1 \\ -16 & 1 \\ -28 & 1 \\ -43.5 & 1 \end{bmatrix}, Y = \begin{bmatrix} 8 \\ 10 \\ 14 \\ 17 \end{bmatrix}$$

it follows that

$$\hat{a} = \begin{bmatrix} a \\ b \end{bmatrix} = (B^T B)^{-1} B^T Y = \begin{bmatrix} -0.25361 \\ 6.258339 \end{bmatrix}$$

Therefore, the GM(1, 1) ordinality response sequence of the distortion date sequence is

$$\hat{q}^{(1)}(k + 1) = 27.667e^{0.25361k} - 24.667$$
$$\hat{q}(k + 1) = \hat{q}^{(1)}(k + 1) - \hat{q}^{(1)}(k)$$

That is,

$$\hat{q}(k + 1) = 27.667e^{0.25361k} - 24.667e^{0.25361(k-1)} = 6.1998e^{0.25361k}$$

Thus, we can obtain a simulated sequence for $Q^{(0)}$ as follows:

$$\hat{Q}^{(0)} = (\hat{q}(1), \hat{q}(2), \hat{q}(3), \hat{q}(4), \hat{q}(5))$$
$$= (6.1998, 7.989, 10.296, 13.268, 17.098)$$

From

$$\varepsilon(k) = q(k) - \hat{q}(k), \quad k = 1, 2, 3, 4, 5$$

we obtain the error sequence as follows:

$$\varepsilon^{(0)} = (\varepsilon(1), \varepsilon(2), \varepsilon(3), \varepsilon(4), \varepsilon(5))$$

$$= (-3.1998, 0.011, -0.296, 0.732, -0.098)$$

And from

$$\Delta_k = \left| \frac{\varepsilon(k)}{q(k)} \right|; \quad k = 1, 2, 3, 4, 5$$

it follows that the sequence of relative errors is

$$\Delta = (\Delta_2, \Delta_3, \Delta_4, \Delta_5) = (0.1\%, 2.96\%, 5.1\%, 0.6\%)$$

From this sequence, we calculate the average relative error

$$\overline{\Delta} = \frac{1}{4} \sum_{k=2}^{5} \Delta_k = 2.19\%$$

With $1 - \overline{\Delta} = 97.81\%$ as the average relative accuracy, and $1 - \Delta_5 = 99.4\%$. Therefore, we can use

$$\hat{q}(k + 1) = 6.1998e^{0.25361k}$$

to carry out our predictions. Because

$$\hat{q}(5 + 1) = \hat{q}(6) \approx 22, \, \hat{q}(6) - \hat{q}(5) \approx 22 - 17 = 5$$

we predict that in five years, counting from the time of the last drought in 2021, there might be a drought. In order to improve the accuracy of our prediction, we can take several different abnormal values to build various models to make predictions.

9.4 Wave Form Forecasting

When the available data sequence vibrates widely with large magnitudes, it is often difficult, if not impossible, to find an appropriate simulation model. In this case, one can consider making use of the pattern of fluctuation of the data to predict the future development of the wavy movement. This kind of prediction is known as a wave form forecasting (Deng, 1985; Liu et al., 2017).

Definition 9.4.1 Let $X = (x(1), x(2), \dots, x(n))$ be the sequence of raw data, then

$$x_k = x(k) + (t - k)[x(k + 1) - x(k)]$$

is known as a k-piece zigzagged line of the sequence X, and

$$\{x_k = x(k) + (t - k)[x(k + 1) - x(k)] \quad k = 1, 2, \ldots, n - 1\}$$

the zigzagged line, still denoted by using X.

Definition 9.4.2 Assume that X is a zigzagged line, let

$$\sigma_{\max} = \max_{1 \le k \le n} \{x(k)\} \text{ and } \sigma_{\min} = \min_{1 \le k \le n} \{x(k)\}.$$

Then.

(1) For any $\forall \xi \in [\sigma_{\min}, \sigma_{\max}]$, $X = \xi$ is known as the ξ-contour (line); and
(2) The solutions $(t_i, x(t_i))(i = 1, 2, \ldots)$ of system of equations

$$\begin{cases} X = \{x(k) + (t - k)[x(k + 1) - x(k)]|k = 1, 2, \ldots, n - 1\} \\ X = \xi \end{cases}$$

is called the ξ-contour points. The ξ-contour point is the intersection of the zigzagged line X and the ξ-contour line.

Proposition 9.4.1 *If on the ith segment of X there is a ξ-contour point, then the coordinates of this point are given by* $\left(i + \frac{\xi - x(i)}{x(i+1) - x(i)}, \xi\right)$.

Proof The equation of i-piece zigzagged line of the sequence X is as follows:

$$X = x(i) + (t_i - i)[x(i + 1) - x(i)]$$

From

$$\begin{cases} X = x(i) + (t_i - i)[x(i + 1) - x(i)] \\ X = \xi \end{cases}$$

We have

$$t_i = i + \frac{\xi - x(i)}{x(i + 1) - x(i)}$$

Definition 9.4.3 Let $X_\xi = (P_1, P_2, \ldots, P_m)$ be the sequence of ξ-contour points of X such that point P_i is located on the ith segment. Let

$$q(i) = t_i + \frac{\xi - x(t_i)}{x(t_i + 1) - x(t_i)}, \quad i = 1, 2, \ldots, m$$

Then $Q^{(0)} = (q(1), q(2), \ldots, q(m))$ is known as the ξ-contour time moment sequence. By establishing a GM(1,1) model using this ξ-contour moment sequence, one can produce the predicted values for future ξ-contour time moments:

$$\hat{q}(m+1), \hat{q}(m+2), \ldots, \hat{q}(m+k).$$

Definition 9.4.4 The lines $X = \xi_i (i = 0, 1, 2, \ldots, s)$, where $\xi_0 = \sigma_{min}$, $\xi_1 = \frac{1}{s}(\sigma_{max} - \sigma_{min}) + \sigma_{min}, \ldots \xi_i = \frac{i}{s}(\sigma_{max} - \sigma_{min}) + \sigma_{min}, \ldots, \xi_{s-1} = \frac{s-1}{s}(\sigma_{max} - \sigma_{min}) + \sigma_{min}, \xi_s = \sigma_{max}$ are known as equal time distanced contours. When taking contour lines, one needs to make sure that the corresponding contour moments satisfy the conditions for establishing valid GM(1,1) models.

Definition 9.4.5 Let $X = \xi_i (i = 1, 2, \ldots, s)$ be s different contours,

$$Q_i^{(0)} = (q_i(1), q_i(2), \ldots, q_i(m_1)), \quad i = 1, 2, \ldots, s,$$

stand for the sequence of ξ_i-contour time moments, and

$$\hat{q}_i(m_i + 1), \hat{q}_i(m_i + 2), \ldots, \hat{q}_i(m_i + k_i), \quad i = 1, 2, \ldots, s$$

the GM(1,1) predicted ξ_i-contour time moments. If there are $i \neq j$ such that

$$\hat{q}_i(m_i + l_i) = \hat{q}_j(m_j + l_j),$$

then these values are known as a pair of invalid moments.

Proposition 9.4.2 *Let $\hat{q}_i(m_i + j)$, $j = 1, 2, \ldots, k_i$, $i = 1, 2, \ldots, s$, be the GM(1,1) predicted ξ_i-contour time moments. After deleting all invalid predictions, order the rest in terms of their magnitudes as follows:*

$$\hat{q}(1) < \hat{q}(2) < \cdots < \hat{q}(n_s),$$

where $n_s \leq k_1 + k_2 + \cdots + k_s$. If $X = \xi_{\hat{q}(k)}$ is the contour line corresponding to $\hat{q}(k)$. Then the predicted wavy curve of $X^{(0)}$ is given below:

$$X = \hat{X}^{(0)} = \{\xi_{\hat{q}(k)} + [t - \hat{q}(k)][\xi_{\hat{q}(k+1)} - \xi_{\hat{q}(k)}] \mid k = 1, 2, \ldots, n_s\}.$$

9.5 System Forecasting

9.5.1 The Five-Step Modeling Process

Generally, when studying a system one should first establish a mathematical model through which the overall functionality of the system, abilities of coordination, incidence relations, causal relations, and dynamic relationships between different parts can be quantitatively investigated. This kind of study has to be guided by an early qualitative analysis, and there must be close connection between the quantitative and qualitative studies. As for the development of the system's model, one generally goes through the following five steps: development of thoughts, analysis of relevant factors, quantification, dynamics, and optimization. This is the so-called five-step modeling (Deng, 1985).

Step 1: Develop thoughts and form concepts. Through an initial qualitative analysis, one clarifies his goal, possible paths and specific procedures, and then verbally and precisely describes the desired outcomes. This is the initial language model of the problem (see Fig. 9.3).

Step 2: Examine all the factors involved in the language model and their mutual relationships in order to pinpoint the causes and conclusions. Then, construct a line-drawing to depict the causal relationships (Fig. 9.3). Each pair (or a group) of causes and effect form a link. A system might be made up of many such links. At the same time, a quantity can be a cause of a link and also a consequence of another link. When several of these links are connected, one obtains a line drawing of many links that organically form the system of our concern (Fig. 9.4).

Step 3: Quantitatively study each causality link and obtain an approximate quantitative relationship, which is a quantified model.

Step 4: For each link, collect additional input–output data, on which dynamic GM models are established. Such dynamic models are higher level quantitative models. They can further reveal the relationships between input and output, and their laws of transformation. They are the foundation of systems analysis and optimization.

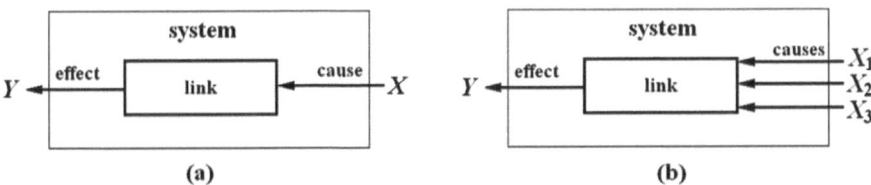

Fig. 9.3 Depicted causal relationships

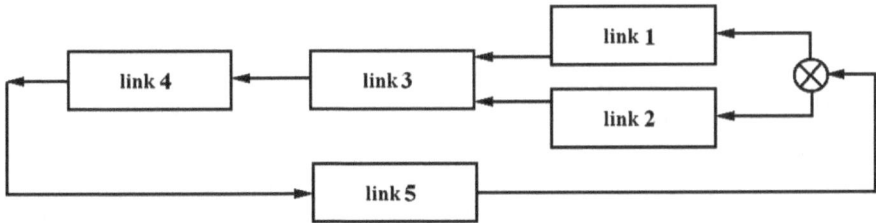

Fig. 9.4 Line drawing of an abstract system

Step 5: Systematically investigate the established dynamic models by adjusting their structures, mechanisms, and parameters, in order to arrive at the purpose of optimizing the outcome and realizing the desired conclusions. Models obtained in this way are known as optimal models.

The procedure of five-step modeling is such a holistic process that at five different stages five different kinds of models are established: language models, network models, quantified models, dynamic models, and optimized models. In the entire process of modeling, the conclusions of the next level should be repeatedly fed back so that the modeling exercise itself becomes a feedback system making the model system as perfect as possible.

9.5.2 System Models for Prediction

For a system with many mutually related factors and many autonomous controlling variables, no single model can reflect adequately the development and change of the system. To effectively study such a system and to predict its future behaviors, one should consider establishing a system of models (Deng, 1985; Liu et al., 2020).

Definition 9.5.1 Assume that

$$X_i^{(0)} = \left(x_i^{(0)}(1), x_i^{(0)}(2), \ldots, x_i^{(0)}(n)\right), \quad i = 1, 2, \ldots, m,$$

are sequences of raw data for the state variables of a system, and

$$U_j^{(0)} = \left(u_j^{(0)}(1), u_j^{(0)}(2), \ldots, u_j^{(0)}(n)\right), \quad j = 1, 2, \ldots, s,$$

are sequences of data of the control variables. Then the following

$$x_1^{(0)} = a_{11}z_1^{(1)} + a_{12}x_2^{(1)} + \cdots + a_{1m}x_m^{(1)} + b_{11}u_1^{(1)} + b_{12}u_2^{(1)} + \cdots + b_{1s}u_s^{(1)}$$
$$x_2^{(0)} = a_{21}x_1^{(1)} + a_{22}z_2^{(1)} + \cdots + a_{2m}x_m^{(1)} + b_{21}u_1^{(1)} + b_{22}u_2^{(1)} + \cdots + b_{2s}u_s^{(1)}$$
$$\cdots \quad \cdots \quad \cdots \quad \cdots \quad \cdots \quad \cdots \quad \cdots \quad \cdots \quad \cdots \quad \cdots \quad \cdots \quad \cdots$$
$$x_m^{(0)} = a_{m1}x_1^{(1)} + a_{m2}x_2^{(1)} + \cdots + a_{mm}z_m^{(1)} + b_{m1}u_1^{(1)} + b_{m2}u_2^{(1)} + \cdots + b_{ms}u_s^{(1)}$$
$$\frac{du_1^{(1)}}{dt} = c_1 u_1^{(1)} + d_1, \quad \frac{du_2^{(1)}}{dt} = c_2 u_2^{(1)} + d_2, \ldots, \quad \frac{du_s^{(1)}}{dt} = c_s u_s^{(1)} + d_s$$

are known as system models for prediction. As a matter of fact, each system model for prediction consists of m DGM(1,$m + s$) and s EGM(1,1) models. If we write the previous system models for prediction using the terminology of matrices, we have

$$\begin{cases} X^{(0)} = AX^{(1)} + BU^{(1)} \\ \quad U^{(0)} = CU^{(1)} + D \end{cases}$$

where $X^{(1)} = (x_1^{(1)}, x_2^{(1)}, \cdots, x_m^{(1)})^T$, $U^{(1)} = (u_1^{(1)}, u_2^{(1)}, \ldots, u_s^{(1)})^T$, $A = [a_{kl}]_{m \times m}$, $B = [b_{pq}]_{m \times s}$, $C = diag[c_j]_{s \times s}$, and $D = [d_j]_{s \times 1}$.

X is known as the state vector, U the control vector, A the state matrix, B the control matrix, C the development matrix, and D the grey effect vector.

Proposition 9.5.1 *For the previous system models for prediction, the time response sequences are given as follows:*

$$\hat{x}_i^{(0)}(k) = a_{i1}x_i^{(1)}(k) + a_{i2}x_2^{(1)}(k) + \cdots + a_{im}x_m^{(1)}(k) + b_{i1}u_1^{(1)}(k)$$
$$+ b_{i2}u_2^{(1)}(k) + \cdots + b_{is}u_s^{(1)}(k), \quad i = 1, 2, ..., m$$

$$\hat{u}_j^{(0)}(k) = (1 - e^{c_j})(u_j^{(0)}(1) - \frac{d_j}{c_j})e^{-c_j(k-1)}, \quad j = 1, 2, ..., s$$

9.6 Practical Applications

Example 9.6.1 Let us look at a wavy curve prediction for the (synthetic) stock index of Shanghai stock exchange. Using the stock index data of the stock index weekly closes of Shanghai stock exchange, the time series plot from February 21, 1997, through to October 31, 1998, is shown in Fig. 9.5 (Dang & Liu, 2009).

Let us take

$$\xi_1 = 1140, \xi_2 = 1170, \xi_3 = 1200, \xi_4 = 1230, \xi_5 = 1260$$
$$\xi_6 = 1290, \xi_8 = 1350, \xi_7 = 1320, \xi_9 = 1380.$$

Then the corresponding ξ_i-contour time moment sequences are given below:

Fig. 9.5 Shanghai stock exchange index (Feb. 21, 1997, to Oct. 31, 1998)

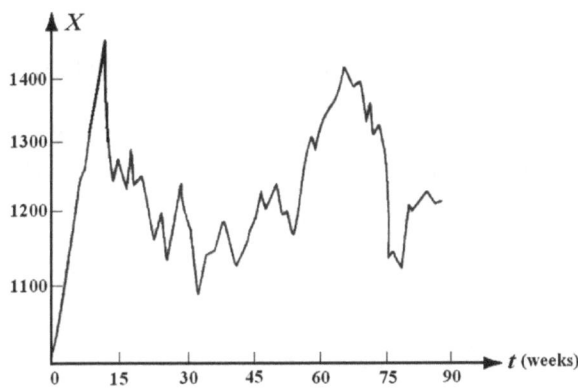

(1) For $\xi_1 = 1140$,

$$Q_1^{(0)} = \{q_1(k)\}_1^7 = (4.4, 31.7, 34.2, 41, 42.4, 76.8, 78.3)$$

(2) For $\xi_2 = 1170$,

$$Q_2^{(0)} = \{q_2(k)\}_2^{12} = (5.2, 19.8, 23, 25.6, 26.9, 31.2, 34.8, 39.5, 44.6, 76, 76.2, 79.2)$$

(3) For $\xi_3 = 1200$,

$$Q_3^{(0)} = \{q_3(k)\}_3^{11} = (5.9, 19.5, 24.8, 25.2, 26.5, 30.3, 46.2, 53.4, 55.4, 75.5, 79.7)$$

(4) For $\xi_4 = 1230$,

$$Q_4^{(0)} = \{q_4(k)\}_4^{10} = (6.5, 19.2, 28.3, 29.5, 49.7, 50.8, 56.2, 76.4, 82.9, 85)$$

(5) For $\xi_5 = 1260$,

$$Q_5^{(0)} = \{q_5(k)\}_5^7 = (7, 14.2, 16.5, 16.4, 18.8, 56.7, 75.2)$$

(6) For $\xi_6 = 1290$,

$$Q_6^{(0)} = \{q_6(k)\}_6^5 = (8.3, 13.4, 16.9, 56.2, 74.6)$$

(7) For $\xi_7 = 1320$,

$$Q_7^{(0)} = \{q_7(k)\}_7^6 = (8.8, 12.8, 60.2, 71.8, 72.7, 73.6)$$

(8) For $\xi_8 = 1350$,

$$Q_8^{(0)} = \{q_8(k)\}_8^6 = (9.6, 12.5, 61.8, 69.8, 70.9, 71.8)$$

(9) For $\xi_9 = 1380$,

$$Q_9^{(0)} = \{q_9(k)\}_9^4 = (10.8, 12.4, 64.1, 69)$$

Applying the 1-AGO on $Q_i^{(0)} (i = 1, 2, \ldots, 9)$ produces $Q_i^{(1)} (i = 1, 2, \ldots, 9)$, whose EGM(1,1) response sequences are respectively given by:

$\hat{q}_1^{(1)}(k + 1) = 113.91e^{0.215k} - 109.51, \hat{q}_2^{(1)}(k + 1) = 98.58e^{0.159k} - 93.83,$

$\hat{q}_3^{(1)}(k + 1) = 102.08e^{0.166k} - 96.18, \hat{q}_4^{(1)}(k + 1) = 151.66e^{0.160k} - 145.16,$

$\hat{q}_5^{(1)}(k + 1) = 13e^{0.435k} - 6, \hat{q}_6^{(1)}(k + 1) = 21.94e^{0.539k} - 13.64,$

$\hat{q}_7^{(1)}(k + 1) = 185.08e^{0.192k} - 176.28, \hat{q}_8^{(1)}(k + 1) = 193.19e^{0.186k} - 182.57,$

Fig. 9.6 The predicted wavy curve of Shanghai stock exchange index (Nov. 1998 to March 2000)

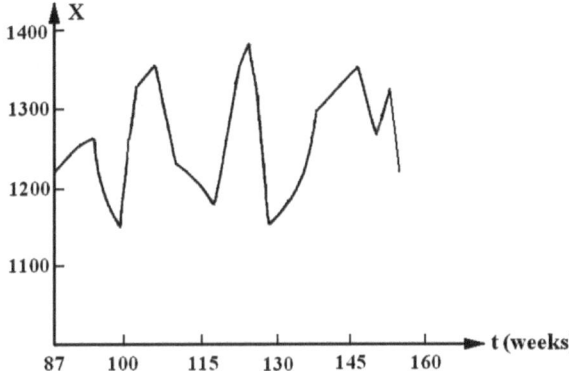

$$\hat{q}_9^{(1)}(k + 1) = 45.22e^{0.490k} - 35.39$$

By letting $\hat{q}_i(k + 1) = \hat{q}_i^{(1)}(k + 1) - \hat{q}_i^{(1)}(k)$, we obtain the following ξ_i-contour prediction sequences, $i = 1, 2, \ldots, 9$,

$$\hat{Q}_1^{(0)} = (\hat{q}_1(12), \hat{q}_1(13)) = (99.8, 127.7)$$

$$\hat{Q}_2^{(0)} = (\hat{q}_2(13), \hat{q}_2(14), \hat{q}_2(15)) = (96.8, 116.7, 131.4)$$

$$\hat{Q}_3^{(0)} = (\hat{q}_3(12), \hat{q}_3(13), \hat{q}_3(14)) = (95.7, 114.2, 133.8)$$

$$\hat{Q}_4^{(0)} = (\hat{q}_4(11), \hat{q}_4(12), \hat{q}_4(13)) = (110.9, 134.2, 152.8)$$

$$\hat{Q}_5^{(0)} = (\hat{q}_5(8), \hat{q}_5(9)) = (94.2, 148.8)$$

$$\hat{Q}_6^{(0)} = (\hat{q}_6(6)) = (135.5)$$

$$\hat{Q}_7^{(0)} = (\hat{q}_7(7), \hat{q}_7(8), \hat{q}_7(9)) = (101.9, 123.4, 149.5)$$

$$\hat{Q}_8^{(0)} = (\hat{q}_8(7), \hat{q}_8(8), \hat{q}_8(9)) = (105, 119.8, 144.6)$$

$$\hat{Q}_9^{(0)} = (\hat{q}_9(5)) = (122.3)$$

Based on these predictions, we construct the predicted wavy curve for the Shanghai stock exchange index for the time period from November 1998 to the end of 1999 (see Fig. 9.6).

References

Deng, J. L. (1985). *Grey control system*. Wuhan: Press of Huazhong University of Science and Technology.

Deng, J. L. (1990). *Grey system tutorial*. Wuhan: Press of Huazhong University of Science and Technology.

Dang, Y. G., & Liu, S. F. (2009). *On grey prediction and decision-making models.* Beijing: Science Press.

Liu, S. F. (2021). *Grey systems theory and its applications.* Beijing: Science Press.

Liu, S. F., & Guo, T. B. (1991). *Grey systems theory and its application.* Kaifeng: Press of Henan University.

Liu, S. F., Yang, Y. J., & Forrest, J. (2017). *Grey data analysis.* Singapore: Springer-Verlag.

Liu, S. F. Jian, L. R., & Mi, C. M. (2020). *Menagemeng forecasting and decision-making Methods.* Beijing: Science Press.

Chapter 10
Grey Models for Decision-Making

10.1 Introduction

Deciding on what actions to take based on actual circumstances and pre-determined goals is known as decision-making. The essential meaning of decision-making is to make a decision or to choose a course of actions. Decision-making not only plays an important part in various kinds of management activities, but also appears throughout every person's daily life. The concept of decision-making can be divided into two categories: general and specific. In the general category, each decision-making stands for an entire process of activities, including posting questions, collecting data, establishing a goal, making, analyzing, and evaluating a plan of action, implementing the plan, feedback, and modifying the plan. In the specific category, decision-making only represents the step of choosing a specific plan of action out of the entire decision-making process. Also, some scholars understand decision-making as choosing and picking a plan of action under uncertain conditions. In this case, the choice can be most likely influenced by the decision maker's prior experience, attitude, and willingness to take a certain amount of risk. Grey decision-making is about making a decision using decision models that involve grey elements or that combine general decision model and grey systems models. Its focus of study is on the problem of choosing a specific plan.

In this chapter, we define an event as the problem waiting to be resolved, the event needing to be handled, and the current state of a system's behavior. Events are where we begin our investigation.

Definition 10.1.1 Events, countermeasures, objectives, and effects are known as the four key elements of decision-making (Deng, 1986).

Definition 10.1.2 The totality of all events within the range of a research is known as the set of events of the study, denoted $A = \{a_1, a_2, \ldots, a_n\}$, where a_i, $i = 1, 2, 3, \ldots, n$, stands for the ith event. The totality of all possible countermeasures is known as the set of countermeasures, denoted $B = \{b_1, b_2, \ldots, b_m\}$ with b_j, $j = 1, 2, \ldots, m$, be the jth countermeasure.

Definition 10.1.3 The Cartesian product $A \times B = \{(a_i, b_j) | a_i \in A, b_j \in B\}$ of the event set A and the countermeasure set B is known as the set of decision schemes, written as $S = A \times B$, where each ordered pair $s_{ij} = (a_i, b_j)$, for any $a_i \in A$, $b_j \in B$, is known as a decision scheme.

For example, in the decision-making on what to plant in agriculture, weather conditions can be used as the set of events, with a normal year denoted as a_1, a drought year as a_2, and a flood year as a_3. Then, the set of events is

$$A = \{a_1, \ a_2, \ a_3\}$$

Different strains of crops can be seen as countermeasures, with corn denoted as b_1, Chinese sorghum as b_2, soybeans as b_3, sesame b_4, potatoes and yams as b_5, ...; then the countermeasure set is given as

$$B = \{b_1, b_2, b_3, b_4, b_5, \cdots\}$$

Therefore, the set of decision scheme is

$$S = A \times B = \{s_{11}, s_{12}, \ldots, s_{15}, \ldots, s_{21}, \ldots, s_{25}, \ldots, s_{31}, \ldots, s_{35}, \ldots\}$$

where $s_{ij} = (a_i, b_j)$.

Here, events and countermeasures are simple. Therefore, the constructed decision schemes are relatively simple, too. In practical decision-making, events are often complicated, consisting of many kinds of simple events, so the countermeasures are complicated, too. Hence, the resultant decision schemes can be extremely complicated.

Let us continue to use the previous agricultural decision-making example. The set of events is the organic body consisting of weather, soil, irrigation, fertilizer, agricultural chemicals, work force, and technology. The countermeasures are not simply the individual strains of crops, but various proportional combinations of many different strains of crops. Let us define a_1 as an event characterized by a normal year, loam, 50% effective irrigation area, sufficient fertilizer and agricultural chemicals, sufficient work force, and medium level of technology. Additionally, let us define a_2 as an event characterized by a drought year, black earth, 50% effective irrigation area, sufficient fertilizer and work force, lack of agricultural chemicals, and medium level of technology. Then, we have the set of events:

$$A = \{a_1, a_2, \ldots\}$$

Let us write b_1 as the countermeasure including 30% corn + 10% Chinese sorghum + 10% soybeans + 15% sesame + 15% potatoes and yams + 10% others. Also, let us write b_2 as the countermeasure including 10% corn + 20% Chinese sorghum + 30% soybeans + 30% sesame + 10% others. Then, we have the countermeasure set:

$$B = \{b_1, b_2, \ldots\}$$

Now, the decision scheme $s_{11} = (a_1, b_1)$ is that, under the conditions of a normal year, loam, 50% effective irrigation area, with sufficient fertilizer and agricultural chemicals, sufficient workforce, and medium level of technology, we should plant 30% corn, 10% Chinese sorghum + 20% soybeans + 15% sesame + 15% potatoes and yams + 10% others.

Let us look at the example of teaching scheduling. The collection of all course offerings of a fixed semester at a certain school can be seen as the set of events; all teaching faculty of this school, and various teaching methods, such as laboratory, interns, and multimedia, are seen as the set of countermeasures. Based on the circumstances, one teacher can teach several courses, or several teachers teach one course together. The work load could be 100% teaching, or 60% teaching, 20% laboratory, 10% interns, and 10% multimedia and others.

For a given decision scheme $s_{ij} \in S$, evaluating the effects under a set of predetermined objectives and deciding on what to take and what to let go based on evaluation is the decision-making we discuss in this chapter. In the following sections, we will study several different kinds of grey decision-making methods (Deng, 1986; Liu & Guo, 1991).

10.2 Grey Target Decisions

Definition 10.2.1 Let $S = \{s_{ij} = (a_i, b_j) | a_i \in A, b_j \in B\}$ be a set of decision schemes, $u_{ij}^{(k)}$ the effect value of decision scheme s_{ij} with respect to objective k, and R the set of all real numbers. Then $u_{ij}^{(k)}: S \mapsto R$, defined by $s_{ij} \mapsto u_{ij}^{(k)}$, is known as the effect mapping of S with respect to object k.

Definition 10.2.2 If $u_{ij}^{(k)} = u_{ih}^{(k)}$, then we say that the countermeasures b_j and b_h of event a_i are equivalent with respect to objective k, written as $b_j \cong b_h$; and the set

$$B_i^{(k)} = \{b | b \in B, b \cong b_h\}$$

is known as the effect equivalence class of countermeasure b_h of event a_i with respect to objective k.

Definition 10.2.3 If k is such an objective that the greater the effect value is the better, and $u_{ij}^{(k)} > u_{ih}^{(k)}$, then we say that the countermeasure b_j is superior to b_h in terms of event a_i with respect to objective k, written as $b_j \succ b_h$. The set $B_{ih}^{(k)} = \{b | b \in B, b \succ b_h\}$ is known as the superior set of countermeasure b_h of event a_i with respect to objective k.

Similarly, we can define the concept of superior classes of countermeasures for situations where the closer to a fixed moderate value the effect value is the better, and where the smaller the effect value is the better.

Definition 10.2.4 If $u_{ij}^{(k)} = u_{ih}^{(k)}$, then events a_i and a_j are said to be equivalent in terms of the countermeasure b_h with respect to objective k, written $a_i \cong a_j$. The set

$$A_{jh}^{(k)} = \{a | a \in A, \quad a \cong a_i\}$$

is known as the effect equivalence class of events of the countermeasure b_h with respect to objective k.

Definition 10.2.5 If k is such an objective that the greater the effect value is the better, and $u_{ih}^{(k)} > u_{jh}^{(k)}$, then we say that event a_i is superior to event a_j in terms of countermeasure b_h with respect to objective k, denoted $a_i \succ a_j$. The set

$$A_{jh}^{(k)} = \{a | a \in A, a \succ a_j\}$$

is known as the superior class of event a_j in terms of countermeasure b_h with respect to objective k.

Similarly, the concept of superior classes can be defined for situations where the closer to a fixed moderate value the effect value is the better, and where the smaller the effect value is the better.

Definition 10.2.6 If $u_{ij}^{(k)} = u_{hl}^{(k)}$, then scheme s_{ij} is equivalent to scheme s_{hl} under objective k, denoted $s_{ij} \cong s_{hl}$. The set

$$S^{(k)} = \{s | s \in S, s \cong s_{hl}\}$$

is known as the effect equivalence class of scheme s_{hl} under objective k.

Definition 10.2.7 If k is such an objective that the greater the effect value is the better, and $u_{ij}^{(k)} > u_{hl}^{(k)}$, then scheme s_{ij} is said to be superior to scheme s_{hl} under objective k, denoted $s_{ij} \succ s_{hl}$. The set

$$S_{hl}^{(k)} = \{s | s \in S, s \succ s_{hl}\}$$

is known as the effect superior class of scheme s_{hl} under objective k.

Similarly, the concept of superior classes for scheme effects can be defined for scenarios where the closer to a fixed moderate value the effect value of a scheme is the better, and where the smaller the effect value of the scheme is the better.

Proposition 10.2.1 Assume that $S = \{s_{ij} = (a_i, b_j) | a_i \in A, b_j \in B\} \neq \emptyset$ and $U^{(k)} = \{u_{ij}^{(k)} | a_i \in A, b_j \in B\}$ is the set of effects under objective k, and $\{S^{(k)}\}$ the set of effect equivalence classes of schemes under objective k. Then the mapping $u^{(k)} : \{S^{(k)}\} \rightarrow U^{(k)}$, defined by $S^{(k)} \mapsto u_{ij}^{(k)}$, is bijective.

Definition 10.2.8 Let $d_1^{(k)}$ and $d_2^{(k)}$ be the upper and lower threshold values of the decision effects of s_{ij} under objective k. Then $S^1 = \left\{ r \middle| d_1^{(k)} \leq r \leq d_2^{(k)} \right\}$ is known as the one-dimensional grey target of objective k, $u_{ij}^{(k)} \in [d_1^{(k)}, d_2^{(k)}]$ a satisfactory effect under objective k, the corresponding s_{ij} a desirable scheme with respect to objective k, and b_j a desirable countermeasure of event a_i with respect to objective k.

Proposition 10.2.2 Assume that $u_{ij}^{(k)}$ stands for the effect value of scheme s_{ij} with respect objective k. If $u_{ij}^{(k)} \in S^1$, that is, s_{ij} is a desirable scheme with respect to objective k. Then for any $s \in S_{ij}^{(k)}$, s is also a desirable scheme. That is, when s_{ij} is desirable, all schemes in its effect superior class are desirable.

The discussion above applies to cases involving a single objective. Nevertheless, grey targets of decision-making with multi-objectives can also be addressed.

Definition 10.2.9 Assume that $d_1^{(1)}$ and $d_2^{(1)}$ are the threshold values of decision effects of objective 1, $d_1^{(2)}$ and $d_2^{(2)}$ the threshold values of decision effects of objective 2. Then

$$S^2 = \left\{ (r^{(1)}, r^{(2)}) \middle| d_1^{(1)} \leq r^{(1)} \leq d_2^{(1)}, d_1^{(2)} \leq r^{(2)} \leq d_2^{(2)} \right\}$$

is known as a grey target of two-dimensional decision-making. If the effect vector of scheme s_{ij} satisfies $u_{ij} = \left\{ u_{ij}^{(1)}, u_{ij}^{(2)} \right\} \in S^2$, then s_{ij} is seen as a desirable scheme with respect to objectives 1 and 2, and b_j a desirable countermeasure for event a_i with respect to objectives 1 and 2.

Definition 10.2.10 Assume that $d_1^{(1)}, d_2^{(1)}; d_1^{(2)}, d_2^{(2)}; \ldots; d_1^{(s)}, d_2^{(s)}$ are respectively the threshold values of decision effects under objectives 1, 2, ..., s. Then the following region of the s–dimensional Euclidean space

$$S^s = \left\{ (r^{(1)}, r^{(2)}, \ldots, r^{(s)}) \middle| d_1^{(1)} \leq r^{(1)} \leq d_2^{(1)}, d_1^{(2)} \leq r^{(2)} \leq d_2^{(2)}, \ldots, d_1^{(s)} \leq r^{(s)} \leq d_2^{(s)} \right\}$$

is known as a grey target of an s–dimensional decision-making. If the effect vector of scheme s_{ij} satisfies

$$u_{ij} = (u_{ij}^{(1)}, u_{ij}^{(2)}, \ldots, u_{ij}^{(s)}) \in S^s$$

where $u_{ij}^{(k)}$ stands for the effect value of the scheme s_{ij} with respect to objective k, $k = 1, 2, \ldots, s$, then s_{ij} is known as a desirable scheme with respect to objectives 1, 2, ..., s, and b_j a desirable countermeasure of event a_i with respect to objectives 1, 2, ..., s.

Intuitively, the grey targets of a decision-making essentially represent the location of satisfactory effects in terms of relative optimization. In many practical circumstances, it is impossible to obtain the absolute optimization so that people are happy if they can achieve a satisfactory outcome. Of course, based on the need, one can gradually shrink the grey targets of his decision-making to a single point in order to obtain the ultimate optimal effect, where the corresponding scheme is the most desirable, and the corresponding countermeasure the optimal countermeasure.

Definition 10.2.11 The following equation

$$R^s = \left\{ (r^{(1)}, r^{(2)}, \ldots, r^{(s)}) \middle| (r^{(1)} - r_0^{(1)})^2 + (r^{(2)} - r_0^{(2)})^2 + \cdots + (r^{(s)} - r_0^{(s)})^2 \le R^2 \right\}$$

is known as an s-dimensional spherical grey target centered at $r_0 = (r_0^{(1)}, r_0^{(2)}, \ldots, r_0^{(s)})$ with radius R. The vector $r_0 = (r_0^{(1)}, r_0^{(2)}, \ldots, r_0^{(s)})$ is seen as the optimum effect vector.

For $r_1 = (r_1^{(1)}, r_1^{(2)}, \ldots, r_1^{(s)}) \in R$,

$$|r_1 - r_0| = \left[(r_1^{(1)} - r_0^{(1)})^2 + (r_1^{(2)} - r_0^{(2)})^2 + \cdots + (r_1^{(s)} - r_0^{(s)})^2 \right]^{1/2}$$

is known as the bull's-eye distance of vector r_1. The values of this distance reflect the superiority of the corresponding decision effect vectors.

Definition 10.2.12 Let s_{ij} and s_{hl} be two different schemes, and $u_{ij} = (u_{ij}^{(1)}, u_{ij}^{(2)}, \ldots, u_{ij}^{(s)})$ and $u_{hl} = (u_{hl}^{(1)}, u_{hl}^{(2)}, \ldots, u_{hl}^{(s)})$ their effect vectors, respectively. If

$$\left| u_{ij} - r_0 \right| \ge \left| u_{hl} - r_0 \right| \tag{10.1}$$

then scheme s_{hl} is said to be superior to s_{ij}, denoted $s_{hl} \succ s_{ij}$. When the equal sign in Eq. (10.1) holds true, schemes s_{ij} and s_{hl} are said to be equivalent, written $s_{hl} \cong s_{ij}$.

If for $i = 1, 2, \ldots, n$ and $j = 1, 2, \ldots, m$, $u_{ij} \ne r_0$ always holds true, then the optimum scheme does not exist, and the event does not have any optimum countermeasure. If the optimum scheme does not exist, however, there are h and l such that for any $i = 1, 2, \ldots, n$ and $j = 1, 2, \ldots, m$, $|u_{hl} - r_0| \le |u_{ij} - r_0|$ holds true, that is, for any $s_{ij} \in S$, $s_{hl} \succ s_{ij}$ holds, then s_{hl} is known as a quasi-optimum scheme, a_h a quasi-optimum event, and b_l a quasi-optimum countermeasure.

Theorem 10.2.1 Let $S = \{s_{ij} = (a_i, b_j) | a_i \in A, b_j \in B\}$ be a set of schemes, and

$$R^s = \left\{ (r^{(1)}, r^{(2)}, \ldots, r^{(s)}) \middle| (r^{(1)} - r_0^{(1)})^2 + (r^{(2)} - r_0^{(2)})^2 + \cdots + (r^{(s)} - r_0^{(s)})^2 \le R^2 \right\}$$

an s-dimensional spherical grey target. The S becomes an ordered set with "superiority" as its order relation $\prec$.

Theorem 10.2.2 There must be quasi-optimum scheme in the set of decision schemes of $(S, \succ)$.

Proof This is a restatement of Zorn's Lemma in set theory.

Example 10.2.1 Consider event a_1 of reconstructing an old building. There are three possibilities: $b_1 = $ renovate the building completely; $b_2 = $ tear down the building and reconstruct another; and $b_3 = $ simply maintain what the building is by fixing up minor problems. Let us make a grey target decision using three objectives: cost, functionality, and construction speed.

Solution Let us denote the cost as objective 1, the functionality as objective 2, and the construction speed as objective 3. Then, we have the following three decision schemes:

$$s_{11} = (a_1, b_1) = (\text{reconstruction, renovation}),$$
$$s_{12} = (a_1, b_2) = (\text{reconstruction, new building}), \text{ and}$$
$$s_{13} = (a_1, b_3) = (\text{reconstruction, maintenance}).$$

Evidently, different decision schemes with respect to different objectives have different effects; and the standards for measuring the effects are also accordingly different. For instance, regarding cost, the lesser the better; for functionality, the higher the better; and for speed, the faster the better. Let us divide the effects of the decision schemes into three classes: good, okay, and poor.

The effect vectors of the decision schemes are respectively defined as follows:

$$u_{11} = \left(u_{11}^{(1)}, u_{11}^{(1)}, u_{11}^{(3)} \right) = (2, 2, 2),$$
$$u_{12} = \left(u_{12}^{(1)}, u_{12}^{(2)}, u_{12}^{(3)} \right) = (3, 1, 3), \text{ and}$$
$$u_{13} = \left(u_{13}^{(1)}, u_{13}^{(2)}, u_{13}^{(3)} \right) = (1, 3, 1).$$

Let the bull's eye be located at $r_0 = (1, 1, 1)$ and compute the bull's-eye distances

$$|u_{11} - r_0| = \left[\left(u_{11}^{(1)} - r_0^{(1)} \right)^2 + \left(u_{11}^{(2)} - r_0^{(2)} \right)^2 + \left(u_{11}^{(3)} - r_0^{(3)} \right)^2 \right]^{1/2}$$
$$= \left[(2-1)^2 + (2-1)^2 + (2-1)^2 \right]^{1/2} = 1.73$$

$$|u_{12} - r_0| = \left[\left(u_{12}^{(1)} - r_0^{(1)} \right)^2 + \left(u_{12}^{(2)} - r_0^{(2)} \right)^2 + \left(u_{12}^{(2)} - r_0^{(3)} \right)^2 \right]^{1/2}$$
$$= \left[(3-1)^2 + (1-1)^2 + (3-1)^2 \right]^{1/2} = 2.83$$

$$|u_{13} - r_0| = \left[\left(u_{13}^{(1)} - r_0^{(1)}\right)^2 + \left(u_{13}^{(2)} - r_0^{(2)}\right)^2 + \left(u_{13}^{(3)} - r_0^{(3)}\right)^2\right]^{1/2}$$

$$= \left[(1-1)^2 + (3-1)^2 + (1-1)^2\right]^{1/2} = 2$$

where $|u_{11} - r_0|$ is the smallest. So, the effect vector $u_{11} = (2, 2, 2)$ of the decision scheme s_{11} enters the grey target. Hence, renovation is a satisfactory decision.

10.3   Other Approaches to Grey Decision

10.3.1   Grey Relational Decision

The bull's-eye distance between a decision effect vector and the center of the target measures the superiority of the scheme in comparison with other schemes. At the same time, the grey relational degree between the effect vector of a decision scheme and the optimum effect vector can be seen as another way to evaluate the superiority of a decision scheme.

Definition 10.3.1 Let $S = \{s_{ij} = (a_i, b_j) | a_i \in A, b_j \in B\}$ be a set of decision schemes, and $u_{i_0 j_0} = \{u_{i_0 j_0}^{(1)}, u_{i_0 j_0}^{(2)}, \ldots, u_{i_0 j_0}^{(s)}\}$ the optimum effect vector. If the decision scheme corresponding to $u_{i_0 j_0}$ satisfies $u_{i_0 j_0} \notin S$, then $u_{i_0 j_0}$ is known as an imagined optimum effect vector, and $s_{i_0 j_0}$ the imagined optimum scheme.

Proposition 10.3.1 Let S be the same as above and the effect vector of scheme s_{ij} is $u_{ij} = \{u_{ij}^{(1)}, u_{ij}^{(2)}, \ldots, u_{ij}^{(s)}\}$, for $i = 1, 2, \ldots, n$, $j = 1, 2, \ldots, m$.

(1)   When k is an objective such that the greater its effect value is the better, let
$u_{i_0 j_0}^{(k)} = \max_{1 \leq i \leq n, 1 \leq j \leq m} \{u_{ij}^{(k)}\}$;

(2)   When k is an objective such that the closer to a fixed moderate value u_0 its effect value is the better, let $u_{i_0 j_0}^{(k)} = u_0$; and

(3)   When k is an objective such that the smaller its effect value is the better, let $u_{i_0 j_0}^{(k)} = \min_{1 \leq i \leq n, 1 \leq j \leq m} \{u_{ij}^{(k)}\}$, then $u_{i_0 j_0} = \{u_{i_0 j_0}^{(1)}, u_{i_0 j_0}^{(2)}, \ldots, u_{i_0 j_0}^{(s)}\}$ is the imagined optimum effect vector.

Proposition 10.3.2 Assume the same as in Proposition 10.3.1 and let $u_{i_0 j_0} = \{u_{i_0 j_0}^{(1)}, u_{i_0 j_0}^{(2)}, \ldots, u_{i_0 j_0}^{(s)}\}$ be the imagined optimum effect vector, ε_{ij} the grey absolute relational degree between u_{ij} and $u_{i_0 j_0}$, for $i = 1, 2, \ldots, n$, $j = 1, 2, \ldots, m$. If for any $i \in \{1, 2, \ldots, n\}$ and $j \in \{1, 2, \ldots, m\}$ satisfying $i \neq i_1$ and $j \neq j_1$, $\varepsilon_{i_1 j_1} \geq \varepsilon_{ij}$ always holds true, then $u_{i_1 j_1}$ is a quasi-optimum effect vector and $s_{i_1 j_1}$ a quasi-optimum decision scheme.

Grey relational decisions can be made by following the following steps:

Step 1: Determine the set of events $A = \{a_1, a_2, \ldots, a_n\}$ and the set of counter-measures $B = \{b_1, b_2, \ldots, b_m\}$. And then construct the set of decision schemes $S = \{s_{ij} = (a_i, b_j) | a_i \in A, b_j \in B\}$.

Step 2: Choose the objectives $1, 2, \ldots, s$, for the decision-making.

Step 3: Compute the effect values $u_{ij}^{(k)}$ of the individual decision scheme s_{ij}, $i = 1, 2, \ldots, n$, $j = 1, 2, \ldots, m$, with respect to objective k, obtained in the decision effect sequence $u^{(k)}$

$$u^{(k)} = (u_{11}^{(k)}, u_{12}^{(k)}, \ldots, u_{1m}^{(k)}; u_{21}^{(k)}, u_{22}^{(k)}, \ldots, u_{2m}^{(k)}; \ldots; u_{n1}^{(k)}, u_{n2}^{(k)}, \ldots, u_{nm}^{(k)});$$
$$k = 1, 2, \ldots, s.$$

Step 4: Compute the average image of the decision effect sequence $u^{(k)}$ with respect to objective k, which is still written the same as

$$u^{(k)} = (u_{11}^{(k)}, u_{12}^{(k)}, \ldots, u_{1m}^{(k)}; u_{21}^{(k)}, u_{22}^{(k)}, \ldots, u_{2m}^{(k)}; \ldots; u_{n1}^{(k)}, u_{n2}^{(k)}, \ldots, u_{nm}^{(k)});$$
$$k = 1, 2, \cdots, s$$

Step 5: Based on the results of Step 4, write out the effect vector $u_{ij} = \left\{ u_{ij}^{(1)}, u_{ij}^{(2)}, \ldots, u_{ij}^{(s)} \right\}$ of decision scheme s_{ij}, for $i = 1, 2, \ldots, n$, $j = 1, 2, \cdots, m$.

Step 6: Compute the imagined optimum effect vector $u_{i_0 j_0} = \left\{ u_{i_0 j_0}^{(1)}, u_{i_0 j_0}^{(2)}, \ldots, u_{i_0 j_0}^{(s)} \right\}$.

Step 7: Calculate the grey absolute relational degree ε_{ij} between u_{ij} and $u_{i_0 j_0}$, $i = 1, 2, \ldots, n$, $j = 1, 2, \ldots, m$.

Step 8: From $\max_{1 \le i \le n, 1 \le j \le m} \{\varepsilon_{ij}\} = \varepsilon_{i_1 j_1}$, the quasi-optimum effect vector $u_{i_1 j_1}$ and the quasi-optimum decision scheme $s_{i_1 j_1}$ are obtained.

Example 10.3.1 Let us look at grey relational decision-making regarding the evaluation of looms.

Solution Let us denote the event of evaluating loom models by a_1. Then the event set is $A = \{a_1\}$. There are three loom models under consideration: Model 1: purchase projectile loom, which is treated as countermeasure b_1; Model 2: select air jet loom, which is treated as countermeasure b_2; Model 3: choose rapier loom, which is treated as countermeasure b_3. Thus, the set of countermeasure is $B = \{b_1, b_2, b_3\}$, and the set of decision schemes is $S = \left\{ s_{ij} = (a_i, b_j) | a_i \in A, b_j \in B \right\} = \{s_{11}, s_{12}, s_{13}\}$.

Now, let us determine the objectives. According to the functionality of looms, eleven objectives are chosen. The weft-insertion rate (m/min) of the looms is objective 1. The efficiency of the looms is objective 2. The total investment (in ten thousand US\$) on the looms is objective 3. The total energy cost (W/a) is objective 4. The total area (m^2) of the land to be occupied by the looms is objective 5. The total manpower (person) is objective 6. The quantity of weft yarn waste (cm/weft) is objective 7. The cost of replacement parts (ten thousand Yuan/a) is objective 8. Noise (dB) is

Table 10.1 Objective values for the looms

Model	Projectile loom	Air jet loom	Rapier loom
Weft-insertion rate (m/min)	1000	1200	800
Efficiency (%)	92	90	92
Total investment (10 K US$)	880	336	612
Total energy consumption (W/a)	374	924	816
Total land needed (m²)	1760	1092	2124
Total manpower (person)	18	22	24
Quantity of weft yarn waste (cm/weft)	5	6	10
Cost of parts (10 K ¥/a)	37	35	75
Noise (dB)	85	91	91
Quality	Best	Good	Fine
Adaptability	Good	Better	Best

objective 9. The quality of the produced fabric is objective 10. And, the adaptability of the type of loom is objective 11.

Under the assumptions that the above-mentioned three loom models produce the same kind of grey fabric meeting the same set of requirements, and that these looms will produce the same amount of annual output, let us conduct the associated computations for the said loom models. Our quantitative calculations lead to relevant values for the objectives, some of which determined from the literature and field investigations (see Table 10.1).

In the following equations, we compute decision effect sequences $U^k(k = 1, 2, \ldots, 11)$ with respect to the objectives.

For objective 1, we have $U^{(1)} = (u_{11}^{(1)}, u_{12}^{(1)}, u_{13}^{(1)}) = (1000, 1200, 800)$.

For objective 2, we have $U^{(2)} = (u_{11}^{(2)}, u_{12}^{(2)}, u_{13}^{(2)}) = (92, 90, 92)$.

For objective 3, we have $U^{(3)} = (u_{11}^{(3)}, u_{12}^{(3)}, u_{13}^{(3)}) = (880, 336, 612)$.

For objective 4, we have $U^{(4)} = (u_{11}^{(4)}, u_{12}^{(4)}, u_{13}^{(4)}) = (374, 924, 816)$.

For objective 5, we have $U^{(5)} = (u_{11}^{(5)}, u_{12}^{(5)}, u_{13}^{(5)}) = (1760, 1092, 2124)$.

For objective 6, we have $U^{(6)} = (u_{11}^{(6)}, u_{12}^{(6)}, u_{13}^{(6)}) = (18, 22, 24)$.

For objective 7, we have $U^{(7)} = (u_{11}^{(7)}, u_{12}^{(7)}, u_{13}^{(7)}) = (5, 6, 10)$.

For objective 8, we have $U^{(8)} = (u_{11}^{(8)}, u_{12}^{(8)}, u_{13}^{(8)}) = (37, 35, 75)$.

For objective 9, we have $U^{(9)} = (u_{11}^{(9)}, u_{12}^{(9)}, u_{13}^{(9)}) = (85, 91, 91)$.

For objective 10, we have $U^{(10)} = (u_{11}^{(10)}, u_{12}^{(10)}, u_{13}^{(10)}) = (\text{best, good, fine})$.

For objective 11, we have $U^{(11)} = (u_{11}^{(11)}, u_{12}^{(11)}, u_{13}^{(11)}) = (\text{good, better, best})$.

Quantify the last two qualitative objectives as follows:

$$U^{(10)} = (u_{11}^{(10)}, u_{12}^{(10)}, u_{13}^{(10)}) = (9, 8, 7)$$
$$U^{(11)} = (u_{11}^{(11)}, u_{12}^{(11)}, u_{13}^{(11)}) = (8, 7, 9)$$

We now compute the average images of the decision effect sequences for each of the objectives:

$U^{(1)} = (1, 1.2, 0.8); U^{(2)} = (1.01, 0.98, 1.01); U^{(3)} = (1.44, 0.55, 1.01)$

$U^{(4)} = (0.53, 1.31, 1.16); U^{(5)} = (1.06, 0.66, 1.28); U^{(6)} = (0.84, 1.03, 1.13)$

$U^{(7)} = (0.71, 0.86, 1.43); U^{(8)} = (0.76, 0.71, 1.53); U^{(9)} = (0.96, 1.02, 1.02)$

$U^{(10)} = (1.13, 1, 0.87);$ and $U^{(11)} = (1, 0.87, 1.13)$

We also compute the effect vectors U_{ij} of decision schemes $s_{ij}, i = 1, j = 1, 2, 3$:

$$U_{11} = (u_{11}^{(1)}, u_{11}^{(2)}, \ldots, u_{11}^{(11)})$$
$$= (1, 1.01, 1.44, 0.53, 1.06, 0.84, 0.71, 0.76, 0.96, 1.13, 1),$$
$$U_{12} = (u_{12}^{(1)}, u_{12}^{(2)}, \ldots, u_{12}^{(11)})$$
$$= (1.2, 0.98, 0.55, 1.31, 0.66, 1.03, 0.86, 0.71, 1.02, 1, 0.87), \text{ and}$$
$$U_{13} = (u_{13}^{(1)}, u_{13}^{(2)}, \ldots, u_{13}^{(11)})$$
$$= (0.8, 1.01, 1.01, 1.16, 1.28, 1.13, 1.43, 1.53, 1.02, 0.87, 1.13)$$

According to the principle of constituting optimum reference sequences, from the average images of the decision effect sequences of the objectives, it follows that:

For objective 1, the greater the effect value is the better, so $U_{i_0 j_0}^{(1)} = \max\{u_{ij}^{(1)}\} = u_{12}^{(1)} = 1.2$;

For objective 1, the higher the effect value is the better, so $U_{i_0 j_0}^{(2)} = \max\{u_{ij}^{(2)}\} = u_{11}^{(2)} = 1.01$;

For objective 3, the smaller effect value is the better, so $U_{i_0 j_0}^{(3)} = \min\{u_{ij}^{(3)}\} = u_{12}^{(3)} = 0.55$;

For objective 4, the smaller effect value is the better, so $U_{i_0 j_0}^{(4)} = \min\{u_{ij}^{(4)}\} = u_{11}^{(4)} = 0.53$;

For objective 5, the smaller effect value is the better, so $U_{i_0 j_0}^{(5)} = \min\{u_{ij}^{(5)}\} = u_{11}^{(5)} = 0.66$;

For objective 6, the smaller effect value is the better, so $U_{i_0 j_0}^{(6)} = \min\{u_{ij}^{(6)}\} = u_{11}^{(6)} = 0.84$;

For objective 7, the smaller effect value is the better, so $U_{i_0 j_0}^{(7)} = \min\{u_{ij}^{(7)}\} = u_{11}^{(7)} = 0.71$;

For objective 8, the smaller effect value is the better, so $U_{i_0 j_0}^{(8)} = \min\left\{u_{ij}^{(8)}\right\} = u_{12}^{(8)} = 0.71$;

For objective 9, the smaller effect value is the better, so $U_{i_0 j_0}^{(9)} = \min\left\{u_{ij}^{(9)}\right\} = u_{11}^{(9)} = 0.96$;

For objective 10, the higher effect value is the better, so $U_{i_0 j_0}^{(10)} = \max\left\{u_{ij}^{(10)}\right\} = u_{11}^{(10)} = 1.13$; and

For objective 11, the higher effect value is the better, so $U_{i_0 j_0}^{(11)} = \max\left\{u_{ij}^{(11)}\right\} = u_{13}^{(11)} = 1.13$.

That is, we obtain the following optimum reference sequence:

$$U_{i_0 j_0} = (u_{i_0 j_0}^{(1)}, u_{i_0 j_0}^{(2)}, \ldots, u_{i_0 j_0}^{(11)})$$
$$= (1.2, 1.01, 0.55, 0.53, 0.66, 0.84, 0.71, 0.71, 0.96, 1.13, 1.13)$$

From u_{ij} and $u_{i_0 j_0}$, we compute the absolute grey relational degrees:

$$\varepsilon_{11} = 0.628, \quad \varepsilon_{12} = 0.891, \quad \varepsilon_{13} = 0.532$$

From the definition of grey relational decision-making, it follows that because $\max\left\{\varepsilon_{ij}\right\} = \varepsilon_{12} = 0.891$, U_{12} is the quasi-optimum vector and s_{12} the quasi-optimum decision scheme. That is to say, in terms of producing general grey fabric, the air jet loom is the best choice among the available loom models.

10.3.2   Grey Development Decision

Grey development decision-making is done based on the development tendency or the future behaviors of the decision scheme of concern. It does not necessarily place specific emphasis on the current effect of the scheme. Instead it focuses more on the change of the decision effect over time. This method of decision-making can be and has been employed for long-term planning as well as the decision-making of large scale engineering projects and urban planning. It looks at problems from the angle of development while attempting to make feasible arrangements and avoiding repetitious constructions so that great savings of capital and manpower can be achieved. What we have discussed earlier are static decision schemes with a fixed time moment. Because we now involve the concept of time, as time moves, constantly changing decision effects are considered.

Definition 10.3.2 Assume that $A = \{a_1, a_2, \ldots, a_n\}$ is a set of events, $B = \{b_1, b_2, \ldots, b_m\}$ a set of countermeasures, and $S = \{s_{ij} = (a_i, b_j) | a_i \in A, b_j \in B\}$ the set of decision schemes. Then,

$$u_{ij}^{(k)} = (u_{ij}^{(k)}(1), u_{ij}^{(k)}(2), \ldots, u_{ij}^{(k)}(h))$$

is known as the decision effect time series of scheme s_{ij} with respect to objective k.

Definition 10.3.3 Let the decision effect time series of the scheme s_{ij} with respect to objective k be

$$u_{ij}^{(k)} = (u_{ij}^{(k)}(1), u_{ij}^{(k)}(2), \ldots, u_{ij}^{(k)}(h))$$

$\hat{a}_{ij}^{(k)} = \left[a_{ij}^{(k)}, b_{ij}^{(k)} \right]^T$ the least squares estimate of the parameters of the EGM(1,1) model of $u_{ij}^{(k)}$. Then the inverse accumulation restoration of the EGM(1,1) time response of $u_{ij}^{(k)}$ is given by

$$\hat{u}_{ij}^{(k)}(l+1) = \left[1 - \exp(a_{ij}^{(k)}) \right] \cdot \left[u_{ij}^{(k)}(1) - \frac{b_{ij}^{(k)}}{a_{ij}^{(k)}} \right] \exp\left(-a_{ij}^{(k)} \cdot l \right)$$

Assume that the restored sequence through inverse accumulation of the EGM(1,1) time response of the decision effect time series of the scheme s_{ij} with respect to objective k is

$$\hat{u}_{ij}^{(k)}(l+1) = \left[1 - \exp(a_{ij}^{(k)}) \right] \cdot \left[u_{ij}^{(k)}(1) - \frac{b_{ij}^{(k)}}{a_{ij}^{(k)}} \right] \exp\left(-a_{ij}^{(k)} \cdot l \right)$$

When objective k satisfies that the greater the effect value is the better, if

(1) $\max_{1 \leq i \leq n, 1 \leq j \leq m} \left\{ -a_{ij}^{(k)} \right\} = -a_{i_0 j_0}^{(k)}$, then $s_{i_0 j_0}$ is known as the optimum scheme of development coefficients with respect to objective k;

(2) $\max_{1 \leq i \leq n, 1 \leq j \leq m} \left\{ \hat{u}_{ij}^{(k)}(h+l) \right\} = \hat{u}_{i_0 j_0}^{(k)}(h+l)$, then $s_{i_0 j_0}$ is known as the optimum scheme of predictions with respect to objective k.

Similarly, the concepts of optimum schemes of development coefficients and predictions can be defined for cases of objectives satisfying that the smaller the effect value is the better, and that the closer to a moderate value the effect value is the better, respectively. In particular, for objectives satisfying that the smaller the effect value is the better, one only needs to replace "max" in the items (1) and (2) above by "min"; if k is an objective satisfying that the closer to a fixed moderate value the effect value is the better, one can determine the moderate value of the development coefficients or predicted values at first; then define the optimum scheme based on the distances of the development coefficients or predicted values to the moderate value.

In practical applications, one may face the scenarios that either both the optimum scheme of development coefficients and predictions are the same, or that they are different. Even so, the following theorem tells us that eventually these optimum schemes would converge into one.

Theorem 10.3.1 Assume that k is such an objective that the greater its effect value is the better, $s_{i_0 j_0}$ is the optimum scheme of development coefficients, that is, $-a_{i_0 j_0}^{(k)} = \max_{1 \le i \le n, 1 \le j \le m} \left\{ -a_{ij}^{(k)} \right\}$, and $\hat{u}_{i_0 j_0}^{(k)}(h + l + 1)$ is the predicted value for the decision effect of $s_{i_0 j_0}$. Then there must be $l_0 > 0$ such that

$$\hat{u}_{i_0 j_0}^{(k)}(h + l_0 + 1) = \max_{1 \le i \le n, 1 \le j \le m} \left\{ \hat{u}_{ij}^{(k)}(h + l_0 + 1) \right\}$$

That is, in a sufficiently distant future, $s_{i_0 j_0}$ will also be the optimum scheme of predictions.

Proof See Liu and Lin (2006, pp. 340–341) for details.

Similar results hold true for those objectives satisfying either that the smaller the effect value is the better or that the closer to a fixed moderate value the effect value is the better.

At this junction, careful readers might have noticed that Theorem 10.3.1 does not state the case that there are some increasing and decreasing sequences among decision effect time series at the same time. As a matter of fact, for objectives satisfying that the greater the effect value is the better, there is no need to consider decreasing decision effect time series. For objectives satisfying that the smaller the effect value is the better, all increasing decision effect time series are deleted in advance in all discussions. As for objectives satisfying that the closer to a moderate value the effect value is the better, one can consider only either increasing or decreasing decision effect time series depending on the circumstances involved.

10.3.3 Grey Clustering Decision

Grey cluster decision is useful for synthetic evaluations of objects with respect to several different criteria so that decisions can be made about whether or not an object meets the given standards for inclusion in or exclusion from a set. This method has often been employed for classification decision-making regarding objects or people. For instance, school students can be classified based on their individual capabilities to receive information, to comprehend what is provided, and to grow so that different teaching methods can be applied and different students can be enrolled in different programs. As a second example, based on different sets of criteria, comprehensive evaluations can be done for general employees, technicians, and administrators respectively so that decisions can be made regarding who is qualified for his/her job, who is ready for a promotion, and so on.

Definition 10.3.4 Assume that there are n objects to make decisions on, m criteria, s different grey classes, the quantified evaluation value of object i with respect to criterion j is x_{ij}, $f_j^k(*)$ are the possibility functions of the kth grey class with respect to the jth criterion, and w_j is the synthetic decision-making weight of criterion j such that $\sum_{j=1}^m w_j = 1$, $i = 1, 2, \ldots, n, j = 1, 2, \ldots, m, k = 1, 2, \ldots, s$. Then

$$\sigma_i^k = \sum_{j=1}^m f_j^k(x_{ij})w_j$$

is known as the decision coefficient for the object i to belong to grey class k; $\sigma_i = (\sigma_i^1, \sigma_i^2, \ldots, \sigma_i^s)$ is known as the decision coefficient vector of object i, $i = 1, 2, \ldots, n$; and $\sum = (\sigma_i^k)_{n \times s}$ the decision coefficient matrix. If $\max_{1 \leq k \leq s}\{\sigma_i^k\} = \sigma_i^{k^*}$, then the decision is that the object i belongs to grey class k^*.

In practical applications, it is quite often the case that many objects belong to the same decision grey class at the same time, while there is a constraint on how many objects are allowed in the grey class. When this occurs, we can further determine individual objects' precedence in grey class k^* on the basis of the size of integrate clustering coefficients.

10.4 Multi-attribute Intelligent Grey Target Decision Model

In this section, we will study a new decision model, which is constructed on the basis of four new functions of uniform effect measures. This new decision model sufficiently considers the two different scenarios of whether or not the effect values of the objectives actually hit the targets with very clear physics significance. First, a grey target is defined as a satisfying region, which a decision maker wants to reach, with an inside ideal point across multiple objectives. To facilitate the uniform distance measure of a decision strategy to the pre-defined grey target, four kinds of measure procedures are designed including the effect measures for benefit-type objectives and cost-type objectives, the lower effect measure for moderate-type objectives, and the upper effect measure for moderate-type according to three types of decision objective including benefit objective, cost objective, and non-monotonic objective with a most preferred middle value. Then, a matrix of synthetic effect measures can be easily obtained based on the uniform distance measure of a decision strategy to the grey target over different objectives. Based upon the obtained matrix information, different decision strategies can be evaluated easily and comprehensively. The proposed method has a clear physical meaning as missing target, hitting target as well as hitting performance (Liu et al., 2013; Liu et al., 2017; Liu, 2021).

10.4.1 The Uniform Effect Measure

Definition 10.4.1

(1) Let k be a benefit type objective, that is, for k the larger the effect value is the better, and the decision grey target of objective k is $u_{ij}^{(k)} \in [u_{i_0 j_0}^{(k)}, \max_i \max_j \{u_{ij}^{(k)}\}]$, that is, $u_{i_0 j_0}^{(k)}$ stands for the threshold effect value of objective k. Then

$$r_{ij}^{(k)} = \frac{u_{ij}^{(k)} - u_{i_0 j_0}^{(k)}}{\max\limits_i \max\limits_j \{u_{ij}^{(k)}\} - u_{i_0 j_0}^{(k)}} \tag{10.2}$$

is referred to as the effect measure of a benefit-type objective.

(2) Let k be a cost-type objective, that is, for k the smaller the effect value is the better, and the decision grey target of objective k is $u_{ij}^{(k)} \in [\min_i \min_j \{u_{ij}^{(k)}\}, u_{i_0 j_0}^{(k)}]$, that is, $u_{i_0 j_0}^{(k)}$ stands for the threshold effect value of objective k. Then

$$r_{ij}^{(k)} = \frac{u_{i_0 j_0}^{(k)} - u_{ij}^{(k)}}{u_{i_0 j_0}^{(k)} - \min\limits_i \min\limits_j \{u_{ij}^{(k)}\}} \tag{10.3}$$

is referred to as the effect measure of cost-type objective.

(3) Let k be a moderate-value type objective, that is, for η_k the closer to a moderate value A the effect value is the better, and the decision grey target of objective η_k is $u_{ij}^{(k)} \in [A - u_{i_0 j_0}^{(k)}, A + u_{i_0 j_0}^{(k)}]$, that is, both $A - u_{i_0 j_0}^{(k)}$ and $A + u_{i_0 j_0}^{(k)}$ are respectively the lower and upper threshold effect values of objective k. Then,

(i) When η_k,

$$r_{ij}^{(k)} = \frac{u_{ij}^{(k)} - A + u_{i_0 j_0}^{(k)}}{u_{i_0 j_0}^{(k)}} \tag{10.4}$$

is referred to as the lower effect measure of moderate-value type objective.

(ii) When $u_{ij}^{(k)} \in [A, A + u_{i_0 j_0}^{(k)}]$,

$$r_{ij}^{(k)} = \frac{A + u_{i_0 j_0}^{(k)} - u_{ij}^{(k)}}{u_{i_0 j_0}^{(k)}} \tag{10.5}$$

is referred to as the upper effect measure of moderate-value type objective.

The effect measures of benefit-type objectives reflect the degrees of both how close the effect sample values are to the maximum sample values and how far away they are from the threshold effect values of the objectives. Similarly, the effect measures of cost-type objectives represent how close the effect sample values are to the minimum effect sample values and how far away the effect sample values are from the threshold effect values of the objectives; the lower effect measures of moderate-value type objectives indicate how far away the effect sample values that are smaller than the moderate value A are from the lower threshold effect value, and the upper effect measures indicate how far away the effect sample values that are greater than the moderate value A are from the upper threshold effect values of the objectives.

For situations of missing targets, there are the following four different possibilities:

(1) The effect value of a benefit-type objective is smaller than the threshold value $u_{i_0 j_0}^{(k)}$, that is, $u_{ij}^{(k)} < u_{i_0 j_0}^{(k)}$;

(2) The effect value of a cost-type objective is greater than the threshold value $u_{i_0 j_0}^{(k)}$, that is, $u_{ij}^{(k)} > u_{i_0 j_0}^{(k)}$;

(3) The effect value of a moderate-value type objective is smaller than the lower threshold effect value $A - u_{i_0 j_0}^{(k)}$, that is, $u_{ij}^{(k)} < A - u_{i_0 j_0}^{(k)}$; and

(4) The effect value of a moderate-value type objective is greater than the upper threshold effect value $A + u_{i_0 j_0}^{(k)}$, that is, $u_{ij}^{(k)} > A + u_{i_0 j_0}^{(k)}$.

In order for the effect measures of each type of objective to satisfy the condition of normality, that is, $r_{ij}^{(k)} \in [-1, 1]$, without loss of generality, we can assume that:

For a benefit-type objective, $u_{ij}^{(k)} \geq -\max_i \max_j \left\{ u_{ij}^{(k)} \right\} + 2u_{i_0 j_0}^{(k)}$;

For a benefit-type objective, $u_{ij}^{(k)} \leq -\min_i \min_j \left\{ u_{ij}^{(k)} \right\} + 2u_{i_0 j_0}^{(k)}$;

For cases where the effect value of a moderate-value type objective is smaller than the lower threshold effect value $A - u_{i_0 j_0}^{(k)}$, $u_{ij}^{(k)} \geq A - 2u_{i_0 j_0}^{(k)}$; and

For cases where the effect value of a moderate-value type objective is greater than the upper threshold effect value $A + u_{i_0 j_0}^{(k)}$, $u_{ij}^{(k)} \leq A + 2u_{i_0 j_0}^{(k)}$.

With these assumptions, we have the proposition below.

Proposition 10.4.1 The effect measures $r_{ij}^{(k)} (i = 1, 2, \ldots, n; \ j = 1, 2, \ldots, m; \ k = 1, 2, \ldots, s)$, as defined in Definition 10.4.1, satisfy the following properties:

(1) $r_{ij}^{(k)}$ is non-dimensional; (2) the more ideal the effect, the larger $r_{ij}^{(k)}$ is; and (3) $r_{ij}^{(k)} \in [-1, 1]$.

Definition 10.4.2 $r_{ij}^{(k)} (i = 1, 2, \ldots, n; j = 1, 2, \ldots, m; k = 1, 2, \ldots, s)$, as defined in Definition 10.4.1, is called uniform effect measure of decision scheme s_{ij}.

For decision scheme s_{ij} of hitting the target, $r_{ij}^{(k)} \in [0, 1]$; and for decision scheme s_{ij} of missing the target, $r_{ij}^{(k)} \in [-1, 0]$.

Definition 10.4.3 For a given set S, define $R^{(k)} = \left(r_{ij}^{(k)} \right)_{n \times m}$ as the matrix of uniform effect measure of S with respect to objective k. For $s_{ij} \in S$, $r_{ij} = (r_{ij}^{(1)}, r_{ij}^{(2)}, \cdots, r_{ij}^{(s)})$ is known as the vector of uniform effect measure of the decision scheme s_{ij}.

10.4.2 The Weighted Synthetic Effect Measure

Definition 10.4.4 Assume that η_k stands for the decision weight of objective k, $k = 1, 2, \ldots, s$, satisfying $\sum_{k=1}^{s} \eta_k = 1$, then $\sum_{k=1}^{s} \eta_k \cdot r_{ij}^{(k)}$ is called a weighted synthetic effect measure of the decision scheme s_{ij}, which is still denoted as $r_{ij} = \sum_{k=1}^{s} \eta_k \cdot r_{ij}^{(k)}$; and $R = (r_{ij})_{n \times m}$ is known as the matrix of weighted synthetic effect measures.

In the case of weighted synthetic effect measures, $r_{ij} \in [-1, 0]$ belongs to the decision scheme s_{ij} of missing the target, while $r_{ij} \in [0, 1]$ belongs to the decision scheme s_{ij} of hitting the target. For the decision scheme of hitting the target, we can further compare the superiority of events a_i, countermeasures b_j, and decision schemes s_{ij} respectively by using the magnitudes of the weighted synthetic effect measures, $i = 1, 2, \ldots, n$, $j = 1, 2, \ldots, m$.

Definition 10.4.5 (1) If $\max_{1 \leq j \leq m}\{r_{ij}\} = r_{ij_0}$, then b_{j_0} is known as the optimum countermeasure of event a_i; (2) If $\max_{1 \leq i \leq n}\{r_{ij}\} = r_{i_0 j}$, then a_{i_0} is known as the optimum event corresponding to countermeasure b_j; (3) If $\max_{1 \leq i \leq n, 1 \leq j \leq m}\{r_{ij}\} = r_{i_0 j_0}$, then $s_{i_0 j_0}$ is known as the optimum decision scheme.

The weighted multi-attribute grey target decision can be made by following the steps below:

Step 1: Based on the set $A = \{a_1, a_2, \ldots, a_n\}$ of events and the set $B = \{b_1, b_2, \ldots, b_m\}$ of countermeasures, construct the set of decision schemes $S = \{s_{ij} = (a_i, b_j) | a_i \in A, b_j \in B\}$;

Step 2: Determine the decision objectives $k = 1, 2, \ldots, s$;

Step 3: Determine the decision weights $\eta_1, \eta_2, \cdots, \eta_s$ of the objectives;

Step 4: For each objective $k = 1, 2, \cdots, s$, compute the corresponding observed effect matrix $U^{(k)} = (u_{ij}^{(k)})_{n \times m}$;

Step 5: Determine the threshold effect value of objective $k = 1, 2, \ldots, s$;

Step 6: Calculate the matrix $R^{(k)} = (r_{ij}^{(k)})_{n \times m}$ of uniform effect measures of objective $k = 1, 2, \ldots, s$;

Step 7: From $r_{ij} = \sum_{k=1}^{s} \eta_k \cdot r_{ij}^{(k)}$, compute the matrix of synthetic effect measures $R = (r_{ij})_{n \times m}$; and

Step 8: Determine the optimum decision scheme $s_{i_0 j_0}$.

The proposed model here has a unique feature of clear physical meaning presented as missing target, hitting target and hitting performance of different decision strategies with a pre-defined grey target. The distance of a strategy to the grey target over different objectives is calculated through effect measure functions as follows: the concept of upper effect measure reflects the distance of the observed effect value from the maximum observed effect value; the concept of lower effect measure indicates the distance between the observed effect value from the minimum observed effect value; and the concept of moderate effect measure tells the distance of the observed effect value from the pre-defined most preferred effect value in the middle.

To aggregate the performance of a strategy over different objectives, one can make use of the concept of upper effect measure for benefit objectives where the larger or the more the effect sample values are the better; for cost objectives where the smaller or the fewer the effect sample values are the better, one can utilize the concept of lower effect measure. As for non-monotonic objectives that require "neither too large nor too small" and/or "neither too many nor too few," one can apply the concept of moderate effect measure. The effect measure for benefit and cost type objectives, the lower effect measure for moderate type, and the upper effect measure for moderate type can be further integrated as uniform effect measures by incorporating weight information over different objectives. The value of uniform effect measures is located in the interval of $[-1, 1]$ and has a crystal physical meaning: if a strategy hits the target, the value will be positive and the larger the closer to the ideal point in the grey target; if a strategy misses the target, the value will be negative. The new model has been applied to the selection of the supplier of a key component used in the production of large commercial aircrafts and this application confirmed its feasibility.

Example 10.4.1 Let us look at the selection of the supplier of a key component used in the production of large commercial aircrafts (Liu et al., 2010; Liu et al., 2013; Liu, 2021).

In China, the production of large commercial aircrafts is managed using the model of main manufacturers—suppliers, where a great amount of key components comes from international suppliers. So, the scientific approach to decision-making regarding the selection of relevant suppliers is a key determinant of the success or failure of the operation. As a typical decision-making problem involved in the production process of sophisticated products, the selection of suppliers is generally accomplished through public bidding. Usually the main manufacturer first lists his demands, then each potential supplier puts together their proposal to outline how they meet the needs of the manufacturer. After collecting the proposals, the manufacturer comprehensively evaluates all the suppliers' submissions to select the optimum proposal and sign the purchase agreement. As for what factors actually affect the manufacturer's decision, it is an extremely complicated matter. In order to arrive at educated and scientifically sound decisions, there is a need to analyze all the involved factors closely and holistically.

During the selection of international suppliers for a specific key component of the production of large commercial aircrafts, there were there suppliers accepted into the second round of the tender. To decide on the eventual supplier, let us go through the following steps.

Table 10.2   The objectives' evaluation system

Objective	Quality	Price	Delivery	Design	Competitiveness
Unit	Qualitative	Million US$	Month	Qualitative	Qualitative
Order #	1	2	3	4	5
Weight	0.25	0.22	0.18	0.18	0.17

Step 1: Establish the sets of events, countermeasures, and situations. Let us define event a_1 as the selection of a supplier for the said component for the production of large commercial aircrafts. So, the set of events is $A = \{a_1\}$. Define the selection of supplier 1, supplier 2, or supplier 3 to be our countermeasures b_1, b_2, and b_3, respectively, so that the set of countermeasures is $B = \{b_1, b_2, b_3\}$. Therefore, our set of situations in this case is $S = \{s_{ij} = (a_i, b_j) | a_i \in A, b_j \in B, i = 1; j = 1, 2, 3\} = \{s_{11}, s_{12}, s_{13}\}$.

Step 2: Determine the decision objectives. Through three rounds of surveys with relevant experts, the following 5 objectives are considered: quality, price, time of delivery, design proposal, and competitiveness.

Among these objectives, competitiveness, quality, and design proposal are qualitative. They are scored by relevant experts' evaluations, and the higher the evaluation scores the better. That is, they are benefit-type objectives. Let us take the threshold value $u_{i_0 j_0}^{(k)} = 9, k = 1, 4, 5$. For the objective of cost, the lower the cost the better. So, it is a cost-type objective. Let us take the threshold value $u_{i_0 j_0}^{(2)} = 15$. The objective of time of delivery is one of moderate-value type. The main manufacturer desires the delivery at the end of the 16th month with 2 months' deviation allowed. That is, $u_{i_0 j_0}^{(3)} = 2$, the lower threshold effect value is $16 - 2 = 14$, and the upper threshold effect value is $16 + 2 = 18$.

Step 3: Determine the decision weights of the objectives. To this end, we apply the Analytic Hierarchy Process (AHP) method (see Table 10.2 for details).

Step 4: Determine the effect sample vectors of each of the objectives:

$$U^{(1)} = (9.5, 9.4, 9), U^{(2)} = (14.2, 15.1, 13.9), U^{(3)} = (15.5, 17.5, 19),$$
$$U^{(4)} = (9.6, 9.3, 9.4), U^{(5)} = (9.5, 9.7, 9.2).$$

Step 5: Assign the threshold effect values for the objectives. Because competitiveness, quality, and design proposal are all benefit-type objectives, let us take the threshold values $u_{i_0 j_0}^{(k)} = 9, k = 1, 4, 5$. Because price is a cost-type objective, let us take the threshold value $u_{i_0 j_0}^{(2)} = 15$. Because time of delivery is a moderate value-type objective and the main manufacturer desires the delivery at the end of the 16th month with a tolerance of ± 2 months, we set $u_{i_0 j_0}^{(3)} = 2$, the lower threshold effect value $16 - 2 = 14$, and the upper threshold effect value $16 + 2 = 18$.

Step 6: Calculate the vectors of uniform effect measures. For the three qualitative objectives, competitiveness, quality, and design proposal, we employ the effect measures of benefit-type. For the objective of price, we utilize the effect measures of cost-type. For the objective of time of delivery, we apply the lower and upper effect measures. Thus, we obtain the following vectors of uniform effect measures:

$$R^{(1)} = [1, 0.8, 0], R^{(2)} = [0.73, -0.09, 1], R^{(3)} = [0.75, 0.25, -0.5],$$
$$R^{(4)} = [1, 0.5, 0.67], \text{ and } R^{(5)} = [0.71, 1, 0.29].$$

Step 7: From $r_{ij} = \sum_{k=1}^{5} \eta_k \cdot r_{ij}^{(k)}$, we compute the following vector of synthetic effect measures:

$$R = [r_{11}, r_{12}, r_{13}] = [0.8463, 0.4852, 0.2999].$$

Step 8: Make the final decision. Because $r_{11} > 0, r_{12} > 0, r_{13} > 0$, it means that all these three suppliers have hit the target. This result implies that it is reasonable for these suppliers to enter the second round of the tender. However, based on $\max_{1 \leq j \leq 3} \{r_{1j}\} = r_{11} = 0.08463$, it follows that the main manufacturer should sign the agreement with supplier 1.

10.5 On Paradox of Rule of Maximum Value and Its Solution

The thought and methodology of statistical decision emerged in Britain in the late eighteenth century (Bayes, 1763). L. J. Savage built the system of Bayesian Decision Theory in the book titled *The foundations of statistics* in 1954. According to Bayesian Decision Theory (Savage, 1954), the reasonable action subject abide by the principles of maximum subjective expected utility in their decision-making. In 1969, R. Nozick published his article on the Newcomb Paradox (Nozick, 1969), which led to a major divergence of views in Bayesian Decision Theory and gave rise to Causal Decision Theory (Lewis, 1973) and Evidential Decision Theory, also known as D-S Theory (Dempster, 1968; Shafer, 1976). A. Gibbard and W. Harper tried to solve the Newcomb Paradox by defining two different expected utilities named as U-utility and V-utility in 1978 (Gibbard and Harper, 1978). Then, in 1981 E. Eells thought that people could solve the Newcomb Paradox by revising the principles of maximum utility (Eells, 1981). J. M. Joyce put forward a general theory of conditional beliefs to illuminate the nature of causal beliefs and their role in rational choice (Joyce, 1999). It was concluded that there is less difference than is usually thought between causal decision theory and evidential decision theory (Joyce, 1999). S. Burgess thought that the resolution of the Newcomb problem was unqualified (Burgess, 2004), and in 2010 D. H. Wolpert and G. Benford proved that two Bayes nets are incompatible based on game theory (Wolpert & Benford, 2010).

In the course of decision-making, people need divide their decision objects into different classes or clusters, then compare and sort the objects in the same class or cluster to help them choose the right object or objects. All types of cluster evaluation models, such as statistical clustering analysis (Tryon, 1939), fuzzy clustering (Bezdek, 1981), and grey clustering evaluation models such as grey variable weight clustering model (Deng, 1986), grey fixed weight clustering evaluation model (Liu, 1993), grey cluster evaluation model using end-point triangular possibility functions (Liu & Zhu, 1993; Liu, 2006), grey cluster evaluation model using center-point triangular possibility functions (Liu et al., 2011, 2012), and grey cluster evaluation model using mixed triangular possibility functions (Liu et al., 2014; Liu, Yang, & Fang, 2015; Liu, Zhang, & Jian, 2015) use the rule of maximum value of component of cluster coefficient vector $\sigma_i = (\sigma_i^1, \sigma_i^2, \ldots, \sigma_i^s)$ as a basis for determining ascription of decision objects.

In cases where more than one object belongs to a class or cluster, people may be confronted with a decision paradox. For example, assume that $\delta_1 = (0.4, 0.35, 0.25)$ and $\delta_2 = (0.41, 0.2, 0.39)$ are the clustering coefficient vectors of objects 1 and 2, respectively. It is demonstrably the case that objects 1 and 2 both belong to class 1, according to the principles of maximum value of clustering coefficient. Also, object 2 is better than object 1 given that $0.41 > 0.4$. However, if we were to consider the values of all the components of δ_1, δ_2 in an integrated manner, object 1 could be perceived as being superior to object 2. This is a paradox.

In this section, we try and find a solution for the decision paradox by using weight vector group with kernel, weighted comprehensive clustering coefficient vector and a two-stages decision model (Liu et al., 2018; Liu et al., 2022).

10.5.1 The Weight Vector Group with Kernel

Clustering coefficient vectors cannot be compared with each other because usually they are not unit vectors. Therefore, firstly all clustering coefficient vectors need to be unitized.

Definition 10.5.1 Assume that $\sigma_i = (\sigma_i^1, \sigma_i^2, \cdots, \sigma_i^s), i = 1, 2, \cdots, n$ are n clustering coefficient vectors, $\delta_i^k = \frac{\sigma_i^k}{\sum_{k=1}^s \sigma_i^k}$, δ_i^k is called unitized clustering coefficient of decision-making object i belonging to class k. Clearly, $\delta_i^k (k = 1, 2, \cdots s)$ satisfy $\sum_{i=1}^s \delta_i^k = 1$.

Definition 10.5.2 $\delta_i = (\delta_i^1, \delta_i^2, \ldots, \delta_i^s); (i = 1, 2, \ldots, n)$ is called unitized clustering coefficient vectors of decision-making object i. The following conclusion about unitized clustering coefficient vector δ_i is also suitable for non-unitized clustering coefficient vector σ_i. Therefore, the "unitized" can be omitted.

Sort the components of δ_i according to their values, that is, $\delta_i^{k_1} \geq \delta_i^{k_2} \geq \cdots \geq \delta_i^{k_l} \geq \cdots \geq \delta_i^{k_s}$.

Definition 10.5.3 Assume that $\max_{1 \leq k \leq s} \{\delta_i^k\} = \delta_i^{k^*}$, then $\delta_i^{k^*}$ is called the maximum component of clustering coefficient vector δ_i.

Given that all the corresponding coefficients of two decision coefficient vectors δ_i, δ_j are equal, then there is no difference between δ_i, δ_j. When two objects i, j belong to a class k^* and the maximum component $\delta_i^{k^*} > \delta_j^{k^*}$, it means that δ_i is better than δ_j by the rule of maximum value; but it is possible to think that δ_j is better than δ_i if we consider the values of all the components of δ_1, δ_2 in an integrated manner. This is a decision paradox of rule of maximum value.

To solve the decision paradox of rule of maximum value, firstly the weight vector group of kernel clustering is defined. The basic step to solve the paradox is to cluster the information which is included in other components around δ_i^k, and supporting objects i come under class k into component k. Then it is necessary to obtain a new decision coefficient vector which contains factors included in other components around δ_i^k.

Definition 10.5.4 Assume that there are s classes of decision-making, and real numbers $w_k \geq 0$, $k = 1, 2, \ldots, s$, let

$$\eta_1 = \frac{1}{\sum_{k=1}^{s} w_k}(w_s, w_{s-1}, w_{s-2}, \ldots, w_1),$$

$$\eta_2 = \frac{1}{w_{s-1} + \sum_{k=2}^{s} w_k}(w_{s-1}, w_s, w_{s-1}, w_{s-2}, \ldots, w_2),$$

$$\eta_3 = \frac{1}{w_{s-1} + w_{s-2} + \sum_{k=3}^{s} w_k}(w_{s-2}, w_{s-1}, w_s, w_{s-1}, \ldots, w_3),$$

$$\cdots,$$

$$\eta_k = \frac{1}{\sum_{i=s-k+1}^{s-1} w_i + \sum_{i=k}^{s} w_i}$$
$$(w_{s-k+1}, w_{s-k+2}, \ldots, w_{s-1}, w_s, w_{s-1}, \ldots, w_k),$$

$$\cdots,$$

$$\eta_{s-1} = \frac{1}{w_{s-1} + \sum_{k=2}^{s} w_k}(w_2, w_3, \ldots, w_{s-1}, w_s, w_{s-1}),$$

$$\eta_s = \frac{1}{\sum_{k=1}^{s} w_k}(w_1, w_2, w_3, \ldots, w_{s-1}, w_s),$$

then $\eta_k (k = 1, 2, \ldots, s)$ is called a weight vector group with kernel.

Note: s-dimensional vector $\eta_k = (\eta_k^1, \eta_k^2, \ldots, \eta_k^s)(k = 1, 2, \ldots, s)$ is the multiplication of scalar $a_k = \frac{1}{\sum_{i=s-k+1}^{s-1} w_i + \sum_{i=k}^{s} w_i}$ with vector ζ_k, where the function of scalar factor a_k is to ensure $\eta_k (k = 1, 2, \ldots, s)$ is a normalized vector. Also, the k-th component of vector factor $\zeta_k (k = 1, 2, \cdots, s)$ is w_s, which is the maximum component of ζ_k. Then the k-th component w_s can be taken as a center, and the other components on both sides of the k-th component w_s descend step by step. The k-th component with the largest contribution for the decision-making object belongs to grey class k, so the k-th component of ζ_k should take the maximum weight w_s. The values of other components are set by the principle which states that "the component which is closest to the k-th component has the largest contribution for object I belonging to class k, so it is given the largest weight; the component which is farthest from the k-th component has the smallest contribution for object I belonging to class k, so it is given the smallest weight".

10.5.2   The Weighted Comprehensive Clustering Coefficient Vector

Definition 10.5.5 Assume there are n decision objects and s different grey classes, then $\omega_i^k = \eta_k \cdot \delta_i^T$ is called the weighted coefficient of kernel clustering for decision-making of object $_i$ about grey class k. And

$$\omega_i = (\omega_i^1, \omega_i^2, \ldots, \omega_i^s); \quad i = 1, 2, \ldots, n$$

is called the weighted comprehensive clustering coefficient vector of object i.

Definition 10.5.6 Let $\max_{1 \le k \le s}\{\omega_{i_1}^k\} = \omega_{i_1}^{k^*}$, $\max_{1 \le k \le s}\{\omega_{i_2}^k\} = \omega_{i_2}^{k^*}$, when $\omega_{i_1} > \omega_{i_2}$, then decision object i_1 is better than decision object i_2 in grey class k^*.

Definition   10.5.7 Let $\max_{1 \le k \le s}\{\omega_{i_1}^k\} = \omega_{i_1}^{k^*}$, $\max_{1 \le k \le s}\{\omega_{i_2}^k\} = \omega_{i_2}^{k^*}$, $\cdots$, $\max_{1 \le k \le s}\{\omega_{i_l}^k\} = \omega_{i_l}^{k^*}$, in other words, objects $i_1, i_2, \ldots, i_l$ all belong to grey class k^*. Also, $\omega_{i_1} > \omega_{i_2} > \cdots > \omega_{i_l}$, and if the number of objects contained in the decision grey class k^* is l_1, then objects $i_1, i_2, \ldots, i_{l_1}$ are called the taken object of grey class k^*, and the rest of the objects are called the candidates of grey class k^*.

The two stages decision model to solve the decision paradox by the weight vector group with kernel and the weighted comprehensive clustering coefficient vector can be constructed step by step as outlined below.

Stage 1

Step 1: Compute normalized clustering coefficient vector δ_i

$$\delta_i = (\delta_i^1, \delta_i^2, \ldots, \delta_i^s); \quad (i = 1, 2, \ldots, n)$$

Step 2: Estimate the distinguishability of the clustering coefficient vectors of objects belonging to class k^*. If the order of priority of the objects i belonging to class k^* is easy to identify, turn to step 6; in cases where the order of priority of the objects belonging to class k^* is difficult to identify, turn to step 3;

Stage 2

Step 3: Set the weight vector group with kernel $(\eta_1, \eta_2, \ldots, \eta_s)$;

Step 4: Calculate the weighted comprehensive clustering coefficient vector of decision object i

$$\omega_i = (\omega_i^1, \omega_i^2, \ldots, \omega_i^s); \quad i = 1, 2, \ldots, n;$$

Step 5: Determine object i belonging to grey class k^* by $\max_{1 \leq k \leq s} \{\omega_i^k\} = \omega_i^{k^*}$;
Step 6: Sort the decision objects which belong to class k^* according to the values of $\delta_{i_1}^{k^*}, \delta_{i_2}^{k^*}, \ldots, \delta_{i_l}^{k^*}$ for case where there are l objects belonging to class k^*.

10.5.3 Several Functional Weight Vector Groups with Kernel

Proposition 10.5.1 Assume that

$$\eta_1 = \frac{2}{s(s+1)}(s, s-1, s-2, \ldots, 1)$$

$$\eta_2 = \left(\frac{1}{\frac{s(s+1)}{2} + (s-2)}\right)(s-1, s, s-1, s-2, \ldots, 2)$$

$$\eta_3 = \left(\frac{1}{\frac{s(s+1)}{2} + (2s-6)}\right)(s-2, s-1, s, s-1, \ldots, 3)$$

$$\cdots,$$

$$\eta_k = \left\{\frac{1}{\frac{s(s+1)}{2} + \left[(k-1)s - \frac{k(k-1)}{2}\right]}\right\}$$

$$(s-k+1, s-k+2, \ldots, s-1, s, s-1, \ldots, k), \ldots,$$

$$\eta_{s-1} = \frac{2}{\frac{s(s+1)}{2} + (s-2)}(2, 3, \ldots, s-1, s, s-1)$$

$$\eta_s = \frac{2}{s(s+1)}(1, 2, 3, \ldots, s-1, s)$$

Then $\eta_k(k = 1, 2, \cdots, s)$ is a weight vector group with kernel.

Proposition 10.5.2 Assume that

$$\eta_1 = \frac{1}{\sum_{k=1}^{s}\frac{1}{2^k}}\left(\frac{1}{2}, \frac{1}{2^2}, \frac{1}{2^3}, \cdots, \frac{1}{2^{s-1}}, \frac{1}{2^s}\right)$$

$$\eta_2 = \left(\frac{1}{\frac{1}{2^2} + \sum_{k=1}^{s-1}\frac{1}{2^k}}\right)\left(\frac{1}{2^2}, \frac{1}{2}, \frac{1}{2^2}, \frac{1}{2^3}, \cdots, \frac{1}{2^{s-1}}\right)$$

$$\eta_3 = \left(\frac{1}{\frac{1}{2^3} + \frac{1}{2^2} + \sum_{k=1}^{s-2}\frac{1}{2^k}}\right)\left(\frac{1}{2^3}, \frac{1}{2^2}, \frac{1}{2}, \frac{1}{2^2}, \frac{1}{2^3}, \cdots, \frac{1}{2^{s-2}}\right)$$

$$\cdots,$$

$$\eta_k = \left\{ \frac{1}{\sum_{i=2}^{k} \frac{1}{2^i} + \sum_{i=1}^{s-k+1} \frac{1}{2^i}} \right\} \left(\frac{1}{2^k}, \frac{1}{2^{k-1}}, \cdots, \frac{1}{2^2}, \frac{1}{2}, \frac{1}{2^2}, \cdots, \frac{1}{2^{s-k+1}} \right)$$

$$\cdots,$$

$$\eta_{s-1} = \frac{1}{\frac{1}{2^2} + \sum_{k=1}^{s-1} \frac{1}{2^k}} \left(\frac{1}{2^{s-1}}, \frac{1}{2^{s-2}}, \cdots, \frac{1}{2^2}, \frac{1}{2}, \frac{1}{2^2} \right)$$

$$\eta_s = \frac{1}{\sum_{k=1}^{s} \frac{1}{2^k}} \left(\frac{1}{2^s}, \frac{1}{2^{s-1}}, \cdots, \frac{1}{2^3}, \frac{1}{2^2}, \frac{1}{2} \right)$$

Then $\eta_k (k = 1, 2, \ldots, s)$ is a weight vector group with kernel.

Proposition 10.5.3 For case s $= 10$, assume that

$$\eta_1 = \frac{1}{5.5}(1, 0.9, 0.8, 0.7, \ldots, 0.1)$$

$$\eta_2 = \frac{1}{6.3}(0.9, 1, 0.9, 0.8, \ldots, 0.2)$$

$$\eta_3 = \frac{1}{6.9}(0.8, 0.9, 1, 0.9, \ldots, 0.3)$$

$$\cdots,$$

$$\eta_k = \frac{1}{1 + \sum_{i=1}^{k} 0.(10 - i) + \sum_{i=k}^{9} 0.i}(0.(10 - k), 0.8, 0.9, 1, 0.9, \ldots, 0.k)$$

$$\cdots,$$

$$\eta_9 = \frac{1}{6.3}(0.2, \ldots, 0.8, 0.9, 1, 0.9)$$

$$\eta_{10} = \frac{1}{5.5}(0.1, \ldots, 0.7, 0.8, 0.9, 1)$$

Then $\eta_k (k = 1, 2, \ldots, s)$ is a weight vector group with kernel.

10.6 Practical Applications

Example 10.6.1 Strategic supplier selection for the C919 cooperative development. C919 is the first large commercial aircraft developed by Commercial Aircraft Corporation of China Ltd (COMAC). Many domestic and overseas suppliers joined the development program. Suppliers A and B took part in the development task of the C919 program for a specific key component. One of the two suppliers, either A or B, should be chosen and confirmed as a strategic supplier according to COMAC's criteria. A dilemma for strategic supplier selection will be presented to demonstrate the feasibility of the two-stage decision model based on the weight vector group with kernel and the weighted comprehensive clustering coefficient vector to solve the selection dilemma (Liu et al., 2018; Liu et al., 2022; Liu et al., 2017; Liu, 2021).

The consulting group collected all data according the evaluation index system, which was determined in advance. Then the clustering coefficient vectors of A and B are defined as follows:

$$\delta_A = (\delta_A^1, \delta_A^2, \delta_A^3, \delta_A^4, \delta_A^5) = (0.246, 0.338, 0.292, 0.124, 0)$$
$$\delta_B = (\delta_B^1, \delta_B^2, \delta_B^3, \delta_B^4, \delta_B^5) = (0.089, 0.352, 0.312, 0.197, 0)$$

Here, classes 1, 2, 3, 4, 5 correspond to 'especially excellent', 'excellent', 'good', 'moderate', and 'poor', respectively.

From $\max_{1 \leq k \leq 5}\{\delta_A^k\} = 0.338 = \delta_A^2$, $\max_{1 \leq k \leq 5}\{\delta_B^k\} = 0.352 = \delta_B^2$, it is known that the two suppliers A and B both belong to class 'excellent'. It seems that B should be selected and confirmed as the strategic supplier if we compare the clustering coefficients δ_A^2 of A belonging to class excellent with δ_B^2 of B belonging to class excellent, because $\delta_A^2 = 0.338 < \delta_B^2 = 0.352$. But we found that the clustering coefficients $\delta_A^1 = 0.246$ of A belonging to class 'especially excellent' is greater than the clustering coefficients $\delta_B^1 = 0.089$ of B belonging to class 'especially excellent' if we compare δ_A and δ_B in an integrated way. Therefore, the values of each component of the clustering coefficient vectors δ_A and δ_B should be integrated by a weight vector group with kernel.

The weight vector group with kernel presented in proposition 2 is used to integrate the values of each component of the clustering coefficient vectors δ_A and δ_B. Notice that $s = 5$. We obtain:

$$\eta_1 = \frac{32}{31}\left(\frac{1}{2}, \frac{1}{2^2}, \frac{1}{2^3}, \frac{1}{2^4}, \frac{1}{2^5}\right), \eta_2 = \frac{16}{19}\left(\frac{1}{2^2}, \frac{1}{2}, \frac{1}{2^2}, \frac{1}{2^3}, \frac{1}{2^4}\right),$$
$$\eta_3 = \frac{4}{5}\left(\frac{1}{2^3}, \frac{1}{2^2}, \frac{1}{2}, \frac{1}{2^2}, \frac{1}{2^3}\right), \eta_4 = \frac{16}{19}\left(\frac{1}{2^4}, \frac{1}{2^3}, \frac{1}{2^2}, \frac{1}{2}, \frac{1}{2^2}\right),$$
$$\eta_5 = \frac{32}{31}\left(\frac{1}{2^5}, \frac{1}{2^4}, \frac{1}{2^3}, \frac{1}{2^2}, \frac{1}{2}\right)$$

Then, from $\omega_j^k = \eta_k \cdot \delta_j^T$, $j = A, B$, we have

$$\omega_A^1 = \eta_1 \cdot \delta_A^T = \frac{32}{31}\left(\frac{1}{2}, \frac{1}{2^2}, \frac{1}{2^3}, \frac{1}{2^4}, \frac{1}{2^5}\right)(0.246, 0.338, 0.292, 0.124, 0)^T = 0.26$$

$$\omega_A^2 = \eta_2 \cdot \delta_A^T = \frac{16}{19}\left(\frac{1}{2^2}, \frac{1}{2}, \frac{1}{2^2}, \frac{1}{2^3}, \frac{1}{2^4}\right)((0.246, 0.338, 0.292, 0.124, 0)^T = 0.27$$

$$\omega_A^3 = \eta_3 \cdot \delta_A^T = 0.23, \omega_A^4 = \eta_4 \cdot \delta_A^T = 0.16, \omega_A^5 = \eta_5 \cdot \delta_A^T = 0.10$$

$$\omega_A = (\omega_A^1, \omega_A^2, \omega_A^3, \omega_A^4, \omega_A^5) = (0.26, 0.27, 0.23, 0.16, 0.10)$$

$$\omega_B^1 = \eta_1 \cdot \delta_B^T = 0.19, \omega_B^2 = \eta_2 \cdot \delta_B^T = 0.25, \omega_B^3 = \eta_3 \cdot \delta_B^T = 0.24,$$

$$\omega_B^4 = \eta_4 \cdot \delta_B^T = 0.19, \omega_B^5 = \eta_5 \cdot \delta_B^T = 0.12$$
$$\omega_B = (\omega_B^1, \omega_B^2, \omega_B^3, \omega_B^4, \omega_B^5) = (0.19, 0.25, 0.24, 0.19, 0.12)$$

When comparing the weighted comprehensive clustering coefficient vectors of ω_A and ω_B, we found that $\omega_A^1 = 0.26 > \omega_B^1 = 0.19$, $\omega_A^2 = 0.27 > \omega_B^2 = 0.25$; at the same time, $\omega_A^4 = 0.16 < \omega_B^4 = 0.19$, $\omega_A^5 = 0.10 < \omega_B^5 = 0.12$.

So, it can be judged that the supplier A is better than vendor B. Supplier A should be selected and confirmed as the strategic supplier. The outcome can provide a basis for COMAC's strategic supplier selection.

We can obtain the same conclusion if the weight vector group with kernel presented in proposition 1 or proposition 3 is used to integrate the values of each component of the clustering coefficient vectors δ_A and δ_B.

It is directed against the decision paradox that the conclusion we arrive at by comparing the maximum components δ_i^k and δ_j^k of δ_i and δ_j is in conflict with the conclusion we arrive at by comparing δ_i and δ_j, in an integrated way. The decision paradox that the value of the maximum component δ_i^k of δ_i is close to the maximum component δ_j^k of δ_j is solved effectively.

References

Bayes, T. (1763). An essay towards solving a problem in the doctrine of chances. *Philosophical Transactions, 53*, 370–418.

Bezdek. J. C. (1981). *Pattern recognition with fuzzy objective function algorithms*. Kluwer Academic Publishers, Doordrecht.

Burgess, S. (2004).The newcomb problem: Anunquali—fledresolution. *Synthesis, 138*, 261—287.

Dempster, A. P. (1968). A generalization of Bayesian inference. *Journal of the Royal Statistical Society, Series B, 30*, 205–247.

Deng, J. L. (1986). *Grey prediction and decision-making*. Press of Huazhong University of Science and Technology.

Eells, E. (1981). Causality, utility, and decision. *Synthese, 48*, 295–329.

Joyce, J. M. (1999).*The foundations of causal decision theory*. Cambridge: Cambridge University Press.

Nozick, R. (1969). Newcomb's problem and two principles of choice. In N. Rescher (Ed.), *Essays in honor of Carl G. Hempel* (pp. 114–115). Synthese Library. Dordrecht, The Netherlands: D. Reidel.

Shafer, G. (1976). *A mathematical theory of evidence*. Princeton: Princeton University Press.

Savage, L. J. (1954). *The foundations of statistics*. New York: Wiley.

Lewis, D. (1973). *Counterfactuals*. Cambridge, MA: Harvard University Press.

Liu, S. F. (1993). Fixed weight grey clustering evaluation analysis. *New methods of grey systems* (pp. 178–184). Agriculture Press.

Liu, S. (2006). On index system and mathematical model for evaluation of scientific and technical strength. *Kybernetes, 35*(7–8), 1256–1264.

Liu, S. F. (2021). *Grey system theory and its application* (9th ed.). Beijing: Science Press.

Liu, S. & Zhu Y. (1993). Study on triangular model and indexes in synthetic evaluation of regional economy. *Transactions of the Chinese Society of Agricultural Engineering, 2*, 8–131.

Liu, S. F., & Lin, Y. (2006). *Grey information: theory and practical applications*. Springer.

Liu, S. F., Yuan, W. F. & Sheng, K. Q. (2010). Multi-attribute intelligent grey target decision model. *Control and Decision, 25*(8), 1159–1163.

Liu, S. F., Xie, N. M., & Forrest, J. (2011). Novel models of grey relational analysis based on visual angle of similarity and nearness. In *Grey Systems: Theory and Application* (Vol. 1, No. 1, pp. 8–18).

Liu, S. F., Fang, Z. G., & Yang, Y. J., et al. (2012). General grey numbers and its operations. In *Grey Systems: Theory and Application* (Vol. 2, No. 3, pp. 341–349).

Liu, S. F., Xu, B., & Forrest, J., et al. (2013). On uniform effect measure functions and a weighted multi-attribute grey target decision model. *The Journal of Grey System, 25*(1), 1–11.

Liu, S. F., Yang, Y. J., Wu, L. F., et al. (2014). *Grey System Theory and Its Application* (7th ed.). Science Press.

Liu, S. F., Yang Y. J., Fang, Z. G., et al. (2015). Grey cluster evaluation models based on mixed triangular whitenization weight functions. In *Grey Systems: Theory and Application* (Vol. 5, No. 3, pp. 410–418).

Liu, S. F., Zhang, S. G., Jian, L. R., et al. (2015). Performance evaluation of large commercial aircraft vendors. *Journal of Grey System, 27*(1), 1–11.

Liu, S. F., Yang Y. J., & Forrest, J. (2017). *Grey data analysis*. Singapore: Springer-Verlag.

Liu, S. F., Zhang, H. Y., & Yang Y. J. (2018). On paradox of rule of maximum value and its solution. *Systems Engineering: Theory and Practice, 38*(7), 1830–1835.

Liu, S. F., Liu, T., Yuan, W. F., Yang, Y. J. (2022). Solving the dilemma in supplier selection by the group of weight vector with kernel. *Grey Systems: Theory and Application, 12*(3), 624–634.

Wolpert, D. H., Benford, G. (2010). What does Newcomb's paradox teach us? Report of Cornell University. arXiv:1003.1343 [cs.GT]

Chapter 11
Grey Control Systems

11.1 Introduction

As a scientific concept, the so-called control stands for a special effect a controlling device exerts on controlled equipment. It is a purposeful, selective and dynamic activity. A control system contains at least three parts, including a controlling device, controlled equipment, and an information path. A control system made up of these three parts is known as an open loop control system, as shown in Fig. 11.1. Each open loop control system is quite elementary in that the input directly controls the output, with no resistance against disturbances.

A control system with a feedback return is known as a closed loop control system, as shown in Fig. 11.2. The closed loop control system materializes its control through the combined effect of the input and the feedback of the output. One of the outstanding characteristics of closed loop systems is their strong ability to assist disturbances, with their outputs constantly vibrating around pre-determined objectives. Therefore, closed loop control systems possess a degree of stability.

A grey control system stands for such a system whose control information is only partially known, and is known as a grey system for short. The control of grey systems is different to that of general white systems, mainly due to the existence of grey elements in such systems. Under such conditions, one first needs to understand the possible connection between the systems' behaviors and the parametric matrices of the grey elements, how the systems' dynamics differ from one moment to the next and, in particular, how to obtain a white control function to alter the characteristics of the systems and to materialize control of the process of change of the systems. Grey control contains not only the general situation of systems involving grey parameters, but also the construction of controls based on grey systems analysis, modeling, prediction, and decision-making. Grey control thinking can reveal the essence of the problems at hand and help materialize the purpose of control.

© The Author(s), under exclusive license to Springer Nature Singapore Pte Ltd. 2022   277
S. Liu, *Grey Systems Analysis*, Series on Grey System,
https://doi.org/10.1007/978-981-19-6160-1_11

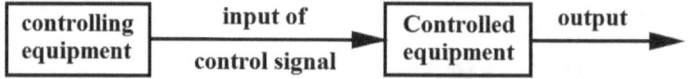

Fig. 11.1 Open loop control system

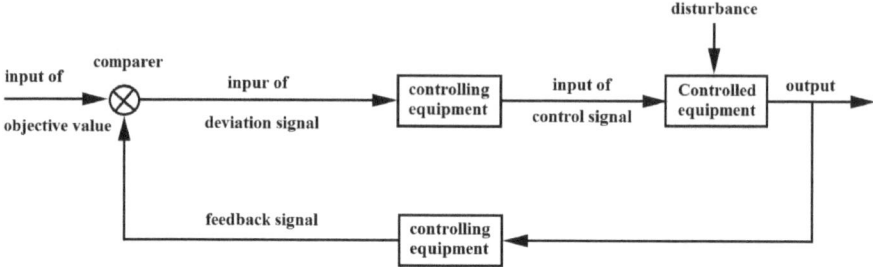

Fig. 11.2 Closed loop control system

11.2 Controllability and Observability of Grey System

The concepts of controllability and observability are two fundamental structural characteristics of systems seen from the angle of control and observation. This section focuses on the problems of controllability and observability of grey linear systems.

Definition 11.2.1 Assume that $U = [u_1, u_2, \ldots, u_s]^T$ is a control vector, $X = [x_1, x_2, \ldots, x_n]^T$ a state vector, and $Y = [y_1, y_2, \ldots, y_m]^T$ the output vector. Then

$$\begin{cases} \dot{X} = A(\otimes)X + B(\otimes)U \\ Y = C(\otimes)X \end{cases} \tag{11.1}$$

is known as the mathematical model of a grey linear control system, where $A(\otimes) \in G^{n \times n}$, $B(\otimes) \in G^{n \times s}$, $C(\otimes) \in G^{m \times n}$. Correspondingly, $A(\otimes)$ is known as the grey state matrix, $B(\otimes)$ the grey control matrix, and $C(\otimes)$ the grey output matrix.

In some studies, to emphasize the fact that U, X, and Y change the dynamic characteristics of the system over time, we also respectively write the control vector, state vector, and the output vector as $U(t)$, $X(t)$, and $Y(t)$.

The first group of equations

$$\dot{X}(t) = A(\otimes)X(t) + B(\otimes)U(t) \tag{11.2}$$

in the mathematical model of grey linear control systems in Eq. (11.1) is known as the state equation, while the second group of equations

$$Y(t) = C(\otimes)X(t) \tag{11.3}$$

is known as the output equation.

Definition 11.2.2 For a given precision and an objective vector $J = [j_1, j_2, \ldots, j_m]^T$, with a controlling device and a control vector $U(t)$ such that the output of the system can reach objective J while satisfying the required precision through controlling the input, then the system is said to be controllable.

Definition 11.2.3 For a given time moment t_0 and a pre-determined precision, if there is $t_1 \in (t_0, \infty)$ such that based on the system's output $Y(t)$, $t \in [t_0, t_1]$, one can measure the system's state $X(t)$ within the required precision, then the system is said to be observable within the time interval $[t_0, t_1]$. If for any t_0, t_1, the system is observable within the interval $[t_0, t_1]$, then the system is said to be observable.

According to control theory, it follows that whether or not a grey system is controllable or observable is determined by whether or not the controllability matrix and the observability matrix, made up of $A(\otimes)$, $B(\otimes)$, are of full rank. That is, the following result holds true.

Theorem 11.2.1 For the system in Eq. (11.1), define

$$L(\otimes) = [B(\otimes) \ A(\otimes)B(\otimes) \ A^2(\otimes)B(\otimes) \ \cdots \ A^{n-1}(\otimes)B(\otimes)]^T$$

$$D(\otimes) = [C(\otimes) \ C(\otimes)A(\otimes) \ C(\otimes)A^2(\otimes) \ \cdots \ C(\otimes)A^{n-1}(\otimes)]^T$$

Then the following hold true (Su & Liu, 2008):

(1) When $rank(L(\otimes)) = n$, the system is controllable; and
(2) When $rank(D(\otimes)) = n$, the system is observable.

Based on this result, the following four theorems can be established.

Theorem 11.2.2 For the system in Eq. (11.1), if the grey control matrix $B(\otimes) \in G^{n \times n}$ satisfies $B(\otimes) = diag[\otimes_{11}, \otimes_{22}, \ldots, \otimes_{nn}]$, where each grey entry along the diagonal is non-zero, then the system is controllable.

Theorem 11.2.3 For the system in Eq. (11.1), if the grey output matrix $C(\otimes) \in G^{n \times n}$ satisfies $C(\otimes) = diag[\otimes_{11}, \otimes_{22}, \ldots, \otimes_{nn}]$, where each grey entry along the diagonal is non-zero, then the system is observable.

Theorem 11.2.4 For the system in Eq. (11.1), if the control matrix $B(\otimes) \in G^{n \times n}$ satisfies $B(\otimes) = diag[\otimes_{11}, \otimes_{22}, \ldots, \otimes_{mm}, 0, \ldots, 0]$ with $rank B(\otimes) = m < n$, and the grey state matrix $A(\otimes)_{n \times n} = diag[0, \ldots, 0, \otimes_{m+1,1}, \otimes_{m+2,2}, \ldots, \otimes_{n,n-m}]$ with $rank A(\otimes) = n - m < n$, then the system is controllable.

Theorem 11.2.5 For the system in Eq. (11.1), if the grey output matrix $C(\otimes) \in G^{m \times n}$ satisfies $C(\otimes) = diag[\otimes_{11}, \otimes_{22}, \ldots, \otimes_{mm}]$ with $rank C(\otimes) = m < n$ and the grey state matrix.

$$A(\otimes) = \begin{pmatrix} 0\cdots 0 & \otimes_{1,m+1} & 0 & \cdots & 0 \\ 0\cdots 0 & 0 & \otimes_{2,m+2} & \cdots & 0 \\ \vdots \cdots \vdots & \vdots & \vdots & \cdots & \vdots \\ 0\cdots 0 & 0 & 0 & \cdots & \otimes_{n-m,n} \\ 0\cdots 0 & 0 & 0 & \cdots & 0 \\ \vdots \cdots \vdots & \vdots & \vdots & \cdots & \vdots \\ 0\cdots 0 & 0 & 0 & \cdots & 0 \end{pmatrix}, \; \mathrm{rank}\, A(\otimes) = n - m < \mathrm{n},$$

then the system is observable.

11.3   Transfer Functions of Grey System

The concept of transfer functions stands for a fundamental relationship between the input and output of time invariant, linear grey control systems. Its rich connection with the expressions of the systems' state spaces can be described by using the concepts of controllability and observability.

11.3.1   Grey Transfer Function

Definition 11.3.1 Assume that the mathematical model of an nth order linear system with grey parameters is given as follows:

$$\otimes_n \frac{d^n x}{dt^n} + \otimes_{n-1} \frac{d^{n-1} x}{dt^{n-1}} + \cdots + \otimes_0 x = \otimes \cdot u(t) \tag{11.4}$$

After applying Laplace transform to both sides of this equation, we obtain

$$G(s) = \frac{X(s)}{U(s)} = \frac{\otimes}{\otimes_n s^n + \otimes_{n-1} s^{n-1} + \cdots \otimes_1 s + \otimes_0} \tag{11.5}$$

where $L(x(t)) = X(s)$ and $L(u(t)) = U(s)$. Equation (11.5) is known as a grey transfer function, which is the ratio of the Laplace transform of the response $x(t)$ of the nth order grey linear control system and the Laplace transform of the driving term $u(t)$. In fact, the transfer function represents a fundamental relationship between the input and output of a first order grey linear control system. From the following theorem, it follows that each nth order grey linear system can be reduced to an equivalent first order grey linear system.

Theorem 11.3.1 For an nth order grey linear system as shown in Eq. (11.4), there is an equivalent first order grey linear system.

Proof Assume that the given nth order grey linear system is

$$\otimes_n \frac{d^n x}{dt^n} + \otimes_{n-1} \frac{d^{n-1} x}{dt^{n-1}} + \cdots + \otimes_0 x = \otimes \cdot u(t)$$

Let

$$x = x_1, \quad \frac{dx}{dt} = \frac{dx_1}{dt} = x_2, \quad \frac{d^2 x}{dt^2} = \frac{dx_2}{dt} = x_3, \ldots, \quad \frac{d^{n-1} x}{dt^{n-1}} = \frac{dx_{n-1}}{dt} = x_n.$$

Therefore, we have

$$\frac{dx_n}{dt} = -\frac{\otimes_0}{\otimes_n} x_1 - \frac{\otimes_1}{\otimes_n} x_2 - \frac{\otimes_2}{\otimes_n} x_3 - \cdots - \frac{\otimes_{n-1}}{\otimes_n} x_n + \frac{\otimes}{\otimes_n} u(t)$$

and the nth order system is reduced to the following first order system

$$\dot{X}(t) = A(\otimes) X(t) + B(\otimes) U(t)$$

where $X(t) = [x_1, x_2, \ldots, x_n]^T$, $U(t) = u(t)$,

$$A(\otimes) = \begin{bmatrix} 0 & 1 & 0 & \cdots & 0 \\ 0 & 0 & 1 & \cdots & 0 \\ \cdots & \cdots & \cdots & & \cdots \\ 0 & 0 & \cdots & \cdots & 1 \\ -\frac{\otimes_0}{\otimes_n} & -\frac{\otimes_1}{\otimes_n} & \cdots & \cdots & -\frac{\otimes_{n-1}}{\otimes_n} \end{bmatrix}, \text{ and } B(\otimes) = \begin{bmatrix} 0 \\ 0 \\ \vdots \\ 0 \\ \frac{\otimes}{\otimes_n} \end{bmatrix}.$$

This ends the proof.

11.3.2 Transfer Functions of Typical Links

A grey control system that is symbolically written in an equation is also known as a grey link. When the transfer function of a link is known, from the relationship $X(s) = G(s) \cdot U(s)$ and the Laplace transform of the driving term, one can obtain the Laplace transform of the response. Then, by using the inverse Laplace transform, one can produce the response $x(t)$. The relationship between the driving and response terms is depicted in Fig. 11.3.

In the following definition, let us look at the transfer functions of several typical links.

Fig. 11.3 The driving and response terms

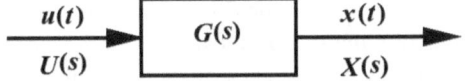

$u(t)$ $G(s)$ $x(t)$

$U(s)$ $X(s)$

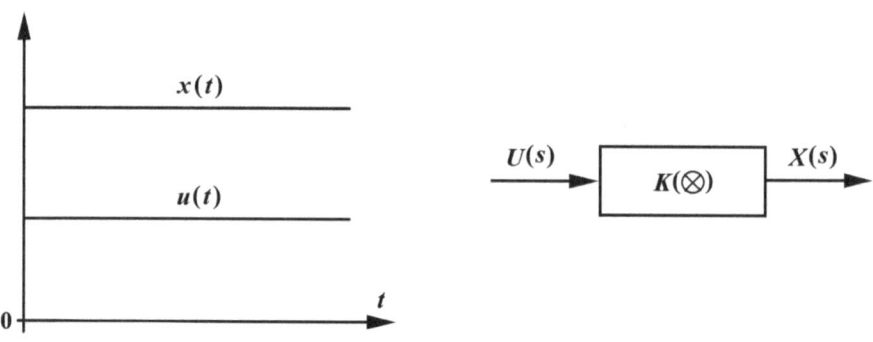

Fig. 11.4 The grey proportional link

Definition 11.3.2 The link between driving term $u(t)$ and response term $x(t)$ satisfying

$$x(t) = K(\otimes)u(t) \tag{11.6}$$

is known as a grey proportional link, where $K(\otimes)$ is the grey magnifying coefficient of the link.

Proposition 11.3.1 The transfer function of a grey proportional link is

$$G(s) = K(\otimes) \tag{11.7}$$

The characteristics of a grey proportional link are that when a jump occurs in the driving quantity, the response value changes proportionally. This kind of change and relationship between the drive and response are depicted in Fig. 11.4.

Definition 11.3.3 When driven by a unit jump, if the response is given by

$$x(t) = K(\otimes)(1 - e^{-tT}) \tag{11.8}$$

then the link is known as a grey inertia link, where T stands for a time constant of the link.

Proposition 11.2.2 The transfer function of a grey inertial link is given by

$$G(s) = \frac{K(\otimes)}{T \cdot s + 1} \tag{11.9}$$

The characteristics of a grey inertia link are that when a jump occurs in the driving quantity, the response can reach a new state of balance only after a period of time. Figure 11.5 provides a block diagram and the curve of change of the response of a grey inertia link when $\tilde{K}(\otimes) = 1$.

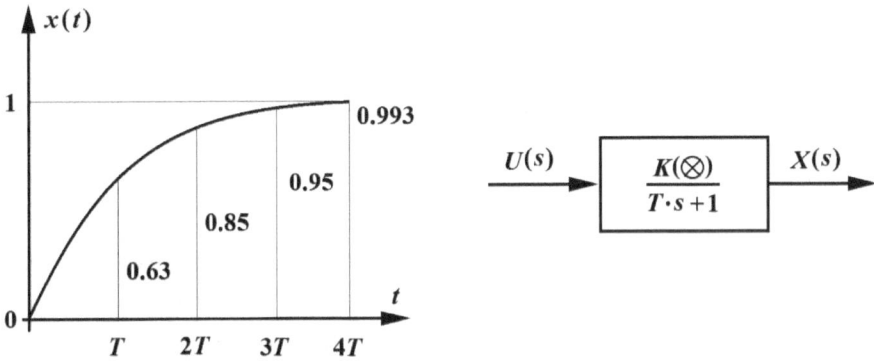

Fig. 11.5 The grey inertia link

Definition 11.3.4 When the drive and response are related as follows, the link is known as grey integral link:

$$x(t) = \int K(\otimes)u(t)dt \qquad (11.10)$$

Proposition 11.3.3 The transfer function of a grey integral link is given below:

$$G(s) = \frac{K(\otimes)}{s} \qquad (11.11)$$

For a grey integral link, when the drive is a jump function, its response is $x(t) = K(\otimes)ut$, as shown in Fig. 11.6.

Definition 11.3.5 If the response and the drive are related as follows, the link is known as a grey differential link:

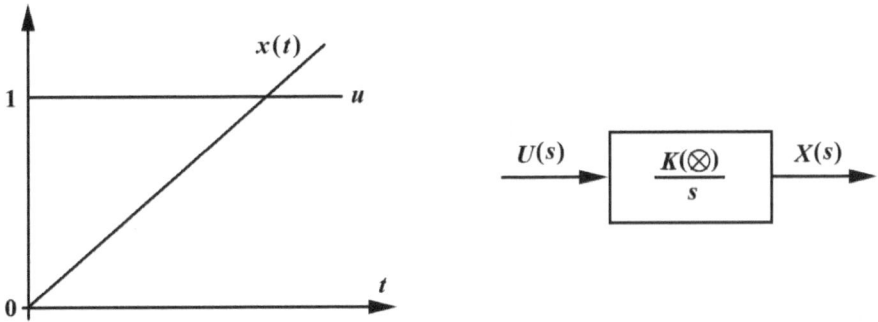

Fig. 11.6 The grey integral link

$$x(t) = K(\otimes)\frac{du(t)}{dt} \tag{11.12}$$

Proposition 11.3.4 The transfer function of a grey differential link is given as follows:

$$G(s) = K(\otimes)s \tag{11.13}$$

The characteristics of a grey differential link are that when the drive stands for a jump, the response becomes an impulse with an infinite amplitude.

Definition 11.3.6 If the drive and response are related as follows, the link is known as a grey postponing link, where $\tau(\otimes)$ is a grey constant:

$$x(t) = u(t - \tau(\otimes)) \tag{11.14}$$

Proposition 11.3.5 The transfer function of a grey postponing link is given below:

$$G(s) = e^{-\tau(\otimes)s} \tag{11.15}$$

For a grey postponing link, when the drive is a jump function, it takes some time for the response to react accordingly. For details, see Fig. 11.7.

The figure above represents some typical links met in practical applications. Many complicated devices and systems can be treated as combinations of these typical links. For instance, when the grey proportional link is combined with a grey differential link, one can obtain a grey proportional differential link. When a grey integral link is connected with grey postponing link, one establishes a grey integral postponing link. Along the same lines, multi-layered combinations can be developed for practical purposes. One of the purposes of studying grey transfer functions is that we can investigate the stabilities and other properties of systems by looking at the extreme values of relevant transfer functions.

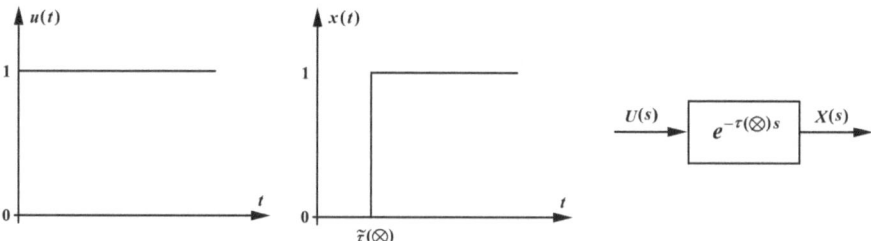

Fig. 11.7 The grey postponing link

11.3.3 Matrices of Grey Transfer Functions

Matrices of grey transfer functions can be employed to express a fundamental relationship between the multi-inputs and multi-outputs of grey linear control systems. In particular, for the following grey linear control system

$$\begin{cases} \dot{X}(t) = A(\otimes)X(t) + B(\otimes)U(t) \\ Y(t) = C(\otimes)X(t) \end{cases}$$

Employing Laplace transforms produces

$$\begin{cases} sX(s) = A(\otimes)X(s) + B(\otimes)U(s) \\ Y(s) = C(\otimes)X(s) \end{cases}$$

and

$$\begin{cases} (sE - A(\otimes))X(s) = B(\otimes)U(s) \\ Y(s) = C(\otimes)X(s) \end{cases}$$

If $(sE - A(\otimes))$ is invertible, then we can further obtain

$$\begin{cases} X(s) = (sE - A(\otimes))^{-1}B(\otimes)U(s) \\ Y(s) = C(\otimes)X(s) \end{cases}$$

That is, we have $Y(s) = C(\otimes)(sE - A(\otimes))^{-1}B(\otimes)U(s)$.

Definition 11.3.7 The $m\,n$ matrix below is known as the matrix of grey transfer functions:

$$G(s) = C(\otimes)(sE - A(\otimes))^{-1}B(\otimes) \tag{11.16}$$

Definition 11.3.8 For an nth order grey linear system, if the state grey matrix $A(\otimes)$ of the corresponding equivalent first order system is non-singular, then

$$\lim_{s \to 0} G(s) = -C(\otimes)A(\otimes)^{-1}B(\otimes) \tag{11.17}$$

is known as a grey gain matrix. If the grey gain matrix $-C(\otimes)A(\otimes)^{-1}B(\otimes)$ is used to replace the transfer function $G(s)$, then the system is reduced into a proportional link. Because $Y(s) = G(s)U(s)$, when $m = s = n$, if $G(s)$ is non-singular, we have the following:

$$U(s) = G(s)^{-1}Y(s) \tag{11.18}$$

Definition 11.3.9 The following matrix is known as a grey structure matrix:

$$G(s)^{-1} = B(\otimes)^{-1}(sE - A(\otimes))C(\otimes)^{-1} \qquad (11.19)$$

When the grey structure matrix is known, to make the output vector $Y(s)$ meet or close to meet a certain expected objective $J(s)$, one can determine the system's control vector $U(s)$ through $G^{-1}(s) \cdot J(s)$. Additionally, we can also discuss the controllability and observability of systems by using matrices of grey transfer functions.

11.4   Robust Stability of Grey System

Stability is a fundamental structural characteristic of systems. It stands for an important mechanism for a system to sustain itself and is a prerequisite for the system to operate smoothly. This is why stability is studied in systems control theory and it is a key objective in relevant engineering designs. Each physical system has to be stable before it can be employed in practical applications.

The stability of grey systems focuses on the investigations of informational changes. It also focuses on whether or not the grey system of concern stays stable or can recover to its stability when the whitenization value of a grey parameter moves within the field of discourse. The existence of grey parameters complicates the study of grey systems stability, and puts them at the center of attention of control theory and control engineering.

In grey systems modeling, there is a distinction between having a postponing term and not having such a term; there is also a difference between having a random term and not having such a term. Ordinarily, grey systems without involving any random and postponing term are known as grey systems; those involving postponing terms without any random terms are grey postponing systems, and those involving random terms are known as grey stochastic systems. In this section, we will study the problem of robust stability of these three kinds of systems.

11.4.1   Robust Stability of Grey Linear Systems

The study of systems' stability is often limited to systems without the effect of any external input. This kind of system is known as an autonomous system. A simple grey linear autonomous system can be written as follows:

$$\begin{cases} \dot{x}(t) = A(\otimes)x(t) \\ x(t_0) = x_0, \quad \forall t \geq t_0 \end{cases} \qquad (11.20)$$

where $x \in R^n$ stands for the state vector, and $A(\otimes) \in G^{n \times n}$ is the matrix of grey coefficients.

Definition 11.4.1 If $A(\tilde{\otimes})$ is a whitenization matrix of the grey matrix $A(\otimes)$, then

$$\begin{cases} \dot{x}(t) = A(\tilde{\otimes})x(t) \\ x(t_0) = x_0 \end{cases} \tag{11.21}$$

is referred to as a whitenization system of the system in Eq. (11.20).

Ordinarily, we assume that the matrix $A(\otimes)$ of grey coefficients of the system in Eq. (11.20) has a continuous matrix cover:

$$A(D) = [L_a, U_a] = \{A(\tilde{\otimes}) : \underline{a}_{ij} \leq \tilde{\otimes} \leq \overline{a}_{ij}, i, j = 1, 2, \ldots, n\},$$

where $U_a = (\overline{a}_{ij})$, $L_a = (\underline{a}_{ij})$.

Definition 11.4.2 If any whitenization system of the system in Eq. (11.20) is stable, then the system in Eq. (11.20) is referred to as robust stable.

The ordinary concept of a system's (robust) stability represents the (robust) asymptotic stability of the system.

Theorem 11.4.1 If there is positive definite matrix P such that

$$PL_a + L_a^T P + 2\lambda_{\max}(P)||U_a - L_a||I_n < 0$$

then the system in Eq. (11.20) is robust stable (Su & Liu, 2009).

Proof Let us take the Lyapunov function $V(x) = x^T P x$. For any whitenization matrix $A(\tilde{\otimes}) \in A(D)$, let us compute the derivative of $V(x)$ with respect to t along the trajectory of the whitenization system and obtain

$$\begin{aligned} \dot{V}(x) &= 2x^T P A(\tilde{\otimes})x = x^T \left(PL_a + L_a^T P \right)x + 2x^T P \Delta A x \\ &\leq x^T \left(PL_a + L_a^T P + 2\lambda_{\max}(P)||U_a - L_a||I_n \right)x < 0, \quad \forall x \neq 0 \\ &\leq x^T \left(PL_a + L_a^T P + 2\lambda_{\max}(P)||U_a - L_a||I_n \right)x < 0, \quad \forall x \neq 0 \end{aligned}$$

This implies that the system in Eq. (11.20) is robust stable. QED. If in Theorem 11.4.1 we let $P = I_n$, then we have the result shown below.

Corollary 11.4.1 If

$$||U_a - L_a|| < -\lambda_{\max}(\frac{L_a + L_a^T}{2}) \tag{11.22}$$

holds true, then the system in Eq. (11.20) is robust stable. If we employ another form of decomposition $A(\tilde{\otimes}) = U_a - \Delta A$ of the whitenization matrix $A(\tilde{\otimes})$ to study the robust stability of the system in Eq. (11.20), then much like in Theorem 11.4.1 and Corollary 11.4.1 we can obtain the following results.

Theorem 11.4.2 If there is a positive definite matrix P such that

$$PU_a + U_a^T P + 2\lambda_{\max}(P)||U_a - L_a||I_n < 0$$

then the system in Eq. (11.20) is robust stable (Su & Liu, 2009).

Corollary 11.4.2 If

$$||U_a - L_a|| \leq -\lambda_{\max}(\frac{U_a + U_a^T}{2}) \tag{11.23}$$

holds true, then the system in Eq. (11.20) is robust stable. Both Corollaries 11.4.1 and 11.4.2 respectively provide us a meaning result, because $U_a - L_a$ in fact stands for the matrix of disturbance errors of the system in Eq. (11.20); Eqs. (11.22) and (11.23) indicate that when the norm of the disturbance error matrix varies within the range of $(0, \lambda)$, the system in Eq. (11.20) will always be stable, where $\lambda = \max\left\{-\lambda_{\max}\left(\frac{L_a + L_a^T}{2}\right), -\lambda_{\max}\left(\frac{U_a + U_a^T}{2}\right)\right\}$.

Theorem 11.4.3 If $L_a + L_a^T + \lambda_{\max}[(U_a - L_a) + (U_a - L_a)^T]I_n < 0$, then the system in Eq. (11.20) is robust stable; if $U_a + U_a^T - \lambda_{\max}[(U_a - L_a) + (U_a - L_a)^T]I_n > 0$, then the system is instable (Su & Liu, 2009).

Example 11.4.1 Let us consider the robust stability problem of the following 2-dimensional grey linear system:

$$\dot{x}(t) = \begin{pmatrix} [-2.3, -1.8] & [0.6, 0.9] \\ [0.8, 1.0] & [-2.5, -1.9] \end{pmatrix} x(t)$$

Solution Through computations, we have

$$||U_a - L_a|| = 0.8072 < -\lambda_{\max}(\frac{L_a + L_a^T}{2}) = 1.3000$$

$$L_a + L_a^T + \lambda_{\max}[(U_a - L_a) + (U_a - L_a)^T]I_n = -1.7759I_n < 0$$

These inequalities indicate that, by using Corollary 11.4.1 or Theorem 11.4.3, we can conclude that this given system is robust stable.

11.4.2 Robust Stability of Grey Linear Time-Delay Systems

The phenomena of timely postponing are very common. They are often the main reason for causing instability, vibration, and poor performance in systems. Therefore, it is very important to investigate the stability problem of postponing systems. In particular, let us look at the following n-dimensional linear postponing autonomous system:

$$\begin{cases} \dot{x}(t) = Ax(t) + Bx(t - \tau), & \forall t \geq 0, \\ x(t) = \phi(t), & \forall t \in [-\tau, 0] \end{cases} \tag{11.24}$$

where $x(t) \in R^n$ stands for the system's state vector, $A, B \in R^{n \times n}$ the known constant matrices, $\tau > 0$ the amount of time of postponing, and $\phi(t) \in C^n[-\tau, 0]$ the nth dimensional space of continuous functions.

Definition 11.4.3 If at least one of the matrices A, B of constants in the linear postponing system in Eq. (11.24) is grey, then this system is referred to as a grey linear postponing autonomous system, denoted as

$$\begin{cases} \dot{x}(t) = A(\otimes)x(t) + B(\otimes)x(t - \tau), & \forall t \geq 0, \\ x(t) = \phi(t), & \forall t \in [-\tau, 0]. \end{cases} \tag{11.25}$$

In the following equation, we assume that the constant matrices in the system in Eq. (11.25) are all grey and have continuous matrix covers; that is, $A(\otimes), B(\otimes)$ respectively have the following form of matrix covers:

$$A(D) = [L_a, U_a] = \{A(\tilde{\otimes}) : \underline{a}_{ij} \leq \tilde{\otimes} \leq \overline{a}_{ij}, \quad i, j = 1, 2, \ldots, n\},$$
$$B(D) = [L_b, U_b] = \{B(\tilde{\otimes}) : \underline{b}_{ij} \leq \tilde{\otimes} \leq \overline{b}_{ij}, \quad i, j = 1, 2, \ldots, n\}$$

where $U_a = (\overline{a}_{ij}), L_a = (\underline{a}_{ij}), U_b = (\overline{b}_{ij}), L_b = (\underline{b}_{ij})$.

Definition 11.4.4 If $A(\tilde{\otimes}), B(\tilde{\otimes})$ are respectively whitenization matrices of $A(\otimes), B(\otimes)$, then

$$\begin{cases} \dot{x}(t) = A(\tilde{\otimes})x(t) + B(\tilde{\otimes})x(t - \tau), & \forall t \geq 0, \\ x(t) = \phi(t), & \forall t \in [-\tau, 0]. \end{cases} \tag{11.26}$$

is referred to as a whitenization system of the system in Eq. (11.25).

Definition 11.4.5 If any whitenization system of the system in Eq. (11.25) is stable, the system in Eq. (11.25) is referred to as robust stable.

Based on whether or not the robust stability condition of a grey postponing system depends on the amount of postponing, the robust stability condition can be divided into two classes: postponing independent and postponing dependent. In particular, the condition for a robust stable system to be postponing independent is that for any time postponing $\tau > 0$, the system is robustly asymptotic stable. Because this condition does not require the amount of postponing, it is appropriate for the study of the stability problem of postponing systems whose amounts of postponing are uncertain or unknown.

The condition for a robust stable system to be postponing dependent is that for some values of postponing $\tau > 0$, the system is robust stable, while for some other values of the postponing $\tau > 0$, the system is not stable. That is why the system's stability is dependent on the amount of postponing.

Theorem 11.4.4 If there are positive definite matrices P, Q and positive constants ε_1, ε_2 such that the symmetric matrix

$$
\begin{pmatrix}
\Xi & PL_b & P & P \\
L_b^\mathrm{T} P & -Q + \varepsilon_2 ||U_b - L_b||^2 I_n & 0 & 0 \\
P & 0 & -\varepsilon_1 I_n & 0 \\
P & 0 & 0 & -\varepsilon_2 I_n
\end{pmatrix} < 0
$$

where $\Xi = L_a^\mathrm{T} P + PL_a + Q + \varepsilon_1 ||U_a - L_a||^2 I_n$, and I_n stands for the identity matrix (the same symbol will be used for the rest of this chapter), then the system in Eq. (11.25) is robust stable (Su, 2012).

Theorem 11.4.5 If there are positive definite matrices P, Q, N and positive constants ε_1, ε_2 such that the symmetric matrix

$$
\begin{pmatrix}
\Gamma & PL_b & \rho I_n & P & P \\
L_b^\mathrm{T} P & -Q + \varepsilon_2 ||U_b - L_b||^2 I_n & 0 & 0 & 0 \\
\rho I_n & 0 & -\overline{N} & 0 & 0 \\
P & 0 & 0 & -\varepsilon_1 I_n & 0 \\
P & 0 & 0 & 0 & -\varepsilon_2 I_n
\end{pmatrix} < 0
$$

where $\Gamma = L_a^\mathrm{T} P + PL_a + Q + \varepsilon_1 ||U_a - L_a||^2 I_n$ and $\overline{N} = N^{-1}$, $\rho = \sqrt{\tau}$, then the system in Eq. (11.25) is robust stable.

Example 11.4.2 Let us look at the following 2-dimensional grey linear postponing system

$$
\begin{cases}
\dot{x}(t) = A(\otimes)x(t) + B(\otimes)x(t - \tau), & \forall t \geq 0, \\
x(t) = \phi(t), & \forall t \in [-\tau, 0].
\end{cases}
$$

Assume that the upper and lower bound matrices of the continuous matrix covers of the grey constant matrices $A(\otimes)$, $B(\otimes)$ are respectively give as follows:

$$
L_a = \begin{pmatrix} -4.38 & 0.20 \\ 0.19 & -4.33 \end{pmatrix}, \quad U_a = \begin{pmatrix} -4.26 & 0.29 \\ 0.27 & -4.22 \end{pmatrix};
$$

$$
L_b = \begin{pmatrix} -0.93 & 0.21 \\ 0.23 & -0.86 \end{pmatrix}, \quad U_b = \begin{pmatrix} -0.88 & 0.24 \\ 0.26 & -0.82 \end{pmatrix}
$$

According to Theorem 9.10, by using the solver in the LMI (linear matrix inequality) control toolbox, we obtain the behavioral solution as follows:

$$P = \begin{pmatrix} 9.4642 & 0.6983 \\ 0.6983 & 9.8228 \end{pmatrix}, \quad Q = \begin{pmatrix} 28.0088 & -0.0340 \\ -0.0340 & 27.8605 \end{pmatrix},$$

$$\varepsilon_1 = 30.0826, \quad \varepsilon_2 = 30.2461.$$

Now from Theorem 9.11, by using the solver in the LMI (linear matrix inequality) control toolbox, we obtain the behavioral solution below:

$$P = \begin{pmatrix} 6.8592 & 0.5061 \\ 0.5061 & 7.1191 \end{pmatrix}, \quad Q = \begin{pmatrix} 20.2294 & -0.0246 \\ -0.0246 & 20.1920 \end{pmatrix}, \quad N = \begin{pmatrix} 0.0456 & 0 \\ 0 & 0.0456 \end{pmatrix},$$

$$\varepsilon_1 = 21.8024, \quad \varepsilon_2 = 21.9209, \quad \tau = 2.7035$$

These results indicate that the system considered in this example is robust stable. And the maximum allowed postponing length of time as obtained from Theorem 11.4.5 is 2.7035.

11.4.3 Robust Stability of Grey Stochastic Linear Time-Delay System

The mathematical model that describes a stochastic system is generally the Itto stochastic differential equation, where the often seen n-dimensional Itto stochastic differential postponing equation is

$$\begin{cases} dx(t) = Ax(t) + Bx(t - \tau) + [Cx(t) + Dx(t - \tau)]dw(t), & \forall t \geq 0, \\ x(t) = \xi(t), \xi(t) \in L_{F_0}^2([-\tau, 0]; R^n), & \forall t \in [-\tau, 0]. \end{cases} \tag{11.27}$$

where $x(t) \in R^n$ stands for the system's state vector, $A, B, C, D \in R^{n \times n}$ known constant matrices, $\tau > 0$ the time of postponing, and $w(t)$ a 1-dimensional Brownian motion defined on a complete probability space $(\Omega, F, \{F_t\}_{t \geq 0}, P)$. $L_{F_0}^2([-\tau, 0]; R^n)$ stands for the totality of all F_0-measurable stochastic variables $\xi = \{\xi(t) : -\tau \leq t \leq 0\}$ that take values from $C([-\tau, 0]; R^n)$ satisfying $\sup_{-\tau \leq t \leq 0} E|\xi(\theta)|^2 < \infty$, while $C([-\tau, 0]; R^n)$ stands for the totality of continuous functions $\phi : [-\tau, 0] \to R^n$. Under the initial condition $x(t) = \xi(t) \in L_{F_0}^2([-\tau, 0]; R^n)$, the system in Eq. (11.27) has an equilibrium point $x(t; \xi)$, and corresponds to the initial value $\xi(t) = 0, x(t; 0) \equiv 0$.

There are several different concepts of stability for stochastic systems. In the following, we list four of the important stabilities.

Definition 11.4.6 The equilibrium point $x(t) \equiv 0$ of the system in Eq. (11.27) is referred to as stochastically stable, if for each $\varepsilon > 0$, $\lim\limits_{x_0 \to 0} P(\sup\limits_{t > t_0} |x(t; t_0, x_0)| > \varepsilon) = 0$.

Definition 11.4.7 The equilibrium point $x(t) \equiv 0$ of the system in Eq. (11.27) is referred to as stochastically asymptotically stable, if it is stochastically stable and $\lim\limits_{x_0 \to 0} P(\lim\limits_{t \to +\infty} x(t; t_0, x_0) = 0) = 1$.

Definition 11.4.8 The equilibrium point $x(t) \equiv 0$ of the system in Eq. (11.27) is referred to as large-scale stochastically asymptotically stable, if it is stochastically stable and for any t_0, x_0, $P(\lim\limits_{t \to +\infty} x(t; t_0, x_0) = 0) = 1$.

Definition 11.4.9 The equilibrium point $x(t) \equiv 0$ of the system in Eq. (11.27) is referred to as mean square exponential stable, if there are positive constants $\alpha > 0, \beta > 0$ such that $E|x(t; t_0, x_0)|^2 \leq \alpha |x_0|^2 \exp(-\beta t), t > t_0$.

A grey system is stochastic if it involves grey parameters. Concepts related to grey stochastic systems are generally introduced based on relevant concepts of conventional stochastic systems. Considering the problems we will study, let us provide the following definitions.

Definition 11.4.10 If at least one of the matrices A, B, C, D of the stochastic linear postponing system in Eq. (11.27) is grey, then the system is referred to as a grey stochastic linear postponing system, written as follows:

$$\begin{cases} dx(t) = A(\otimes)x(t) + B(\otimes)x(t - \tau) + [C(\otimes)x(t) + D(\otimes)x(t - \tau)]dw(t), \\ \quad \forall t \geq 0, \\ x(t) = \xi(t), \xi(t) \in L^2_{F_0}([-\tau, 0]; R^n), \quad \forall t \in [-\tau, 0]. \end{cases}$$

$$(11.28)$$

In this section, we assume that all the coefficient matrices of the system in Eq. (11.28) are grey with continuous matrix covers. That is, the matrix covers of the grey matrices $A(\otimes)$, $B(\otimes)$, $C(\otimes)$, and $D(\otimes)$ are respectively given as follows:

$$A(D) = [L_a, \ U_a] = \{A(\tilde{\otimes}) = (\tilde{\otimes}_{aij})_{n \times n} : \underline{a}_{ij} \leq \tilde{\otimes}_{aij} \leq \overline{a}_{ij}\},$$
$$B(D) = [L_b, \ U_b] = \{B(\tilde{\otimes}) = (\tilde{\otimes}_{bij})_{n \times n} : \underline{b}_{ij} \leq \tilde{\otimes}_{bij} \leq \overline{b}_{ij}\},$$
$$C(D) = [L_c, \ U_c] = \{C(\tilde{\otimes}) = (\tilde{\otimes}_{cij})_{n \times m} : \underline{c}_{ij} \leq \tilde{\otimes}_{cij} \leq \overline{c}_{ij}\},$$

and

$$D(D) = [L_d, \ U_d] = \{D(\tilde{\otimes}) = (\tilde{\otimes}_{dij})_{n \times n} : \underline{d}_{ij} \leq \tilde{\otimes}_{dij} \leq \overline{d}_{ij}\},$$

where $L_a = (\underline{a}_{ij})_{n \times n}, U_a = (\overline{a}_{ij})_{n \times n}, L_b = (\underline{b}_{ij})_{n \times n}, U_b = (\overline{b}_{ij})_{n \times n}, L_c = (\underline{c}_{ij})_{n \times n}, U_c = (\overline{c}_{ij})_{n \times n}, L_d = (\underline{d}_{ij})_{n \times n}$, and $U_a = (\overline{d}_{ij})_{n \times n}$.

Definition 11.4.11 If $A(\tilde{\otimes})$, $B(\tilde{\otimes})$, $C(\tilde{\otimes})$, and $D(\tilde{\otimes})$ are arbitrary whitenization matrices of the grey matrices $A(\otimes)$, $B(\otimes)$, $C(\otimes)$, and $D(\otimes)$, respectively, then

$$\begin{cases} dx(t) = A(\tilde{\otimes})x(t) + B(\tilde{\otimes})x(t-\tau) + [C(\tilde{\otimes})x(t) + D(\tilde{\otimes})x(t-\tau)]dw(t), \\ \quad \forall t \geq 0, \\ x(t) = \xi(t), \xi(t) \in L^2_{F_0}([-\tau, 0]; R^n), \quad \forall t \in [-\tau, 0]. \end{cases}$$

(11.29)

is referred to as a whitenization system of the system in Eq. (11.28).

Definition 11.4.12 If any whitenization system of the system in Eq. (11.28) is large-scale stochastic asymptotic stable, that is,

$$\lim_{t \to \infty} x(t; \xi) = 0 \ a.s.$$

then the system in equation (11.28) is said to be large scale stochastic robust asymptotic stable.

Definition 11.4.13 If any whitenization system of the system in Eq. (11.28) is mean square exponential stable, that is, there are positive constants r_0 and K such that the equilibrium points of whitenization systems of the system in Eq. (11.28) satisfy

$$E|x(t, \xi)|^2 \leq Ke^{-r_0 t} \sup_{-\tau \leq \theta \leq 0} E|\xi(\theta)|^2, \quad t \geq 0,$$

or equivalently

$$\lim_{t \to \infty} \sup \frac{1}{t} \log E|x(t; \xi)|^2 \leq -r_0,$$

then the system in Eq. (11.28) is said to be mean square exponential robust stable.

Theorem 11.4.6 For the system in Eq. (11.28), if there is a positive definite symmetric matrix Q and there are positive constants ε_i, $i = 1, \ldots, 6$, satisfying $M + N < 0$, then for any initial condition $\xi \in C^p_{F_0}([-\tau, 0]; R^n)$ the following holds true:

$$\lim_{t \to \infty} x(t; \xi) = 0 \ a.s.$$

That is, according to Su (2012), the system in Eq. (11.28) is large-scale stochastic robust asymptotic stable, where

$$M = QL_a + L_a^T Q + (\varepsilon_1 + \varepsilon_2)Q + \varepsilon_1^{-1}\lambda_{\max}(Q) \cdot ||U_a - L_a||^2 I_{nc}$$
$$+ (1 + \varepsilon_4)(1 + \varepsilon_5)L_c^T QL + (1 + \varepsilon_4^{-1})(1 + \varepsilon_5)\lambda_{\max}(Q)||U_c - L_c||^2 I_n$$

and

$$N = \varepsilon_2^{-1}(1 + \varepsilon_3^{-1})\lambda_{\max}(Q)||U_b - L_b||^2 I_n + \varepsilon_2^{-1} \cdot (1 + \varepsilon_3)L_b^T Q L_b$$
$$+ (1 + \varepsilon_5^{-1})(1 + \varepsilon_6)L_d^T Q L_d + (1 + \varepsilon_5^{-1})(1 + \varepsilon_6^{-1})\lambda_{\max}(Q)||U_d - L_d||^2 I_n.$$

Theorem 11.4.7 For the system in Eq. (11.28), if there are positive definite symmetric matrix Q and positive constants ε_i, $i = 1, \ldots, 6$, satisfying $K + L < 0$, then for any initial condition $\xi \in C_{F_0}^p([-\tau, 0]; R^n)$, the following holds true:

$$\lim_{t \to \infty} x(t; \xi) = 0 \ a.s$$

That is, the system in Eq. (11.28) is large-scale stochastic asymptotic stable, where

$$K = QL_a + L_a^T Q + (\varepsilon_1 + \varepsilon_2)Q + [\varepsilon_1^{-1}\lambda_{\max}(Q)\text{trace}(G_a^T G_a)$$
$$+ (1 + \varepsilon_4)(1 + \varepsilon_5)\text{trace}(L_c^T L_c) + (1 + \varepsilon_4^{-1})(1 + \varepsilon_5)\lambda_{\max}(Q)\text{trace}(G_c^T G_c)]I_n,$$

and

$$L = [\varepsilon_2^{-1}(1 + \varepsilon_3^{-1})\lambda_{\max}(Q)\text{trace}(G_b^T G_b) + \varepsilon_2^{-1}(1 + \varepsilon_3)\text{trace}(L_b^T L_b)$$
$$+ (1 + \varepsilon_5^{-1})(1 + \varepsilon_6)\text{trace}(L_d^T L_d) + (1 + \varepsilon_5^{-1})(1 + \varepsilon_6^{-1})\lambda_{\max}(Q)\text{trace}(G_d^T G_d)]I_n.$$

If we let the matrix and constants in Theorems 11.4.6 and 11.4.7 be $\varepsilon_1 = \cdots = \varepsilon_6 = 1$ and $Q = I_n$, then we can obtain the following corollaries, respectively.

Corollary 11.4.3 If the upper and lower bound matrices of the continuous matrix covers of the coefficient matrices of the system in Eq. (11.28) satisfy

$$L_a + L_a^T + 2L_b^T L_b + 4L_c^T L_c + 4L_d^T L_d$$
$$< -(2||U_b - L_b||^2 + ||U_a - L_a||^2 + 4||U_d - L_d||^2 + 4||U_c - L_c||^2 + 2)I_n$$

then the system in Eq. (11.28) is large-scale stochastic asymptotic stable.

Corollary 11.4.4 If the upper and lower bound matrices of the continuous matrix covers of the coefficient matrices of the system in Eq. (11.28) satisfy

$$L_a + L_a^T + [2\text{trace}(L_b^T L_b) + 4\text{trace}(L_c^T L_c) + 4\text{trace}(L_d^T L_d)]I_n$$
$$< -(\text{trace}(G_a^T G_a) + 2\text{trace}(G_b^T G_b) + 4\text{trace}(G_c^T G_c) + 4\text{trace}(G_d^T G_d) + 2)I_n$$

then the system in Eq. (11.28) is large-scale stochastic asymptotic stable.

Theorem 11.4.8 For the system in Eq. (11.28), if there are positive definite symmetric matrix Q and positive constants ε_i, $i = 1, \ldots, 3$, satisfying

$$QL_a + L_a^T Q + (\varepsilon_1 + \varepsilon_2)Q + \varepsilon_1^{-1}\lambda_{\max}(Q)||U_a - L_a||^2 I_n$$
$$< -[(1 + \varepsilon_3)\lambda_{\max}(Q)\text{trace}(M_c^T M_c) + \varepsilon_2^{-1}\lambda_{\max}(Q)\,\text{trace}(M_b^T M_b)$$
$$+ (1 + \varepsilon_3^{-1})\lambda_{\max}(Q)\text{trace}(M_d^T M_d)]I_n$$

then the system in Eq. (11.28) is large-scale stochastic robust asymptotic stable. If in Theorem 11.4.8 we let $\varepsilon_1 = \varepsilon_2 = \varepsilon_3 = 1$ and $Q = I_n$, then we have the corollary below.

Corollary 11.4.5 If the upper and lower bound matrices of the matrix covers of the grey coefficient matrices in the system in Eq. (11.28) satisfy

$$L_a + L_a^T + 2I_n + ||U_a - L_a||^2 I_n + 2\text{trace}(M_c^T M_c)I_n$$
$$< -[\text{trace}(M_b^T M_b) + 2\text{trace}(M_d^T M_d)]I_n$$

then the system in Eq. (11.28) is large-scale stochastic robust asymptotic stable.

Theorem 11.4.9 For the system in Eq. (11.28), if there are positive definite symmetric matrix Q and positive constants ε_i, $i = 1, \ldots, 6$, satisfying $\lambda_{\max}(M) + \lambda_{\max}(N) < 0$, then for any initial condition $\xi \in C_{F_0}^p([-\tau, 0]; R^n)$, the following holds true:

$$E|x(t, \xi)|^2 \le Ke^{-r_0 t} \sup_{-\tau \le \theta \le 0} E|\xi(\theta)|^2, \quad t \ge 0,$$

or equivalently,

$$\lim_{t \to \infty} \sup \frac{1}{t} \log E|x(t; \xi)|^2 \le -r_0.$$

where the matrices M, N are the same as in Theorem 11.4.6, $K = \frac{\tau e^{r_0\tau}\lambda_{\max}(N)+\lambda_{\max}(Q)}{\lambda_{\min}(Q)}$, and r_0 is the unique real root of the following equation $r_0\lambda_{\max}(Q) + \lambda_{\max}(M) + e^{r_0\tau}\lambda_{\max}(N) = 0$, then the system in Eq. (11.28) is mean square exponential robust stable.

11.5 Several Typical Grey Control Models

Grey control stands for the control of essential grey systems, including the situation of general control systems involving grey numbers, by constructing controls through employing the thinkingv methods of grey systems analysis, modeling, prediction, and decision-making.

11.5.1 Control of Redundancy Removal

The dynamic characteristics of grey systems are mainly determined by the matrices $G(s)$ of grey transfer functions. So, to realize effect control over the systems' dynamic characteristics, one of the effective methods is to modify and correct the matrices of transfer functions and the structure matrices (Deng, 1965, 1985).

Definition 11.5.1 Assume that $G^{-1}(s)$ is a system's structure matrix, and $G_*^{-1}(s)$ an objective structure matrix, then

$$\Delta^{-1} = G_*^{-1}(s) - G^{-1}(s) \tag{11.30}$$

is known as a structural deviation matrix. From $G^{-1}(s)Y(s) = U(s)$ and $G_*^{-1}(s) = \Delta^{-1} + G^{-1}(s)$, we obtain $(G_*^{-1}(s) - \Delta^{-1})Y(s) = U(s)$. That is,

$$G_*^{-1}(s)Y(s) - \Delta^{-1}Y(s) = U(s) \tag{11.31}$$

Definition 11.5.2 $-\Delta^{-1}Y(s)$ is referred to as a superfluous term. The control through a feedback of $\Delta^{-1}Y(s)$ to cancel the superfluous term is known as a control of redundancy removal (Deng, 1965). Through the effect of the feedback of $\Delta^{-1}Y(s)$, the system $G^{-1}(s)Y(s) = U(s)$ is reduced to

$$G^{-1}(s)Y(s) + \Delta^{-1}Y(s) = U(s)(G^{-1}(s) + \Delta^{-1})Y(s) = U(s)$$

That is, $G_*^{-1}(s)Y(s) = U(s)$ has already processed the desired objective structure.

The number of entries in the structural deviation matrix Δ^{-1}, used in a control with abandonment, directly affects the number of components in the controlling equipment. So, when considering the economics, reliability, and ease of application of a dynamic system, one must keep the number of elements in the deviation matrix Δ^{-1} as low as possible. That is to say, in the objective structural matrix, one should try to keep the corresponding entries of the original structure matrix. The idea of control with abandonment is depicted in Fig. 11.8.

11.5.2 Grey Relational Control

Definition 11.5.3 Assume that $Y = [y_1, y_2, \ldots, y_m]^T$ stands for the output vector, and $J = [j_1, j_2, \cdots, j_m]^T$ the objective vector. If the components of the control vector $U = [u_1, u_2, \ldots, u_s]^T$ satisfy

$$u_k = f_k(\gamma(J, Y)); \ k = 1, \ 2, \ \ldots, \ s \tag{11.32}$$

where $\gamma(J, Y)$ is the grey relational degree between the output vector Y and the objective vector J, then the system control is known as a grey relational control.

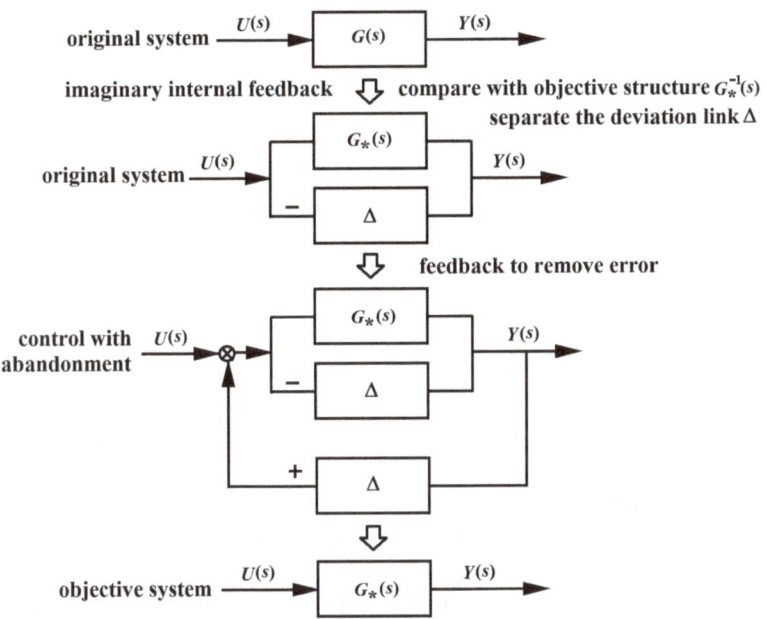

Fig. 11.8 Control of redundancy removal

Fig. 11.9 The grey
relational control system

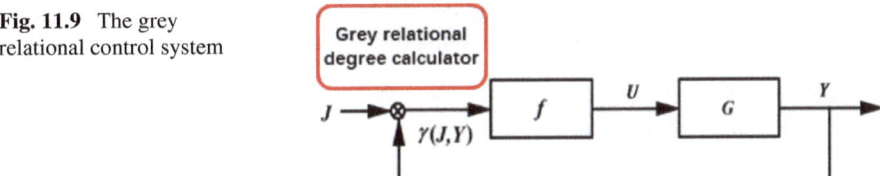

A grey relational control system is obtained by attaching a grey relational controller to the general control system. It determines the control vector U through the grey relational degree of $\gamma(J, Y)$ so that the grey relational degree between the output vector and the objective vector does not go beyond a pre-determined range. The idea of the grey relational control system is depicted in Fig. 11.9.

11.5.3 Control of Grey Prediction

All the various kinds of controls studied earlier are about applying controls after first checking whether or not the system's behavioral sequence satisfies some pre-determined requirements. Such post-event controls evidently suffer from the following weaknesses:

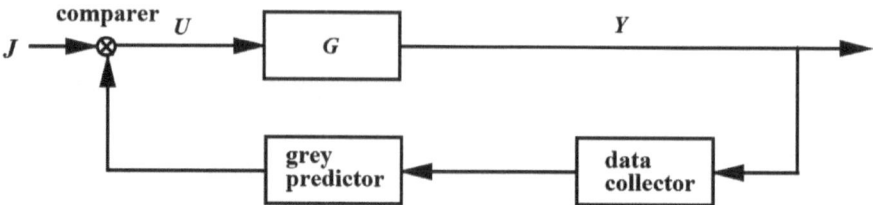

Fig. 11.10 Grey predictive control

(1) Expected future disasters cannot be prevented;
(2) Instantaneous controls cannot be done; and
(3) Adaptability is weak.

The so-called grey predictive control is designed based on the system's future behavioral tendency, which is predicted using the system's behavioral sequences and the patterns discovered from the sequences. This kind of control can be employed to avoid future adverse events from happening; it can be implemented in a timely fashion, and possesses a wide range of applicability.

The idea of a grey predictive control system is graphically shown in Fig. 11.10. Its working principle is that first one must collect and organize the device's behavioral sequence of the output vector Y; secondly, one must use a prediction device to compute the predicted values for the future steps; and lastly, one must compare the predicted values with the objective and determine the control vector U so that the future output vector Y will be as close to the objective J as possible.

Definition 11.5.4 Assume that $j_i(k)$, $y_i(k)$, $u_i(k)$ $(i = 1, 2, \ldots, m)$ are respectively the values of the objective component, output component, and control component at time moment k. For $i = 1, 2, \ldots, m$, let

$$j_i = (j_i(1), j_i(2), \ldots, j_i(n))$$
$$y_i = (y_i(1), y_i(2), \ldots, y_i(n))$$
$$u_i = (u_i(1), u_i(2), \ldots, u_i(n))$$

For the control operator $f : (j_i(\lambda), y_i(\lambda)) \to u_i(k)$,

$$u_i(k) = f(j_i(\lambda), y_i(\lambda)) \tag{11.33}$$

when $k > \lambda$, the system is known as a post-event (or after-event) control; when $k = \lambda$, the system is known as an on-time control; and when $k < \lambda$, the system is known as a predictive control.

Definition 11.5.5 If the control operator f satisfies

$$f(j_i(\lambda), y_i(\lambda)) = j_i(\lambda) - y_i(\lambda) \tag{11.34}$$

That is,

$$u_i(k) = j_i(\lambda) - y_i(\lambda) \tag{11.35}$$

then when $k > \lambda$, the system is known as an error-afterward control; when $k = \lambda$, the system is known as an error-on-time control; and when $k < \lambda$, the system is known as an error-predictive control.

Definition 11.5.6 Let $y_i = (y_i(1), y_i(2), \ldots, y_i(n))(i = 1, 2, \ldots, m)$ stand for a sample of output components and its GM(1,1) response formula be given as follows:

$$\begin{cases} \hat{y}_i^{(1)}(k + 1) = (y_i(1) - \frac{b_i}{a_i})e^{-a_i k} + \frac{b_i}{a_i} \\ \hat{y}_i^{(0)}(k + 1) = \hat{y}_i^{(1)}(k + 1) - \hat{y}_i^{(1)}(k) \end{cases}$$

If the control operator f satisfies

$$u_i(n + k_0) = f(j_i(k), y_i^{(0)}(k)), \quad n + k_0 < k_i, \quad i = 1, 2, \ldots, m \tag{11.36}$$

then the system control is known as a grey predictive control.

In a grey predictive control system, predictions are often done using metabolic models. So, the parameters of the prediction device vary with time. When a new data value output is produced and accepted by the sampling device, an old data value is removed so that a new model is developed. Accordingly, a series of new predicted values are provided. Doing so guarantees the strong adaptability of the system.

Example 11.5.3 Let us look at the EDM (electric discharge machining) grey control system (Yang & Zheng, 1996). The investigation on the control systems of EDM machines has been an important effort in the field of electric discharge machining. Each EDM can be seen as a stochastic time-dependent nonlinear system involving many parameters. Applications mainly include those situations when the conventional controls of linear, constant coefficient systems cannot produce adequate outcomes. The current commonly employed EDM control systems are established based on modern control theory. The frequently applied self-adaptive control systems generally employ mathematical models of approximation with accompanied high costs without actually realizing optimal results. Grey control is not like precise mathematical models based on complete knowledge of a system as addressed in modern control theory. It is also unlike fuzzy control, where the system is treated as a black box as all the information about the internal working of the system is disregarded, which leads to low accuracy controls. The parameters, structures, and other aspects of grey models vary with time. Such dynamic modeling can be highly appropriate for the study of EDM machines with high degrees of uncertainty and produce relatively more satisfactory control effects.

For EDM control systems, the objects of control are EDM machine tools, where outputs need signals from the testing of EDM machine tools as well as the control quantity U, that is, the signals about the control of the EDM machine tools. EDM control systems, in general, mean the control over systems that serve EDM machine tools. For instance, let us look at the traditional gap-voltage feedback servo control system. Due to a lack of linear relationship between the gas voltage, gap size, discharge strength, discharge state, and servo reference voltage, the effect of employing only one gap-voltage feedback servo control system is not very good.

In order to make up for the insufficiency of single loop controls, one can employ double-loop controls with the inner loop being the traditional gap-voltage feedback control and the outer loop being an impulse discharge rate feedback control that instantaneously adjusts the inner loop. The block-design chart of this control system is depicted in Fig. 11.11. Figure 11.11 shows that this control design represents a system of two loops. Based on the collected sequence of gap voltage readings $U_g(K)$, the inner loop employs the GM model to predict the next moment $\widehat{U}_g(K+i+1)$. Here, i stands for the prediction steps, which are then fed into the input end to determine the servo reference voltage value U_s, which is a proportionality coefficient. The outer loop establishes a GM model based on a sequence $Y(K)$ of output values to predict the next steps $\widehat{Y}(K+i+1)$. When these predicted values are compared with requirements Y^*, a sequence $e(K) = \widehat{Y} - Y^*$ of errors is found. These error values are then fed back into the system to adjust the proportionality coefficient K_1 and the servo reference voltage U_s, in order to adjust the inner loop. That is,

$$\Delta U = K_1(Y^* - \widehat{Y}), \ U_s = K_2\widehat{U}_g - \Delta U$$

Therefore, $U_s = K_2\widehat{U}_g - K_1(Y^* - \widehat{Y})$, where parameters K_1, K_2 are determined by experiments.

Example 11.5.4 Let us now look at the grey predictive control for the vibration of a rotor system (Zhu & Zhi, 2002). The theory and methods for active vibration control of rotors have caught more attention in recent years. Many new control theories, such as neural network theory, time-delay theory, self-learning theory, fuzzy theory, and H$^\infty$ theory have gradually been employed in research on active control

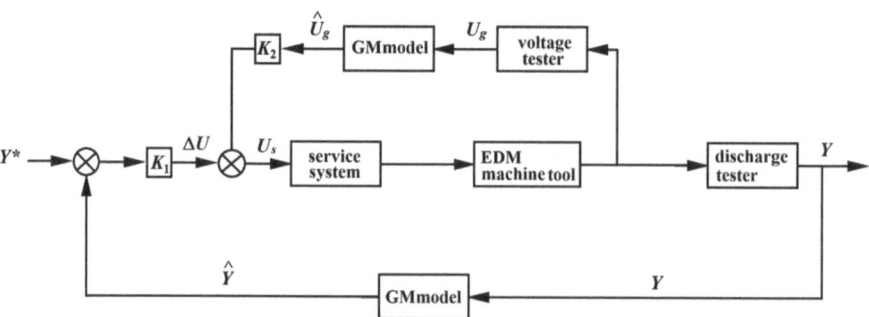

Fig. 11.11 EDM control system

Table 11.1 Sampled data of $I^0(k)$ and $x^0(k)$ when the transducer's sensitivity is 10^4 V/m

$I^0(k)$(A)	0.1	0.125	0.175	0.225	0.325
$x^0(k)$ (dm m)	1.4	1.35	1.2	0.9	0.65

theory of rotors, leading to some good outcomes. For a Jeffcott symmetric rotor follower system with an electromagnetic damper as its executor, we employ control theory and methods of grey predictions to investigate an active amplitude control of vibration. We first establish a grey predictive control module with the GM(1,1) as its main component. In the vibration control system of the rotor, let $I^0(k)$ and $x^0(k)$, $k = 1, 2, \ldots, n$, respectively be electric current inputting into the electromagnetic damper and the corresponding maximum output amplitude of the rotor vibration. By employing the available experimental measurement results from the literature, we obtain a set of data of $I^0(k)$ and the relevant $x^0(k)$, as shown in Table 11.1, when the sensitivity of the transducer is 10^4 V/m.

Based on the mechanism of the GM(1,1) model, we establish the following modification model of the system based on the errors of the grey predictions:

$$\hat{a}^{(0)}(k+1) = -a[x^{(0)}(1) - \beta]e^{-ak} + \delta(k-i)(-a')[q^{(0)}(1) - \beta']e^{-a'k}$$

where $a = 0.1862$; $x^{(0)}(1) = 1.4$; $\beta = 9.3298$; $a' = 0.14$; $q^{(0)}(1) = 0.36$; $\beta' = 3.78$, and

$$\delta(k-i) = \begin{cases} 1 & k \geq i \\ 0 & k < i \end{cases}$$

The design of our grey predictive control of the rotor system is shown in Fig. 11.12.

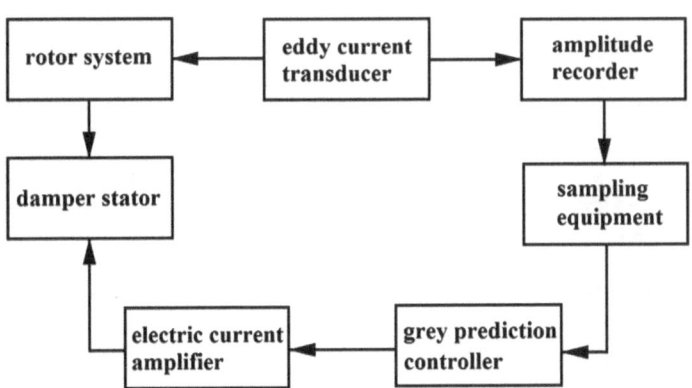

Fig. 11.12 Grey predictive control of the rotor system

Table 11.2 Simulations and actual measurements

I (A)	X_{1m} (m)	X_{2m} (m)	X_{3m} (m)	e_{12} (%)	e_{13} (%)	e_{23} (%)
0.1	1.41×10^{-4}	1.4×10^{-4}	1.4×10^{-4}	0.71	0.71	0
0.125	1.27×10^{-4}	1.227×10^{-4}	1.35×10^{-4}	3.5	-5.93	-9.11
0.175	1.03×10^{-4}	0.949×10^{-4}	1.2×10^{-4}	9	-13.8	-20.92
0.225	0.83×10^{-4}	0.745×10^{-4}	0.9×10^{-4}	11.4	-7.78	-17.22
0.325	0.55×10^{-4}	0.46×10^{-4}	0.65×10^{-4}	19.57	-13.38	-29.23

In this control system, the displacement signal of the rotor system is measured by the eddy current transducer. The sampling equipment collects the data from the amplitude recorder, and through the effect of the grey prediction controller, controlling voltage is produced. This voltage is transformed into a controlling electric current through the current amplifier. Then, when this electric current flows through the stator coil of the electromagnetic damper, an electromagnetic force is created, which in turn controls the amplitude of vibration of the rotor within the expected range so that the system's stability is achieved.

For this grey predictive control system developed for the vibration of the said Jeffcott symmetric rotor follower, our computer simulation, when compared to the physical measurements of the amplitudes, indicates that the maximum amplitudes under the control are only about 7% of those physically observed without the control imposed on the rotor system. Table 11.2 respectively provides the results of the maximum amplitudes of two separate computational simulations and the actual measurements X_{1m}, X_{2m}, and X_{3m}, along with the change in the static electricity i of the electromagnetic damper, when the sensitivity of the transducer is $k_1 = 10^4$ V/m, and the corresponding errors e_{12}, e_{13}, and e_{23} between X_{1m} and X_{2m}, X_{1m} and X_{3m}, and X_{2m} and X_{3m}.

References

Deng, J. L. (1965). A synthesis method of parallel correction devices for multivariable linear systems. *Acta Automatica Sinica, 3*(1), 13–26.

Deng, J. L. (1985). *Grey control system*. Wuhan: Press of Huazhong University of Science and Technology.

Su, C. H., & Liu, S. F. (2008). On the asymptotic stability of grey stochastic linear delay systems. *Control and Decision, 23*(5), 571–574.

Su, C. H., & Liu, S. F. (2009). On robust stability of p-moment index of stochastic system with distributed delay and the interval parameters. *Applied Mathematics and Mechanics, 30*(7), 856–864.

Su, C. H. (2012). Robust stability of grey stochastic time-delay systems with impulsive effects. *Systems science and mathematics, 32*(5), 537–548.

Yang, J., Zheng, L. G., Zhou, J. L, Ni, Y. H. (1996). *Research on EDM grey control system. Electrical Processing*, (5), 22–23+45.

Zhu, X. P., Zhi, X. Z., et al. (2002). Grey predictive control of the vibration of rotor systems. *Mechanical Science and Technology, 21*(1), 97–101. http://igss.nuaa.edu.cn/

Chapter 12
Spectrum Analysis of Sequence Operators

12.1 Introduction

Behavioral prediction of a system under the influence of shocking disturbances has always been a difficult problem. In this case, the available data of the system's behavior can no longer truthfully reflect the law of change of the system. At this situation, if we directly established our model and made our predictions using the severely affected data without first considering the disturbance, then our predictions would most likely fail. This is because the model would not have described the true state of change of the underlying system. Therefore, one of the main tasks of grey forecasting is to uncover the laws of change of certain system variables themselves based on the available data of the system (Liu, 1991).

As usually, a general data sequence composited by various factors of trend and noise (Fig. 12.1a), cycles (Fig. 12.1b), shock disturbance by long-duration impulse (Fig. 12.1c), shock disturbance by transient impulse (Fig. 12.1d), and some factors be ignored (Fig. 12.1e), even some factors joining with noise or be seen as noise (Fig. 12.1a). The evolution rule of data series may change at some points which are called change points (Page, 1955). Before and after the change points, people need to use different models to describe the change rule of data series. The difference may be the change of model form, or the change of one or some parameters in the model (Fig. 12.1f).

It's very difficult to analyze and discriminate the factors of a data sequence in time domain. Thanks to spectrum analysis, we can transfer data in time domain to frequency domain by Fourier transformation. Then analyze and discriminate various factors of a data sequence in frequency domain (Liu et al., 2020).

© The Author(s), under exclusive license to Springer Nature Singapore Pte Ltd. 2022
S. Liu, *Grey Systems Analysis*, Series on Grey System,
https://doi.org/10.1007/978-981-19-6160-1_12

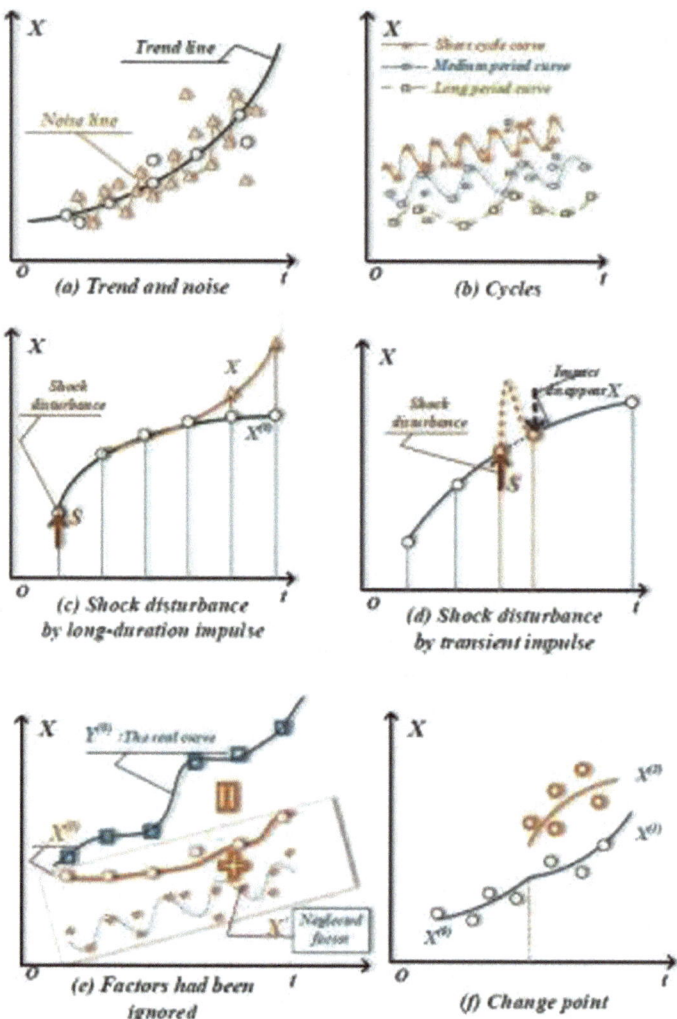

Fig. 12.1 Various factors of a data sequence

12.2 Spectrum Analysis of Time Series Data

Changhai Lin et al. introduced spectrum analysis into grey system theory firstly in 2019 (Lin et al., 2019, 2022).

Generally, system data is presented in the form of time series data. Due to the influence of system disturbance, there will be some deviation between the observed data and the original behavior data series. It is of great significance for people to understand the evolution law of system to analyze and recognize the influence of system disturbance factors correctly. The spectrum analysis of time series data provides us a

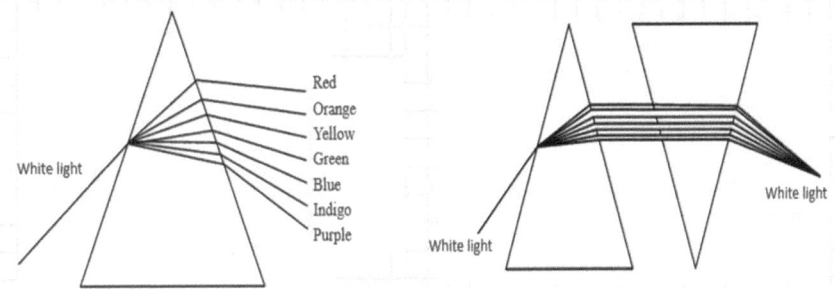

Fig. 12.2 Decomposition and synthesis of white light

new perspective to understand time series data, which is another form of time series data.

Spectral analysis is an important term proposed by Isaac Newton. He first used the concept of spectral analysis in his paper submitted to the Royal Society in 1672 (Newton, 1672). In this paper, Newton mentioned the famous prism experiment. As we all know, prism can decompose sunlight into seven colors. The principle is to decompose the light of different colors from the white light by using the different refractive index of the medium to the light of different colors. In the experiment, Newton also used two positive and negative inverted prisms. Through the first prism, white sunlight was decomposed into different colors of light, while the second inverted prism synthesized different colors of light into white sunlight. In the whole process of decomposition and synthesis, the essence of light has not changed. Prism can be seen as a conversion tool to show the characteristics of light (Newton, 1672) (Fig. 12.2).

The knowledge of physics tells us that every color of light represents a frequency range in visible light. Color analysis of light belongs to the scope of spectrum analysis.Spectrum analysis, as the name implies, its research on the object is carried out in the frequency domain. In the process of system analysis, the data of system behavior observed in the real physical world are mostly time series data recorded based on time, which can be abstracted as a function of time and belong to the scope of time domain. The spectrum analysis of time series data is based on signal decomposition. With the help of the mathematical tool of Fourier transform, the spectrum analysis regards the time series signal as the superposition of sine waves or cosine waves with different periods and amplitudes. The sine wave or cosine wave with different periods and different amplitudes is defined as the frequency content with one amplitude in frequency domain. The conversion process of time series data from time domain to frequency domain can also be expressed as the process of mapping different frequency content of time signal to frequency domain.

The spectrum analysis of time series data is a method of information mining. Information that is not easy to find in time domain analysis can be found through spectrum analysis. By decomposing and analyzing the time series data, we can quantitatively analyze the periodic law contained in the time series data. The magnitude and proportion of different frequency content can be quantitatively analyzed by calculating the frequency amplitude at different frequency points of time series signal.

12.3   Filtering Effect of Mean Operator and Accumulation Operator

The classical model of grey system - mean GM (1,1) is based on accumulation operator and mean operator. The dual effects of accumulation operator and mean operator produce magical effects, so that people can use the mean GM (1,1) to obtain high simulation and prediction accuracy based on few data.

In 1987, Professor Deng Julong studied the grey exponential law of the accumulation operator (Deng, 1987), and found that the the randomness of grey data sequence can be weaken under the action of the accumulation operator and show the variation law of the exponential function. Referring to the digital signal processing (DSP) system, we will study the mean operator and accumulation operator in the frequency domain through Z-transform, as well as the filtering effect of their series action. The contents and main conclusions of this section are based on the research of Lin et al. (2020).

12.3.1   Filtering Effect of Mean Operator

The general 2-term weighted moving average operator can be rewritten into the following form

$$y[n] = b_0 x[n] + b_1 x[n-1] \quad b_0 + b_1 = 1 \tag{12.1}$$

Equation (12.1) can be regarded as the transfer function of DSP signal system, the following Eq. (12.2) can be obtained from Z transformation.

$$Y[z] = b_0 X[z] + b_1 X[z] z^{-1} \tag{12.2}$$

From Eq. (12.2), it is easy to obtain the frequency domain expression of the transfer function of the digital filter system corresponding to the 2-term weighted moving average operators as follows

$$H[z] = \frac{Y[z]}{X[z]} = b_0 + b_1 z^{-1} \tag{12.3}$$

Let $b_0 = b_1 = 0.5$, The frequency domain expression of the transfer function of the digital filter system corresponding to the mean operator can be obtained as

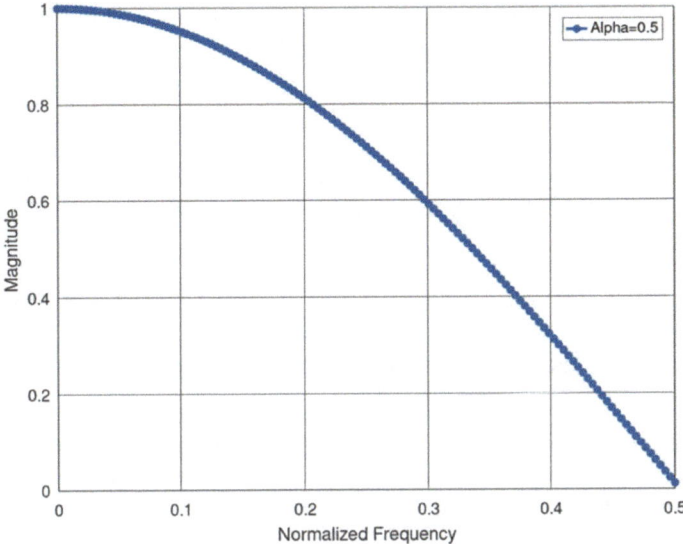

Fig. 12.3 The frequency domain curve of mean operator equivalent filter transfer function

follows

$$H[z] = \frac{Y[z]}{X[z]} = 0.5 + 0.5z^{-1} \tag{12.4}$$

It can be seen from Fig. 12.3 that when the frequency content is 0, the frequency amplitude is 1, and when the frequency content is greater than zero, the frequency amplitude is less than 1. And the higher the frequency content, the smaller the frequency amplitude. That is, the mean operator has the effect of low-pass filtering, the low-frequency part (Evolution Law) of the data remains basically unchanged under the action of the mean operator, and the high-frequency part (fluctuation or disturbance) will be compressed and suppressed. Through spectrum analysis, it is further confirmed that the randomness of grey data sequence can be weaken and the real evolution law will be presented under the action of mean operator.

12.3.2   Filtering Effect of Accumulation Operator

The first order accumulation operator (1-AGO) can be rewritten into the following form

$$y[n] = x[n] + y[n-1] \tag{12.5}$$

Equation (12.5) can be regarded as the transfer function of DSP signal system, the following Eq. (12.6) can be obtained from Z transformation.

$$Y[z] = X[z] + Y[z]z^{-1} \tag{12.6}$$

From Eq. (12.6), it is easy to obtain the frequency domain expression of the transfer function of the digital filter system corresponding to 1-AGO as follows

$$H[z] = \frac{Y[z]}{X[z]} = \frac{1}{1 - z^{-1}} \tag{12.7}$$

From Eq. (12.7) and $z = e^{j\omega}$, it follows that

(1) When $0 \le \omega \le \pi/3$, |H[ω]|>1. The amplitude of output Y[ω] will be greater than the amplitude of input X[ω].The system amplifies the input spectrum.
(2) When $\omega < \pi/3$, $|H[\omega]| < 1$. The amplitude of output Y[ω] will be less than the amplitude of input X[ω]. The system compresses or suppresses the input spectrum.
(3) $z = 1$ is the pole of the transfer function of the digital filter corresponding to 1-AGO.

The frequency domain curve of transfer function of 1-AGO equivalent filter as shown in Fig. 12.4.

The first-order accumulation operator equivalent digital filter belongs to low-pass filter, that is, the low-frequency content (less than a critical frequency) in the input

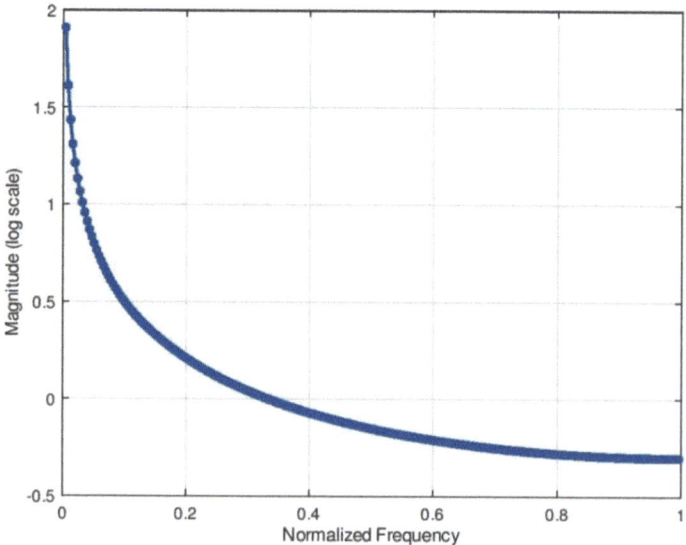

Fig. 12.4 The frequency domain curve of transfer function of 1-AGO equivalent filter

signal can pass through or be amplified. The high frequency content (greater than a critical frequency) in the signal will be compressed or suppressed.

The data fluctuation and random disturbance of general discrete data series belong to high-frequency content. These information will be suppressed during the action of 1-AGO equivalent digital filter. Aperiodic system evolution law belongs to low-frequency signal, which can pass through or be amplified in the process of 1-AGO equivalent digital filter. It is also proved that for general non-negative quasi smooth sequences, the randomness can be reduced by the action of accumulation operator, showing an approximate exponential growth law.

Since the transfer function of digital filter corresponding to 1-AGO has pole, and the frequency content $\omega = 0$ is its pole. This means that when the frequency content is 0, the transfer function of 1-AGO corresponding digital filter has infinite amplification effect. Furthermore, the conclusion of theorem 4.7.3 of this book is confirmed from the mechanism: the function of accumulation operator shouldn't over. That is, if the action sequence of the r-th accumulation operator of $X^{(0)}$ has obvious exponential law, the application of AGO operator will destroy its regularity and turn the exponential law grey.

12.3.3 Filtering Effect of Series Operator

Let the corresponding digital filter transfer functions mean operator (12.4) and accumulation operator (12.7) be denoted by $H_E(z)$ and $H_A(z)$ respectively. According to the transfer function calculation formula of series system, the transfer function of equivalent filter of the series operator of 1-AGO and mean operator can be obtained as follows

$$
\begin{aligned}
H[z] &= H_A(z)H_E(z) \\
&= \frac{1}{1 - z^{-1}} \cdot \left(0.5 + 0.5z^{-1}\right) \\
&= \frac{0.5 + 0.5z^{-1}}{1 - z^{-1}}
\end{aligned}
\tag{12.8}
$$

The frequency domain curve of series equivalent filter of the 1-AGO and mean operator as shown in Fig. 12.5.

As can be seen from the comparison with Fig. 12.4 that The accumulation operator acting alone or the accumulation operator acting in series with the mean operator can produce similar amplification effect on the low-frequency part of the signal. But for the high-frequency part of the data sequence (fluctuation and noise), the series operator

$$
H[z] = H_A(Z)H_E(Z)
$$

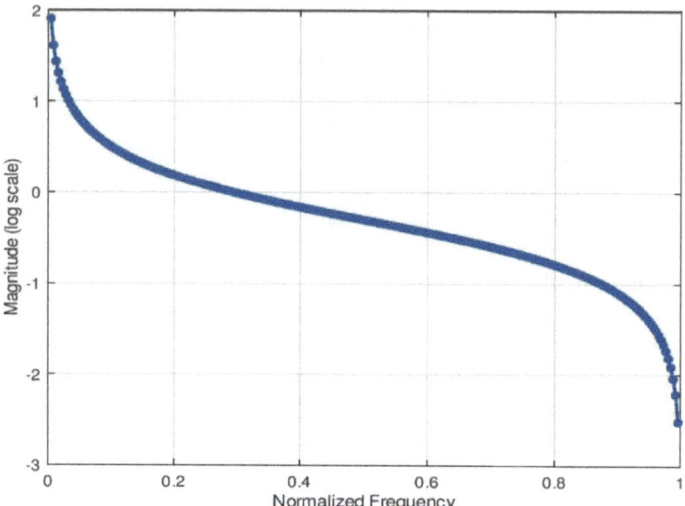

Fig. 12.5 The frequency domain curve of series equivalent filter of the 1-AGO and mean operator

has stronger suppression effect than the accumulation operator acting alone. The signal-to-noise ratio of the series operator sequence is significantly improved. This also proves from the mechanism why the mean GM (1,1) model can obtain high simulation and prediction accuracy based on small data in most cases.

12.4 Spectrum Analysis of Buffer Operator

In order to solve the prediction problem of impact disturbance system, Liu Sifeng proposed the concept of buffer operator, established the axiom system of buffer operator, and designed the widely used average weakening buffer operator (AWBO) (Liu, 1991). Please refer to Chap. 4 of this book for details.

Let the x(k)d in the following AWBO

$$x(k)d = \frac{1}{n-k+1}[x(k) + x(k+1) + \cdots + x(n)]; k = 1, 2, \ldots, n$$

be denoted by y(k). The AWBO can be rewritten into the following formula (12.9)

$$y(k) = \frac{1}{n-k+1}\left[\sum_{i=1}^{n} x(i) - \sum_{i=1}^{k-1} x(i)\right] k = 1, 2, \ldots, n \qquad (12.9)$$

Replace k with k−1, we have

$$y(k-1) = \frac{1}{n-k+2}\left[\sum_{i=1}^{n}x(i) - \sum_{i=1}^{k-2}x(i)\right] k = 2, \ldots, n \qquad (12.10)$$

Eliminate the denominator at the right end of Eqs. (12.9) and (12.10), we have

$$y(k)(n-k+1) = \left[\sum_{i=1}^{n}x(i) - \sum_{i=1}^{k-1}x(i)\right] k = 1, 2, \ldots, n \qquad (12.11)$$

$$y(k-1)(n-k+2) = \left[\sum_{i=1}^{n}x(i) - \sum_{i=1}^{k-2}x(i)\right] k = 2, \ldots, n \qquad (12.12)$$

Subtract Eqs. (12.11 and 12.12), we obtain

$$y(k)(n-k+1) - y(k-1)(n-k+2) = \sum_{i=1}^{k-2}x(i) - \sum_{i=1}^{k-1}x(i) \qquad (12.13)$$

Therefore

$$y(k)(n-k+1) - y(k-1)(n-k+2) = -x(k-1), k = 2, 3, \ldots, n \quad (12.14)$$

The digital signal processing expression corresponding to Eq. (12.14) as follows

$$Y(Z)(n-k+1) - Y(Z)Z^{-1}(n-k+2) = X(Z)Z^{-1} \quad k = 2, 3, \ldots, n \quad (12.15)$$

So, the transfer function of AWBO can be obtained

$$H(Z) = \frac{Y(Z)}{X(Z)} = \frac{Z^{-1}}{\left[(n-k+1) - (n-k+2)Z^{-1}\right]} \qquad (12.16)$$

The actual data simulation results show that the AWBO equivalent digital filter also belongs to low-pass filter. For the low-frequency part of the input signal, the amplitude of the spectrum of the AWBO action sequence is higher than that of the reference, which means that AWBO has amplification effect on the low-frequency content in the sequence. For the high-frequency part of the input signal, the amplitude of the spectrum of the AWBO action sequence is lower than the reference amplitude, indicating that AWBO has the effect of restraining, attenuating or blocking the high-frequency content in the sequence. The high-frequency part of the input signal of the impact disturbance system is mainly composed of impact disturbance components. Therefore, AWBO can weaken the impact disturbance (Lin et al., 2021, 2022).

References

Deng, J. L. (1987). Grey exponential law of accumulating generation: The optimization information problem of grey control systems. *Journal of Huazhong University of Science and Technology, 12*(5), 11–16.

Lin, C. H., Wang, Y., & Liu, S. F. et al. (2019). On spectrum analysis of different weakening buffer operators. *The Journal of Grey System, 31*(4), 111–121.

Lin, C. H., Song, Z. Y., Liu, S. F., et al. (2021). Study on mechanism and filter efficacy of AGO/IAGO. *Grey Systems: Theory and Application, 11*(1), 1–21.

Lin, C. H., Liu, S. F., Fang, Z. G., & Yang, Y. J. (2022). Spectrum analysis of moving average operator and construction of time-frequency hybrid sequence operator. *Grey systems-theory and application, 12*(1), 101–116.

Liu, S. F. (1991). The three axioms of buffer operator and their application. *Journal of Grey System, 3*(1), 39–48.

Liu, S. F., Lin, C. H., Tao, L. Y., et al. (2020). On spectral analysis and new research directions in grey system theory. *The Journal of Grey System, 32*(1), 108–117.

Newton, I. (1672). A letter to the royal society

Page, E. S. (1955). A test for a change in a parameter occurring at an unknown point. *Biometrika, 42*(3–4), 523–527.

Appendix
Introduction to Grey Systems Modeling Software

A.1 Introduction

In 1982, Julong Deng initiated and established grey systems theory. Currently, grey systems theory is widely applied in areas such as social sciences, economics, agriculture, meteorology, military and science, providing solutions to a large number of practical problems and challenges met in everyday life. Various versions of grey systems modeling software have played a very important role in such large scale practical applications of grey systems theory. Along with the rapid development of information technology, high level programming languages have gradually matured, applications of computing packages have been routinized, and grey systems modeling programs have also become sophisticated.

In 1986, Xuemeng Wang and Jiangjun Luo (Wang & Luo, 1986) created their grey systems modeling software using BASIC language and published *Programs of Grey Systems' Prediction, Decision-Making, and Modeling*. In 1991, Xiuli Li and Ling Yang (Liu & Guo, 1991) respectively developed grey modeling software using GWBASIC and Turbo C. In 2001, Xuemeng Wang, Jizhong Zhang, and Rong Wang published the book entitled "Computer Procedures for Grey Systems Analysis and Applications," in which they listed the structure and procedure codes established for grey modeling (Wang et al., 2001). All of these computer software packages were developed on the DOS platform and have become obsolete in the more user-friendly Windows framework.

In 2003, Dr. Bing Liu (Liu et al., 2003) developed the first grey systems modeling software for Windows using VisualBasic6.0. As soon as this package was available, it was most welcomed in the community of scholars and practitioners of the grey systems research, and became the first choice of application in the field of grey systems modeling. With the evolution of software development technology, changes in computer operation, and continued grey theory research progress, some of the weaknesses of such software packages came to the fore, including the following:

S. Liu, *Grey Systems Analysis*, Series on Grey System, https://doi.org/10.1007/978-981-19-6160-1

- Data entry was tedious.

The single worksheet frame limited the operational flexibility on data sequences. It was especially inconvenient when large amounts of data were dealt with in clustering analysis and grey decision-making. Additionally, the only available way for data entry made users feet tired so that the efficiency and accuracy of data entry were greatly affected.

- The classification of the modules was not scientifically sound.

This software system divided modules according to the number of data participating in the modeling process. However, the modules should have been designed according to functions.

- The software system could not show relevant computational procedures.

Most of the users of grey systems theory are scientists and practitioners. The purpose of their use of the software system was mainly their scholarly works. This means that, other than the computational outcomes, such professionals are also very interested in knowing specific procedural details. However, the original software package could only provide the final results of the computation and was unable to reveal relevant computational details.

- The system's capability was disconnected from the most recent research.

Grey systems theory has been an extremely active area of the broader systems science. In particular, in recent years some works with high practical value have appeared. However, the software system was not upgraded along with research progress, causing the software to become obsolete in terms of its functions and capability.

- Problems with the choice of package development.

VisualBasic6.0 is a graphics-based software development tool created by Microsoft Company. Due to its simplicity, functionality, versatility, and other strengths, as soon as it was introduced it was welcomed by many software developers. However, VisualBasic6.0 is an IDE (integrated development environment) based on BASIC, a typical programming language with many known weaknesses that greatly limit its applicability in scientific computing. This is because scientific computing requires high levels of accuracy. Thus, grey systems modeling software developed using VisualBasic6.0 inherently suffers from many weaknesses.

A.2 Software Features and Functions

On one hand, an ideal grey systems modeling software package needs to have the computational power to handle practical models, and on the other hand it has to deal with user confirmation, registration, and other functionalities. The software system accompanying this book sufficiently combines the capabilities of the C/S (client/server) and B/S (Browser/Server) modules, where the C/S part completes computational functions, while the B/S part handles the relevant operations that serve the user and his communication with the server. With a view to improve existing systems, the design of this package focuses more on the reliability, practicality, compatibility, upgradability, accuracy, operational convenience, visual appeal and user friendliness (Zeng et al., 2011). This package has the following characteristics:

- Data entry is convenient and fast.

  For data sequences of the same kind, the package provides a rectangular window into which the user can simply copy the sequences with one operation. For grey clustering and grey decision-making modules that involves large amounts of data, it is evidently inconvenient to employ the traditional way of entering data values. In such instances, the user can enter data in an Excel document and then open the data file into this package system. This software system makes use of the powerful data entry ability of Excel while making data entry convenient for the user.
- Modules are designed according to functionalities.

  In software engineering, a module is a relatively independent system unit of procedures. Each such unit of procedures handles and materializes a relatively independent task. In other words, it contains a group of independent procedures. Each program module has its own external characteristics, such as its own name, label, and interfaces. During the design of this software package, the developer scientifically organized the contents of grey systems theory, defined the relevant functions and related modules.
- This system provides operational details as well as periodic results.

  For modules with complicated computational procedures where intermediate results are also important, the system provides a textbox that can store and show multi-line operational details. The user can monitor data changes in each computational step so that he can further understand how the model operates. Also, the software interface provides relevant information to remind the user of the relevant formulas employed in the model.
- The functionalities of the modules are greatly expanded.

  Based on current practical applications of grey systems theory, combined with the most recent research results, this software system is the most up-to-date system available in the market. It includes: weakening operators (mean weakening buffer operators, geometric mean weakening buffer operators, weighted mean weakening buffer operators), strengthening operators (mean strengthening buffer operators, geometric mean strengthening buffer operators, weighted mean strengthening buffer operators), grey incidence analysis (relative degree of incidence, closeness degree of incidence), clustering analysis (based on center-point triangular whitenization weight functions), grey prediction ($GM(1, n)$ and $DGM(1,1)$ models), grey decision analysis (intelligent grey target decision making), among other contents.

Table A.1 The basic constitution of the grey system modeling software

Grey system modeling software	B/S part	User info
		Statistics
	C/S part	Grey sequence operators
		Grey relational analysis
		Grey clustering evaluation
		Grey forecasting
		Grey decision-making

- The degree of accuracy of the computational results can be adjusted.

  The computation precision of different systems is different. In this software system there is a ComboBox, which can select of computational precision. Therefore, the user can choose the desired degree of accuracy for his work.
- The operation of the software system is convenient and easy to learn.

  This software system is based on the Windows interface using pull-down menus, where the commonly employed modeling techniques of grey systems theory are effectively gathered together. The user only needs to have an elementary under-standing of how a desktop PC works to successfully use this software system. At the same time, this system has a relatively strong ability to locate and correct mistakes. When an illegal operation is performed, the system will provide an accurate and detailed hint.
- This system is developed using Visual C#.

  C# is an object-oriented programming language created by Microsoft and an important part of Microsoft'sv .NET development environment. Also, Microsoft Visual C# is an integrated development environment (IDE) constructed on C# by Microsoft. It is designed for the operation of many application software packages created on the .NET framework. C# possesses powerful capabilities, type safety, object orientation, and other superb functions. It is currently the main development tool of C/S software architecture.

A.3 Main Components

The new edition of the grey system modeling software consists of five modules including grey sequence operators, grey relational analysis models, grey clustering evaluation models, grey forecasting models and grey models for decision-making, given the currently available research on grey systems theory and its practical applications (Table A.1). The software system modules are shown in Tables A.2, A.3, A.4, A.5 and A.6.

Table A.2 Grey sequence operators

Grey sequence operators	Weakening operators	Average weakening buffer operator (AWBO)
		Weighted average weakening buffer operator (WAWBO)
		Geometric average weakening buffer operator (GAWBO)
		Weighted geometric average weakening buffer operator (WGAWBO)
	Strengthening operators	Even strengthening buffer operator (ESBO)
		Average strengthening buffer operator (ASBO)
		Weighted average strengthening buffer operator (WASBO)
	Information mining operators	Accumulating generation operator
		Inverse accumulating generation operator
		Even operator by adjacent neighbor
		Operator of stepwise ratio

Table A.3 Grey relational analysis models

Grey relational analysis models	Deng's model of degree of grey relational model
	Absolute degree of grey relational model
	Relative degree of grey relational model
	Synthetic degree of grey relational model
	Closeness degree of grey relational model
	Similitude degree of grey relational model

Table A.4 Grey clustering evaluation models

Grey clustering evaluation models	Grey clustering model of variable weight
	Grey clustering model of fixed weight
	Grey clustering model using center-point mixed triangular possibility function
	Grey clustering model using end-point mixed triangular possibility function

Table A.5 Grey forecasting models

Grey forecasting models	Singular variable models	Even GM(1,1)
		Original difference GM(1,1)
		Even difference GM(1,1)
		Discrete grey model
		Grey Verhulst model
	Multi-variable models	Model GM(0, N)
		Model GM(1,N)

Table A.6 Grey models for decision-making

Grey models for decision-making	Weighted multi-attribute grey target decision
	Two stages model for decision-making

A.4 Operation Guide

A.4.1 The Confirmation System

To verify legal ownership, the user needs to enter his account number and password before he can actually start using the system. However, if every time the user uses the system he has to confirm his legal ownership of the software package, it will become an annoyance. So, to guarantee the legal ownership of the user and maintain the operational simplicity of the system, the system applies the XML-based client programming technique. When the user attempts to run the program for the first time, the system will prompt him to provide the needed account number and password. The provided data will then be delivered through the internet to the database located at the server to verify their legality. When the user attempts to use the program on different, subsequent occasions, he can directly enter the main interface window without having to enter his account number and password again.

On the first time of confirmation, if the user does not have an account number or password, he needs to click on the "User registration" button (see Fig. A.1) to

Fig. A.1 The confirmation window

register for a free user account (B/S). If the user forgets his password, he can click on the "Recall password" button to retrieve his password. Figure A.2 is the flow chart of confirmation.

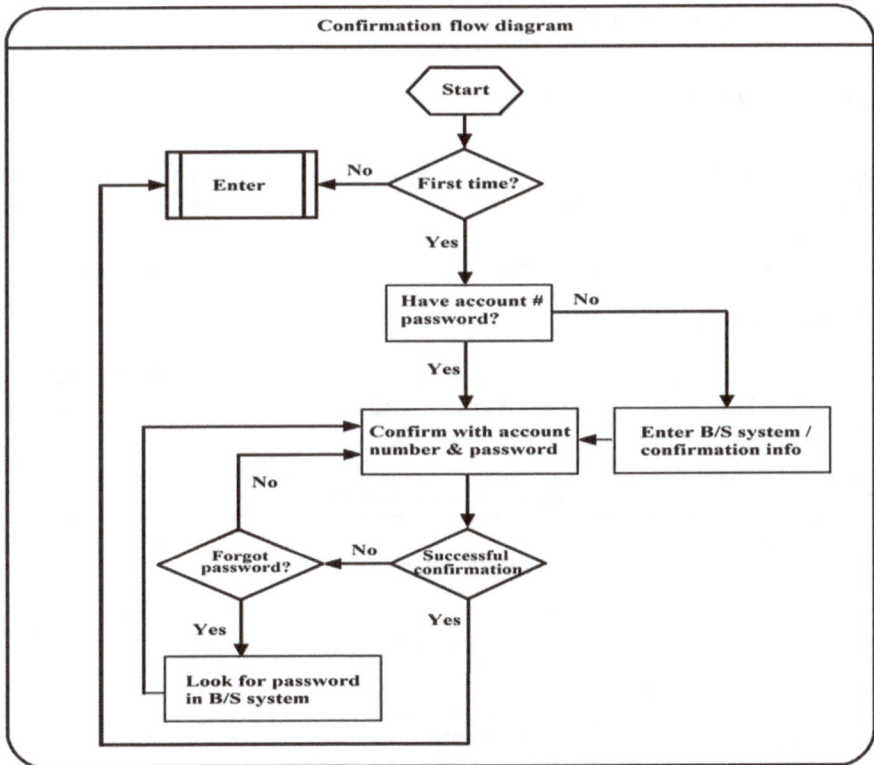

Fig. A.2 The confirmation flow chart

Fig. A.3 The main interface window

A.4.2 Using the Software Package

After successful confirmation of legal ownership, the user will enter the system's main interface window, as shown in Fig. A.3. Various grey systems theory modules (and their sub-modules) are administrated through menus.

Figure A.4 provides the flow chart of various sub-modules of the system.

I Data Entering

Before running the program, one needs to first enter data into the software package and specify the system parameters. As mentioned earlier, there are two ways to input

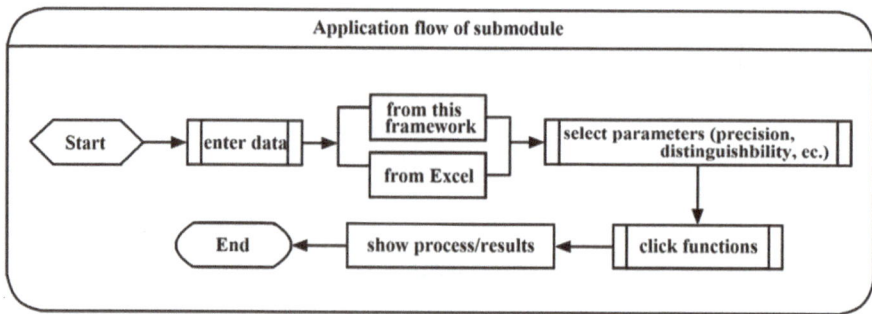

Fig. A.4 Sub-modules

data. One can directly enter data into the provided text box or import data from an external Excel document. For those modules that require large amounts of data, the only way provided for entering data is through importing data from Excel documents. The following sections look at the details of these two data entering methods.

- Enter data directly into the provided text box.

With VisualC#, there are two kinds of controllers available for direct data entry. One is the TextBox controller, and the other the ComboBox controller. The former controller is used to develop the standard Windows' editing controller of the text-box, which is used to acquire the user's input or show what is already stored in the storage space. When entering data into the text-box, right click the mouse inside the text box. When the cursor blinks inside the text box, one can start entering data. The ComboBox of the Windows' window group is mainly used to show data in a down-drop list box. As a default, ComboBox consists of two parts: the top is a text box in which the user is allowed to enter data, and the bottom is a list box where the user can make selections. It is because the ComboBox consists of the text box on the top and the list box at the bottom that it is named a ComboBox. When using the ComboBox to enter data, the user needs to first check whether or not the list box contains the data he wants. If so, he can simply use the mouse to directly make the selection; otherwise, he needs to enter data into the text box on the top. The detailed procedure for entering data in the ComboBox is similar to that of operating the TextBox and is therefore omitted here.

   Note: When entering data using either the TextBox or the ComboBox, the state of entry method needs to be adjusted to half-angle. Data values entered in the state of full angle will be treated by the program as illegal data, which will directly affect the normal operation of the program and potentially lead to unexpected outcomes.

- Import data from an Excel document.

Both the TextBox and ComboBox can only accept small amounts of data values. For entering large sums of information, the use of either the Textbox or the ComboBox is inefficient, and can also lead to errors. To resolve the problem of entering large sums of data values when dealing with grey clustering and grey decision-making, for instance, it is very often the case that large amounts of information are involved, and this software system makes use of the powerful Excel. First enter and edit the needed data in Excel, and then use the provided interface to import the Excel data into the software system. Excel is one of the components of Microsoft Office. It is a tabulated testing and computing software developed for Microsoft Windows and Apple's Macintosh. Its straightforward interface, excellent capabilities of computation and graphics make it the most widely employed PC software used for data analysis. Through Excel, our software package system can conveniently acquire data.

   Each Excel document generally contains three tables, respectively labeled as Sheet1, Sheet 2, and Sheet3. When an Excel document opens, it generally shows Sheet1. When entering data according to the system's requirements, one can directly type in the corresponding values in the relevant rows and columns. Upon finishing data entry into the Excel document, one can employ our system's input function to import the Excel data. When importing an Excel data document, first select the path from which the Excel file is located. As soon as the importing path and location of the file is confirmed, the data will be successfully imported. In fact, the process of importing data connects the Excel file to our system through a specific path so that the data in the Excel file can be mapped into the database controller DataGridView.

DataGridView is a database controller of VisualC#, which can exactly and entirely reveal data from a source file. Through the DataGridView controller of VisualC#, data can be acquired from an Excel file. However, our system does not provide any of the editing capabilities of DataGridView. In other words, if it is found in DataGridView that there is an error in the data, this error cannot be corrected directly within DataGridView. Instead, one has to return to the original Excel file to make the correction and then reimport the entire corrected file back into the grey systems modeling package.

Notes:

- The DataGridView controller does not have any editing capability. To make changes in the data, one has to do it in the original Excel file.
- When entering data into an Excel document, one needs to do so in the mode of "half-angle." All data entered in the mode of "full-angle" will be treated as illegal entries, which will directly affect the normal operation of the grey systems modeling package, and potentially lead to unexpected outcomes.
- The table names of the Excel file have to be the default Sheet1, Sheet2 and/or Sheet3 without any modification, otherwise the import of data will be affected.
- The data entry field of Excel is very large. However, one often needs only a few rows and columns. Make sure that there are no symbols or blank cells accidently entered into other area of the field. Otherwise, the data transfer will be affected.

II Model Computations

(1) Grey sequence operators

Click on "Sequence generation." From the pull-down menu that appears, select the module according to the practical modeling need. The corresponding detailed modeling interface appears. Let us use the "average weakening buffer operator" as an example to illustrate how to apply grey sequence generations. What is shown in Fig. A.5 is the interface of the mean weakening buffer operator.

This interface window contains three main areas: the first shows the original data sequence, which is the area for data entry or importing data; the second area shows the "order and outcome precision," in which it is possible to adjust the order of the operator being applied and the corresponding precision of the computational outputs based on one's modeling needs; and the third the area is where computational results are shown. After the data entry is completed, click on the "mean weakening buffer operator (AWBO)" button. Immediately, the generated sequence will appear in the generated sequence window. Figure A.6 shows a work sheet of an Excel document.

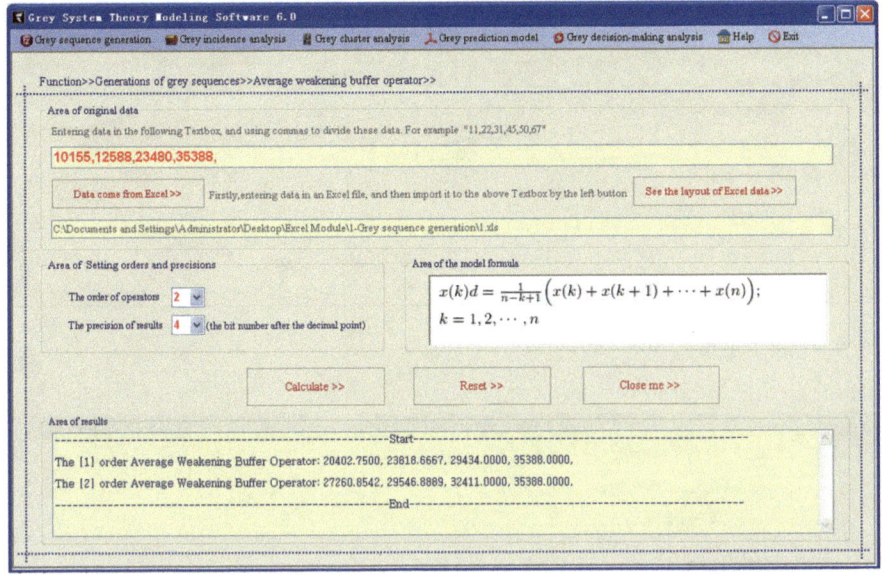

Fig. A.5 The interface of the mean weakening buffer operator

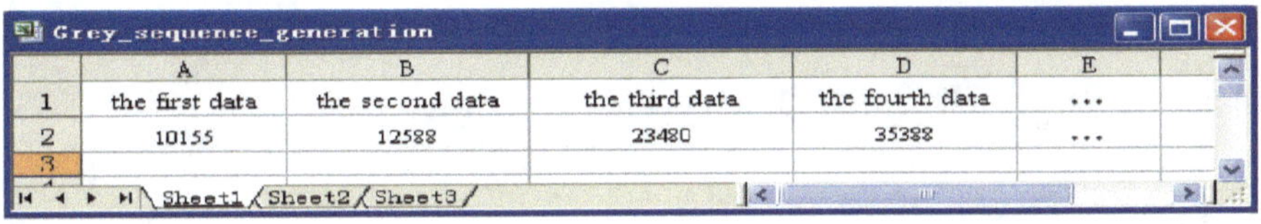

Fig. A.6 The required Excel file format

When applying this Excel capability and importing data from Excel, the user has to follow this shown format exactly.

(2) Grey relational analysis models

Similar to the generation of grey sequences, there are two ways to input data for all parts of relational analysis, so such data entry details are omitted here. However, this is not valid for Deng's degree of grey relation due to the need for a large amount of data. For Deng's degree of grey relation, this software system allows only data entry through Excel documents without the option of direct data entry. Figure A.7 shows the editing format of an Excel document, while Fig. A.8 shows the complete work interface.

(3) Grey clustering evaluation models

Similar to Deng's degree of grey incidence, grey clustering evaluation also requires a large amount of original data. However, in grey clustering evaluation it is possible

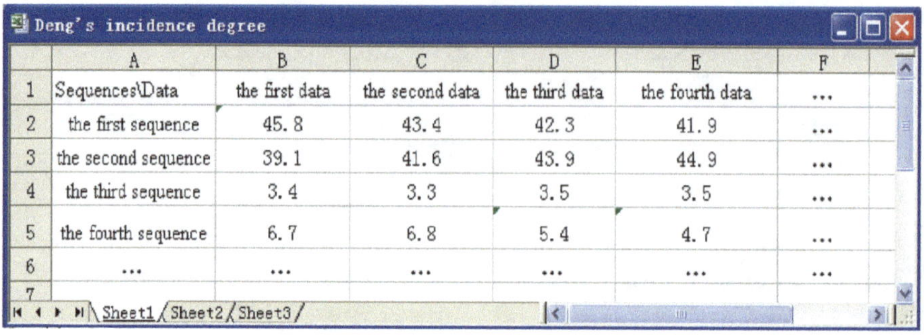

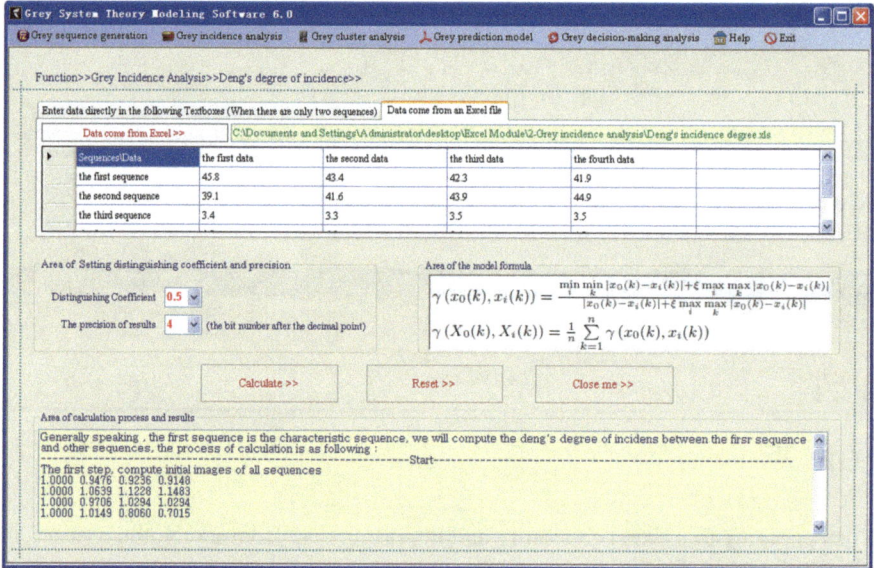

Fig. A.7 The required Excel file format

Fig. A.8 The complete work interface of Deng's degree of grey relation

to have several different types of data, including objective-criterion data, possibility functions, and criteria weights. Therefore, for grey clustering evaluation, the system again provides only one way to enter data, which is by importing Excel files. The key to using this part of the functions is to correctly edit the different types of data in the Excel documents. Sheet1 contains the objective-criteria data (Fig. A.9), Sheet2 the corresponding possibility functions (Fig. A.10), and Sheet3 the weights of the criteria (Fig. A.11).

Figure A.12 shows the operating interface of a grey clustering analysis. As for how to apply grey variable weight clustering and analysis based on center-point mixed

	A	B	C	D	E	F
1		Parameter values	Parameter values	Parameter values	Parameter values	...
2	Object	22.5	4	0	0	...
3	Object	79.37	6	600	0.75	...
4	Object	144	7	300	0.75	...
5	Object	300	6.1	189	12	...
6	Object	456	12	250	12	...
7	Object	189	8	700	1.5	...
8	Object	369	8	1300	2.25	...
9	Object	1127.11	16.2	550	3	...
10	Object	260	11	600	1	...
11	Object	200	8	600	1.25	...
12	Object	475	10	1000	0.75	...
13	Object	314.1	8	900	0.75	...
14	Object	282.8	7.4	1300	0.5	...
15	Object	240	8	1200	0.5	...
16	Object	160	5	1000	0.25	...
17	Object	270	8	1200	0.25	...
18	Object	9	1	200	0	...
19	...	...	...	...	...	...

Sheet1 / Sheet2 / Sheet3

Fig. A.9 The objective-criteria data

	A	B	C	D	E	F
1		the first parameter	the second parameter	the third parameter	the forth parameter	...
2	Whitenization weight function of the first grey class	100,300,-,-	3,10,-,-	200,1000,-,-	0.25,1.25,-,-	...
3	Whitenization weight function of the second grey class	50,150,-,250	2,6,-,10	100,600,-,1100	0,0.5,-,1	...
4	Whitenization weight function of the third grey class	-,-,50,100	-,-,4,8	-,-,300,600	-,-,0.25,0.5	...
5	...	...	...	...	...	...

Sheet1 Sheet2 / Sheet3

Fig. A.10 The corresponding whitenization weight functions

	A	B	C	D	E	F	G	H
1		the first parameter	the second parameter	the third parameter	the forth parameter	...		
2	Weight	0.3	0.25	0.25	0.2	...		

Sheet1 / Sheet2 / Sheet3

Fig. A.11 The weights of the criteria

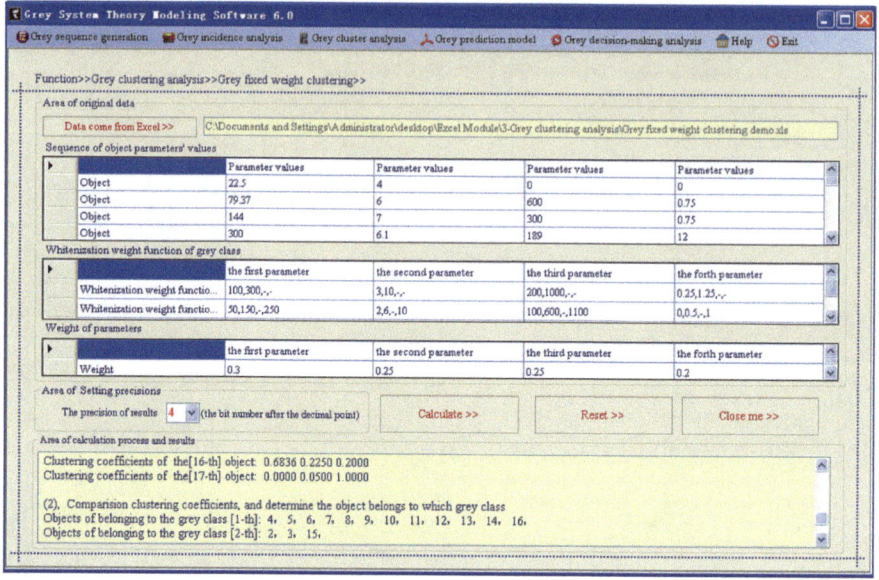

Fig. A.12 The operating interface of a grey clustering evaluation

triangular possibility functions, the operational details are similar and therefore omitted.

(4) Grey prediction models

Grey prediction models stand for an important part of grey systems theory. The operation of each individual prediction model is roughly the same. So, let us use the EGM(1,1) model to illustrate how to use the software system. The main steps include: enter or import data; click the "computation, simulation, prediction" button to compute the model parameters and the simulated values, and select the simulation accuracy; enter the desired number of predicted values, then click "prediction results." Figure A.13 shows the operational interface of the EGM(1,1) model.

(5) Grey decision-making

This part of the software package contains two modules, namely the multi-attribute grey target decision-making model and the two-stage model for decision-making.

The data layout of an intelligent grey target decision-making model is the same as that of any synthesized objective decision-making model, except that there is an additional column of threshold value, as shown in Fig. A.14, where the interval [14, 18] means that the lower effect threshold value is 14 and the upper effect value is 18. Figure A.15 shows the entire operating page of an intelligent grey target decision-making model.

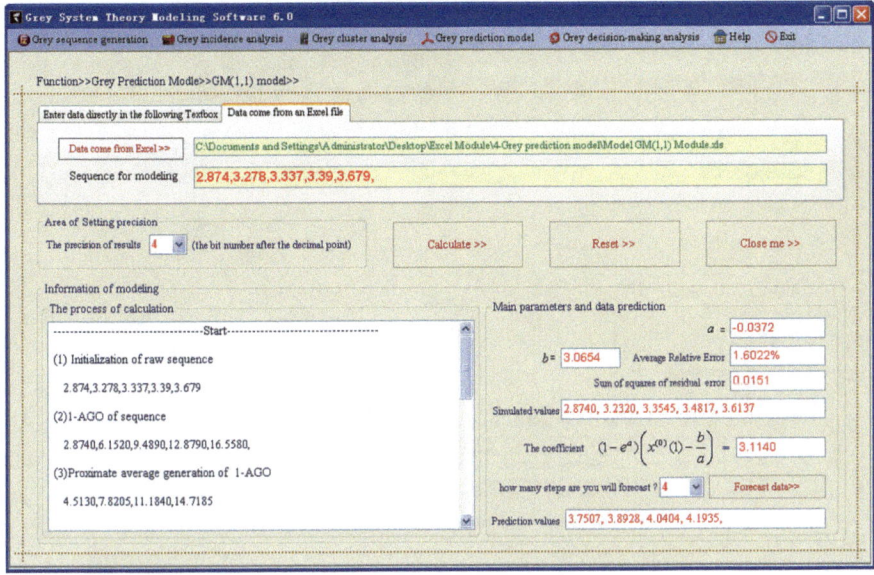

Fig. A.13 The operational interface of the GM(1,1) model

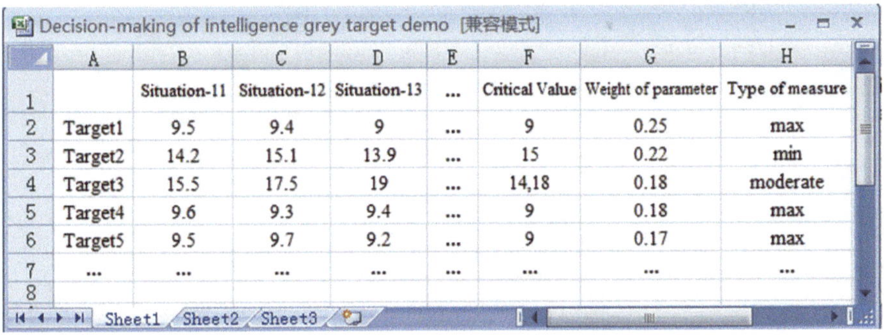

	A	B	C	D	E	F	G	H
1		Situation-11	Situation-12	Situation-13	...	Critical Value	Weight of parameter	Type of measure
2	Target1	9.5	9.4	9	...	9	0.25	max
3	Target2	14.2	15.1	13.9	...	15	0.22	min
4	Target3	15.5	17.5	19	...	14,18	0.18	moderate
5	Target4	9.6	9.3	9.4	...	9	0.18	max
6	Target5	9.5	9.7	9.2	...	9	0.17	max
7	...	...	...	...	...	...	...	...
8								

Decision-making of intelligence grey target demo [兼容模式]

Sheet1 Sheet2 Sheet3

Fig. A.14 The exact layout of data page for multi-attribute grey target decision-making

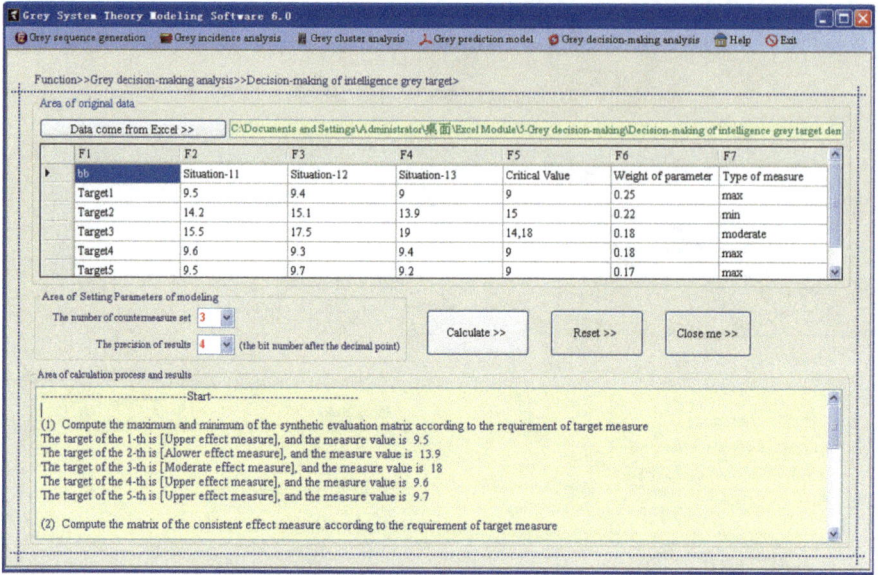

Fig. A.15 The entire operating page of multi-attribute grey target decision-making

Memorabilia of the Establishment and Development of Grey System Theory (1982–2021)

In 1982, professor Julong Deng has published the first paper on Grey System Theory in System and Control Letters.

In 1984, the first national academic conference on grey system theory and applications which chaired by professor Julong Deng was held in Taiyuan.

In 1985, Wuhan grey system consulting department was established.

In 1985, professor Julong Deng has published the first book on Grey System Theory by National Defense Industry Press (In Chinese).

In 1986, the course of grey system theory is included in the postgraduate training plan at both of Huazhong University of Science and Technology and Henan Agricultural University.

In 1987, Wuhan (National) Grey System Research Association which professor Julong Deng served as the president was established.

In 1989, the first Journal of "The Journal of Grey System" which edited by professor Julong Deng was released by "Research Information Ltd" in the UK.

In 1990, a research direction for training doctoral students in the field of grey system theory has set up at system engineering discipline of Huazhong University of Science and Technology.

In 1991, the grey system research office which professor Julong Deng served as the director was established in the Automation Department of Huazhong University of Science and Technology.

In 1996, The 9th national grey system academic conference attended by scholars from both sides of the Taiwan Strait was held in Wuhan.

In 1997, Regional Grey System Society of Taiwan was established.

In 2000, with the approval of the Academic Degrees Committee of the State Council, Nanjing University of Aeronautics and Astronautics has established a doctoral degree authorization point for management science and Engineering which Grey System Theory is listed as the first leading research direction.

In 2000, Institute for Grey Systems Studies was established at Nanjing University of Aeronautics and Astronautics.

In 2002, the first Doctor's Forum on Grey System Theory was held in Nanjing.

In 2004, both the number of publications on grey system theory and the number of citations of IGSS at NUAA are ranked No.1 in web of science.

In 2005, the Grey System Society of China, CSOOPEM, was approved by the Chinese Association for Science and Technology and Ministry of Civil Affairs, China.

In 2006, for the first time, the national grey system academic conference was supported by China Higher Science and Technology Center which professor Tsung-Dao Lee served as the director.

In 2007, the first IEEE International Conference on Grey Systems and Intelligent Services was held in Nanjing.

In 2007, the Technical Committee of IEEE SMC on Grey Systems was established.

In 2008, the course of grey system theory of Nanjing University of Aeronautics and Astronautics was rated as a National Excellence Course.

In 2009, Xuesen Qian, the first winner of the National Highest Science and Technology Award of China, sent professor Sifeng Liu and Dejin Song a letter to encourage their research work.

In 2010, the team with professor Sifeng Liu as the chief expert was rated as a National Excellence Teaching Team of China.

In 2010, the Journal of "Grey Systems-Theory and Application" which edited by professor Sifeng Liu was launched by Emerald Group in the UK.

In 2012, the course of grey system theory of Nanjing University of Aeronautics and Astronautics was selected as a National Excellence Sharing Course of China.

In 2012, the first Workshop of European grey system research collaboration network chaired by professor Yingjie Yang was held at De Montfort University.

In 2013, professor Julong Deng, the founder of Grey System Theory, passed away.

In 2013, the 2013 IEEE International Conference on Grey Systems and Intelligent Services was held in Macau.

In 2013, Professor Sifeng Liu was selected for a Marie Curie International Incoming Fellowship (FP7- PEOPLE- IIF-GA-2013-629051) of the European Union.

In 2014, an international network project entitled "Grey Systems and Its Applications" (IN-2014-020) directed by professor Yingjie Yang was funded by The Leverhulme Trust.

In 2014, a book Series of Grey Systems in Chinese which edited by professor Sifeng Liu were launched by Science Press.

In 2015, the International Association of Grey System and Uncertain Analysis (GSUA) was established.

In 2015, the first International Congress of GSUA was held in Leicester, UK.

In 2017, Polish Scientific Association of Grey Systems which Dr. Rafał Mierzwiak served as the founding president was set up.

In 2017, the 2017 IEEE International Conference on Grey Systems and Intelligent Services was held in Stockholm, Sweden.

In 2017, the book Grey System Theory and Its Applications, which authored by Prof. Sifeng Liu and published by Science Press, is identified as the No.1 top sited books in the pandect of natural science of China.

In 2017, Professor Sifeng Liu has been selected to be one of the top 10 shortlisted promising scientists in the MSCA 2017 Prizes.

In 2018, the course of grey system theory of Nanjing University of Aeronautics and Astronautics was selected as a National Excellence Online Open Course.

In 2018, Grey Systems Society of Pakistan was established.

In 2019, Turkish Association of Grey Systems Theory which professor Erdal Aydemir served as the founding president was established.

In 2019, the 2019 International Congress of GSUA was held in Bangkok, Thailand.

In 2019, Angela Dorothea Merkel, then German Chancellor, praised professor Julong Deng, the founder of grey system theory, and professor Sifeng Liu, a developer of grey system theory.

In 2020, the 2020 International Congress of GSUA was held in Nanjing.

In 2020, the course of grey system theory of Nanjing University of Aeronautics and Astronautics was selected as a National first class offline course.

In 2020, the course of grey system theory of Nanjing University of Aeronautics and Astronautics was selected as a National first class online course.

In 2021, a book Series of Grey Systems which edited by professor Sifeng Liu were launched by Springer-Nature Group.

Sifeng Liu.

Institute for Grey Systems Studies Nanjing University of Aeronautics and Astronautics.

Farewell to Our Tutor

The memorial speech by Sifeng Liu on behalf of students of Prof. Deng.

At 12:15 in the afternoon on June 22, 2013, our most beloved tutor, Professor Deng Julong, saw the end of his eighty-year life journey and left us forever. Professor Deng Julong was a tireless lifelong pioneer and founder of Grey System Theory, so the world has lost a visionary. In recent days, dark clouds gathered and there was a long period of wet weather: it was God's crying for the death of a great scholar.

Prof. Deng graduated from the Department of Electrical Engineering of the Huazhong Institute of Technology in 1955 and then taught at the Department of Automatic Control. In the 1960s, he put forward the idea of control with abandonment. In the 1970s, the method of control with abandonment became a typical control method internationally. In 1982, he pioneered Grey System Theory and created a brand new subject area in the history of science. His academic achievements were highly respected by many in the scientific circles.

In Prof. Deng's academic career of 60 years, there existed neither holidays, weekends nor a line between service and retirement. Prof. Deng was the editor of The Journal of Grey System, an international journal, for 24 years. In his capacity as editor he screened articles, checked the contents of their experiments and edited them in English; he was dedicated and tireless. Until the last moment of his life, he was still working on publishing a scholarly book.

I still remember 1983, when I first participated in a course on Grey System Theory. The mimeographed teaching materials had a blue cover and were presented as a book. It was like finding a treasure, as the book attracted me deeply. The book really inspired me, as I was a young scholar who was going through a period of confusion and lack of direction for academic study. The book shone with sparkles of wisdom and built a lighthouse for a knowledge seeker who was in the mist of trying to find his way in academic research. This book became the light in my life's journey. From then onwards I forged an indissoluble bound with Grey System Theory.

My most unforgettable memory is the first time I joined Prof. Deng's comprehensive course on Grey System Theory in Yu Jiashan's air-raid shelter in 1986. The course deepened my understanding and awareness of many scientific problems in Grey System Theory. In 1995, when I was about 40 years old, I formally became a disciple of Prof. Deng, and from then on I took on the mission of disseminating and developing Grey System Theory.

Today, Grey System Theory is accepted by academics worldwide. A variety of academic works on grey system theory have been published in different languages, including English, Japanese, Korean and Romanian. In 1989, the British journal entitled The Journal of Grey System was launched. In 1997, a Chinese publication named Journal of Grey System was launched in Taiwan. Later in 2004, this same publication started to be published in English. In 2011, Emerald launched a new journal entitled Grey system: Theory and application.

Since November 2007, the biennial IEEE International Conference on Grey System and Intelligent Services has been successfully held in Nanjing, China, attracting scholars from different parts of the world. The fourth IEEE International Conference on Grey System and Intelligent Services will be held at the University of Macau, in November 2013.

Currently, a significant number of scholars from China, United States, England, Germany, Japan, Australia, Canada, Austria, Russia, Turkey, the Netherlands, Iran, and others, have been involved in the research and application of Grey System Theory. Many countries have begun to recruit and cultivate doctoral and other postgraduate students in the area of Grey System Theory. To date, more than 100 students have graduated and received Ph.D.s in this area, and tens of thousands of graduate students have carried out their research on Grey System Theory.

At the Nanjing University of Aeronautics and Astronautics, Grey System Theory has become an important course for undergraduate, masters and doctoral students from many different colleges.

In 2008, Grey System Theory was selected for a national course award in China. In 2013, it was selected as an open learning course, free of charge to all lovers of Grey System Theory.

Currently, there are more than 70 projects on Grey System Theory research and applications funded by the National Natural Science Foundation of China. Many projects have been supported by the European Union, the United Kingdom, USA, Canada, Spain, Romania and other countries.

In 2012, De Montfort University in the UK funded and organized the first European collaboration network for Grey System research, and representatives from 14 European Member States attended the session.

Grey System Theory, as an emerging discipline, has carved its place in science and demonstrated strong vitality.

Farewell to our tutor!

God bless you!

(Prof. Sifeng Liu is a former Ph.D. Student of Prof. Julong Deng. Currently Prof. Liu works at the Nanjing University of Aeronautics and Astronautics. He also serves as the founding director of the Institute for Grey Systems Studies, the founding chair of the IEEE SMC Technical Committee on Grey Systems, and the president of Grey Systems Society of China.)

Bibliography

Abdulshahed, A. M., Longstaff, A. P., & Fletcher, S. (2017). A cuckoo search optimisation-based Grey prediction model for thermal error compensation on CNC machine tools. *Grey Systems: Theory and Application, 7*(2), 146–155.

Andrew, A. M. (2011). Why the world is grey. *Grey Systems: Theory and Application, 1*(2), 112–116.

Amanna, A., Price, M. J., et al. (2011). Grey systems theory applications to wireless communications. *Analog Integrated Circuits and Signal Processing, 69*(2–3), 259–269.

Arasteh, A., Aliahmadi, A., & Omran, M. M. (2014). Application of gray systems and fuzzy sets in combination with real options theory in project portfolio management. *Arabian Journal for Science and Engineering, 39*(8), 6489–6506.

Aydemir, E., Bedir, F., & Ozdemir, G. (2015). Degree of greyness approach for an EPQ model with imperfect items in copper wire industry. *The Journal of Grey System, 27*(2), 13–26.

Bahrami, S., Hooshmand, R. A., & Parastegari, M. (2014). Short term electric load forecasting by wavelet transform and grey model improved by PSO (particle swarm optimization) algorithm. *Energy, 72*, 434–442.

Bai, X., Xu, Y., & Liu, S. F. (2021). Research on the regional leading industry selection of "Kashgar urban agglomerations" based on multi-attribute weighted intelligent grey target decision-making evaluation model. *Grey Systems: Theory and Application, 11*(3), 418–433.

Benitez, C. R. B., Paredes, C. R. B., Lodewijks, G., et al. (2013). Damp trend grey model forecasting method for airline industry. *Expert Systems with Applications, 40*(12), 4915–4921.

Bristow, M., Fang, L. P., & Hipel, K. W. (2012). System of systems engineering and risk management of extreme events: Concepts and case study. *Risk Analysis, 32*(11), 1935–1955.

Chan, J. W. K., & Tong, T. K. L. (2007). Multi-criteria material selections and end-of-life product strategy: Grey relational analysis approach. *Materials & Design, 28*(5), 1539–1546.

Chang, B. R., & Tsai, H. F. (2008). Forecast approach using neural network adaptation to support vector regression grey model and generalized auto-regressive conditional heteroscedasticity. *Expert Systems with Applications, 34*(2), 925–934.

Chang, C. J., Li, D. C., Chen, C. C., et al. (2014). A forecasting model for small non-equigap data sets considering data weights and occurrence possibilities. *Computers & Industrial Engineering, 67*, 139–145.

Chang, S. C., Lai, H. C., & Yu, H. C. (2005). A variable P value rolling grey forecasting model for Taiwan semiconductor industry production. *Technological Forecasting and Social Change, 72*(5), 623–640.

S. Liu, *Grey Systems Analysis*, Series on Grey System, https://doi.org/10.1007/978-981-19-6160-1

Chen, C. I., Chen, H. L., & Chen, S. P. (2008). Forecasting of foreign exchange rates of Taiwan's major trading partners by novel nonlinear Grey Bernoulli model NGBM (1,1). *Communications in Nonlinear Science and Numerical Simulation, 13*(6), 1194–1204.

Chen, C. L., Dong, D. Y., Chen, Z. H., et al. (2008). Grey systems for intelligent sensors and information processing. *Journal of Systems Engineering and Electronics, 19*(4), 659–665.

Chen, J.F., Zhao, S.F., & Yang, L. (2012). On multi-agents decision and simulation of catastrophe insurance based on grey game model. *Soft Science, 26*(7), 131–136.

Chen, M. Y. (1982). Grey dynamic control of boring machine. *Journal of Huazhong University of Science and Technology, 10*(2), 7–11.

Chen, M. Y. (1985). Stability of grey systems and problem of calmness. *Fuzzy Mathematics, 5*(2), 54–58.

Chen, N. L., & Xie, N. M. (2020). Uncertainty representation and information measurement of grey numbers. *Grey Systems: Theory and Application, 10*(4), 495–512.

Chen, R. H., Song Z. Q., & Kang, L. M. (2005). Application of grey system to reservoir evaluation in the seven district of Xinjiang Karamay Oilfield. *Inner Mongolia petroleum chemical industry, 7*, 110–113.

Chen, X.D., Xia, J., Xu, Q. (2009). Self-memory prediction mode with grey differential model. *China Science E: Technological Science, 39*(2), 341–350.

Chen, X. X., & Liu, S. F. (2008). On grey multi-attribute group decision-making with completely unknown weight information. *Management Science of China, 16*(5), 146–152.

Chen, Y. J., Fu, Y. H., & Sun, H. (2007). Grey theory applied in the construction control of steel-pipe-concrete arch bridges. *Journal of Harbing University of Technology, 39*(4), 546–548.

Chen, Y. K., & Tan, X. R. (1995). Grey relational analysis on serum markers of liver fibrosis. *The Journal of Grey System, 7*(1), 63–68.

Cheng, Q. Y., & Qiu, W. H. (2001). Grey system forecast for firing accuracy of gun. *Journal of Systems Science and Systems Engineering, 10*(2), 205–211.

Chiang, J. Y., & Chen, C. K. (2008). Application of grey prediction to inverse nonlinear heat conduction problem. *International Journal of Heat and Mass Transfer, 51*(3), 576–585.

Chiang, Y. M., & Hsieh, H. H. (2009). The use of the Taguchi method with grey relational analysis to optimize the thin-film sputtering process with multiple quality characteristic in color filter manufacturing. *Computers & Industrial Engineering, 56*(2), 648–661.

Chiao, C. H., & Wang, W. Y. (2002). Reliability improvement of fluorescent lamp using grey forecasting model. *Microelectronics Reliability, 42*(1), 127–134.

Chirwa, E. C., & Mao, M. (2006). Application of grey model GM(1, 1) to vehicle fatality risk estimation. *Technological Forecasting and Social Change, 73*(5), 588–605.

Chiu, N. H. (2009). An early software-quality classification based on improved grey relational classifier. *Expert Systems with Applications, 36*(7), 10727–10734.

Chou, J. H., Chen, S. H., & Li, J. J. (2000). Application of the Taguchi-genetic method to design an optimal grey-fuzzy controller of a constant turning force system. *Journal of Materials Processing Technology, 105*(3), 333–343.

Chu, Y. F. (2008). Predictive research on the development scale of Chinese logistics based on grey systems theory. *Management Review, 20*(3), 58–64.

Cristóbal, S., & José, R. (2015). A cost forecasting model for a vessel drydocking. *Journal of Ship Production and Design, 31*(1), 58–62(5)

Cui, J., & Dang, Y. G. (2009). On the GM(1,1) prediction accuracy based on another class of strengthening buffer operators. *Control and Decision, 24*(1), 44–48.

Cui, J., Dang, Y. G., & Liu, S. F. (2008). An improvement method for solving criteria weights based on degrees of grey incidence. *Management Science of China, 16*(5), 141–145.

Cui, J., Dang, Y. G., & Liu, S. F. (2009). Precision analysis of the GM(1,1) model based on a new weakening operator. *Systems Engineering: Theory and Practice, 29*(7), 132–138.

Cui, J., Liu, S., & Xie, N. (2012). Novel grey decision making model and its numerical simulation. *Transactions of Nanjing University of Aeronautics & Astronautics, 29*(2), 112–117.

Cui, L. Z., Liu, S. F., & Wu, Z. P. (2009). Some new weakening buffer operators and their applications. *Control and Decision-Making, 24*(8), 1252–1256.

Cui, L. Z., Liu, S. F., & Wu, Z. P. (2010). On the construction of a new strengthening buffer operator and its application. *Systems Engineering Theory and Practice, 30*(3), 484–489.

Dai, W. Z., & Li, J. F. (2005). Modeling with non-equal-distant GM(1,1). *Systems Engineering: Theory and Practice, 25*(9), 89–93.

Dai, W. Z., & Su, Y. (2012). On the construction of a strengthening buffer operator of new information priorityand its application. *Systems Engineering Theory and Practice, 38*(8), 1329–1334.

Dang, Y. G., Liu, S. F., Liu, B., et al. (2004a). An improvement on the slope degree of grey incidence. *Engineering Science of China, 6*(3), 41–44.

Dang, Y. G., Liu, S. F., Liu, B., et al. (2004b). About weakening buffer operators. *Management Science of China, 12*(2), 108–111.

Dang, Y. G., Liu, S. F., Liu, B., et al. (2004c). Study on multi-criteria decision-making based on interval number incidence. *Journal of Nanjing University of Aeronautics and Astronautics, 36*(3), 403–406.

Delcea, C., Scarlat, E., & Maracine, V. (2012). Grey relational analysis between firm's current situation and its possible causes: A bankruptcy syndrome approach. *Grey Systems: Theory and Application, 2*(2), 229–239.

Delcea, C., Scarlat, E., Liviu-Adrian, C. (2013). Grey relational analysis of the financial sector in Europe. *The Journal of Grey System, 25*(4), 19–30.

Delgado, A., & Romero, I. (2016). Environmental conflict analysis using an integrated grey clustering and entropy-weight method: A case study of a mining project in Peru. *Environmental Modelling & Software, 77*, 108–121.

Deng, J. L. (1982). Grey control SYSTEM. *Journal of Huazhong University of Science and Technology, 10*(3), 9–18.

Deng, J. L. (1985a). The incidence space of grey systems. *Fuzzy Mathematics, 5*(2), 1–10.

Deng, J. L. (1985b). Generation functions of grey systems. *Fuzzy Mathematics, 5*(2), 11–22.

Deng, J. L. (1985c). The GM model of grey systems theory. *Fuzzy Mathematics, 5*(2), 23–32.

Deng, J. L. (1985d). Five kinds of grey predictions. *Fuzzy Mathematics, 5*(2), 33–42.

Deng, J. L. (1985e). Grey situational decision making. *Fuzzy Mathematics, 5*(2), 43–50.

Deng, J. L. (1988). *Grey multi-dimensional planning.* Press of Huazhong University of Science and Technology.

Deng, J. L. (1989). Grey information space. *The Journal of Grey System, 1*(2), 103–118.

Deng, J. L. (1991). Representational forms of information in grey systems theory. *Grey Systems Theory and Practice, 1*(1), 1–12.

Deng, J. L. (1992). Grey causality and information cover. *The Journal of Grey System, 4*(2), 99–119.

Deng, J. L. (1993). Grey differential equation. *The Journal of Grey System, 5*(1), 1–14.

Deng, J. L. (1995). Extent information cover in grey system theory. *The Journal of Grey System, 7*(2), 131–138.

Deng, J. L. (1996). Several problems in grey systems theory and development. In S. F. Liu & Z. X. Xu (Eds.), *New Advances in Grey Systems Research* (pp. 1–10). Press of Huazhong University of Science and Technology.

Deng, J. L., & Ng, D. K. W. (1996). Chaos in Grey Model GM(1, N). *The Journal of Grey System, 8*(1), 1–10.

Deng, J. L., & Zhou, C. S. (1986). Sufficient conditions for the stability of a class of interconnected dynamic systems. *Systems & Control Letters, 5*(2), 105–108.

Dong, F. Y. (2007). Development prediction on the finance of Chinese firms based on newly improved GM(1,1) model. *Management Science of China, 15*(4), 93–97.

Dong, W. J., Liu, S. F., Fang, Z. G., Yang, X. Y., Hu, Q., & Tao, L. Y. (2017). Study of a discrete grey forecasting model based on the quality cost characteristic curve. *Grey Systems: Theory and Application, 7*(3), 376–384.

Dounis, A. I., Tiropanis, P., Argiriou, A., et al. (2011). Intelligent control system for reconciliation of the energy savings with comfort in buildings using soft computing techniques. *Energy and Buildings, 43*, 66–74.

Du, J. L., Liu, S. F., & Liu, Y. (2022). Grey variable dual precision rough set model and its application. *Grey Systems: Theory and Application, 12*(1), 156–173.

Dymova, L., Sevastjanov, P., & Pilarek, M. (2013). A method for solving systems of linear interval equations applied to the Leontief input-output model of economics. *Expert Systems with Applications, 40*(1), 222–230.

E, J. Q., Wang, Y. N., Mei, Z., et al. (2005). Fuzzy adaptive variable weight combination prediction model for energy consumption in copper refining process and its application. *Mining and Metallurgy, 03*, 46–48.

Ejnioui, A., Otero, C. E., & Otero, L. D. (2013). Prioritisation of software requirements using grey relational analysis. *International Journal of Computer Applications in Technology, 47*(2–3), 100–109.

Esme, U., Kulekci, M. K., Ustun, D., et al. (2015). Grey-based fuzzy algorithm for the optimization of the ball burnishing process. *Materials Testing, 57*(7–8), pp. 666–673.

Fan, T. G., Zhang, Y. C., & Quan, X. Z. (2001). Grey model GM(1,1, t) and its chaos feature. *Journal of Grey System, 13*(2), 117–120.

Fang, Z. G., & Liu, S. F. (2003). Grey matrix game models based on pure strategies (1): Construction of standard grey matrix game models. *Journal of Southeast University (natural Science Edition), 33*(6), 796–800.

Fang, Z. G., & Liu, S. F. (2005). A study on the GM(1,1) (GMBIGN(1,1)) model based on interval grey sequences. *Engineering Science of China, 7*(2), 57–61.

Fang, Z. G., Liu, S. F., & Shi, H. X. (2010). *Grey game theory and its applications in economic decision-making.* Taylor & Francis Group.

Farooq, U., Siddiqui, M. A., Gao, L., & Hardy, J. L. (2012). Intelligent transportation systems: An impact analysis for Michigan. *Journal of Advanced Transportation, 46*(1), 12–25.

Feng, Y. C., Li, D., Wang, Y. Z., et al. (2001). A grey model for response time in database system DM3. *Journal of Grey System, 13*(3).

Feng, Z. Y. (1992). Direct grey modeling. *Journal of Applied Mathematics, 15*(3), 345–354.

Gao, W., Feng, X. T. (2004). Study on landslide deformation prediction based on grey evolutionary neural network. *Geotechnical Mechanics, 25*(4), 514–517.

Goel, B., Singh, S., & Sarepaka, R. V. (2015). Optimizing single point diamond turning for monocrystalline germanium using grey relational analysis. *Materials and Manufacturing Processes, 30*(8), 1018–1025.

Golinska, P., Kosacka, M., Mierzwiak, R., et al. (2015). Grey decision making as a tool for the classification of the sustainability level of remanufacturing companies. *Journal of Cleaner Production, 105*, 28–40.

Golmohammadi, D., & Mellat-Parast, M. (2012). Developing a grey-based decision-making model for supplier selection. *International Journal of Production Economics, 137*(2), 191–200.

Guan, Y. Q., & Liu, S. F. (2007). Strengthening buffer operator sequences constructed on fixed points and applications. *Control and Decision, 22*(10), 1189–1193.

Guan, Y. Q., & Liu, S. F. (2008). On matrix of linear buffer operator and its application. *Journal of Applied Mathematics of Universities, 23*(3), 357–362.

Guo, H. (1991). Grey exponential law attributed to mutual complement of series. *The Journal of Grey System, 3*(2), 153–162.

Guo, H., Xiao, X. P., & Forrest, J. (2014). Problem of grey bi-level multi-objective linear programming and its algorithm. *Control and Decision, 29*(7), 1193–1198.

Guo, P. (2007a). MAC control on grey error corrections and application in workloads of machine groups. *Journal of System Simulation, 19*(18), 4326–4330.

Guo, R. K. (2007b). Modeling imperfectly repaired system data via grey differential equations with unequal-gapped times. *Reliability Engineering & System Safety, 92*(3), 378–391.

Guo, X. J., Liu, S. F., & Wu, L. F. (2013). Modeling and algorithm of prediction about pollutants' emission reduction bycombining gray theory and Markov chain. *Research on Computer Application, 30*(12), 3670–3673.

Guo, X. J., Liu, S. F., Wu, L. F., et al. (2015). A multi-variable grey model with a self-memory component and its application on engineering prediction. *Engineering Applications of Artificial Intelligence, 42*, 82–93.

Gurden, S. P., Westerhuis, J. A., Bijlsma, S., et al. (2001). Modelling of spectroscopic batch process data using grey models to incorporate external information. *Journal of Chemometrics, 15*(2), 101–121.

Haken, H. (2011). Book reviews: Grey information: theory and practical applications. *Grey Systems: Theory and Application, 1*(1), 105–106.

Hamzacebi, C., & Es, H. A. (2014). Forecasting the annual electricity consumption of Turkey using an optimized grey model. *Energy, 70*, 165–171.

Hao, Y. H., & Wang, X. M. (2000). Period residual modification of GM(1,1) modeling. *Journal of Grey System, 12*(2), 181–183.

Hao, Y. H., Zhao, J. J., Li, H. M., et al. (2012). Karst hydrological processes and grey system model. *Journal of the American Water Resources Association, 48*(4), 656–666.

He, R. S., & Hang, S. F. (2007). Damage detection by a hybrid real-parameter genetic algorithm under the assistance of grey relation analysis. *Engineering Applications of Artificial Intelligence, 20*(7), 980–992.

He, W. Z., Song, G. X., & Wu, A. D. (2005). A class of schemes for computing the parameters of the GM(1,1) model. *Systems Engineering: Theory and Practice, 25*(1), 69–75.

He, W. Z., & Wu, A. D. (2006). Linear programming methods for estimating the parameters of Verhulst model and applications. *Systems Engineering: Theory and Practice, 26*(8), 141–144.

He, Y., & Bao, Y. D. (1992). Grey Markov prediction model and applications. *Systems Engineering: Theory and Practice, 12*(4), 59–63.

Hipel, K. W. (2011). Book reviews: Grey systems: Theory and applications. *Grey Systems: Theory and Application, 1*(3), 274–275.

Hodzic, M., & Tai, L. C. (2016). Grey Predictor reference model for assisting particle swarm optimization for wind turbine control. *Renewable Energy, 86*, 251–256.

Hsieh, C. H. (2001). Grey data fitting model and its application to image coding. *Journal of Grey System, 13*(1), 245–254.

Hsu, C. C., & Chen, C. Y. (2003a). A modified grey forecasting model for long-term prediction. *Journal of the Chinese Institute of Engineers, 26*(3), 301–308.

Hsu, C. C., & Chen, C. Y. (2003b). Applications of improved grey prediction model for power demand forecasting. *Energy Conversion and Management, 44*(14), 2241–2249.

Hsu, L. C., & Wang, C. H. (2007). Forecasting the output of integrated circuit industry using a grey model improved by the Bayesian analysis. *Technological Forecasting and Social Change, 74*(6), 843–853.

Hsu, Y. T., Yeh, J. (2000). A novel image compression using grey models on a dynamic window. *International Journal of Systems Science, 31*(9), 1125–1141.

Hu, F., Huang, J. G., & Zhang, Q. F. (2007). Research on effect evaluation of underwater navigation devices based on grey systems theory. *Journal of Northwest University of Technology, 25*(3), 411–415.

Hu, W. B., Hua, B., & Yang, C. Z. (2002). Building thermal process analysis with grey system method. *Building and Environment, 37*(6), 599–605.

Hu, X. L., Wu, Z. P., & Han, R. (2013). Analysis on the strengthening buffer operator based on the strictly monotone function. *International Journal of Applied Physics and Mathematics, 3*(2), 132–136.

Hu, Y. C. (2007). Grey relational analysis and radial basis function network for determining costs in learning sequences. *Applied Mathematics and Computation, 184*(2), 291–299.

Huang, F. Y. (1994). Transformation method of grey systems modeling. *Systems Engineering: Theory and Practice, 14*(6), 35–38.

Huang, S. J., & Huang, C. L. (2000). Control of an inverted pendulum using grey prediction model. *IEEE Transactions on Industrial Applications, 36*(2), 452–458.

Huang, J., & Zhong, X. L. (2009). A general accumulation grey predictive control model and its optimal computational scheme. *Systems Engineering: Theory and Practice, 29*(6), 147–156.

Huang, K. Y., & Jane, C. J. (2009). A hybrid model for stock market forecasting and portfolio selection based on ARX, grey system and RS theories. *Expert Systems with Applications, 36*(3), 5387–5392.

Huang, S. J., Chiu, N. H., & Chen, L. W. (2008). Integration of the grey relational analysis with genetic algorithm for software effort estimation. *European Journal of Operational Research, 188*(3), 898–909.

Huang, Y. L., & Fang, B. (2007). Systems' quasi-quantitative incidence matrices based on grey models. *Journal of System Simulation, 19*(3), 474–477.

Huang, Y. P., & Chu, H. C. (1996). Practical consideration for grey modeling and its application to image processing. *The Journal of Grey System, 8*(3), 217–233.

Javed, S. A., & Liu, S. F. (2018). Predicting the research output/growth of selected countries: Application of Even GM (1, 1) and NDGM models. *Scientometrics, 115*, 395–413.

Jia, H. F., & Zheng, Y. Q. (1998). Combined prediction models of grey time series and its application in predicting annual precipitations. *Systems Engineering: Theory and Practice, 18*(8), 42–45.

Jia, Z. Y., Ma, J. W., & Wang, F. J. (2009). Prediction method of assembly product characteristics under the influence of multi part geometric elements. *Journal of Mechanical Engineering, 45*(7), 168–173.

Jiang, S. Q., Liu, S. F., Fang, Z. G., & Liu, Z. X. (2017). Study on distance measuring and sorting method of general grey number. *Grey Systems: Theory and Application, 7*(3), 320–328.

Jiang, S. Q., Liu, S. F., & Liu, Z. X. (2021). General grey number decision-making model and its application based on intuitionistic grey number set. *Grey Systems: Theory and Application, 11*(4), 556–570.

Jiang, Y. Q., Yao, Y., Deng, S. M., et al. (2004). Applying grey forecasting to predicting the operating energy performance of air cooled water chillers. *International Journal of Refrigeration, 27*(4), 385–392.

Jie, J. X., Song, B. F., & Liu, D. X. (2004). A grey incidence method designed for selecting the optimum design of the top layer of aircrafts. *Journal of Systems Engineering, 19*(4), 350–354.

Jin, Y. F., Zhu, Q. S., & Xing, Y. K. (2006). The grey logical method for inserting the missing values in sequential data. *Control and Decisoon, 21*(2), 236–240.

Kadier, A., Abdeshahian, P., Simayi, Y., et al. (2015). Grey relational analysis for comparative assessment of different cathode materials in microbial electrolysis cells. *Energy, 90*, 1556–1562.

Kalsi, N. S., Sehgal, R., & Sharma, V. S. (2013). Grey-based Taguchi analysis for optimization of multi-objective machining process in turning. *Strategic Decision Science, 4*(2), 110–127.

Kahraman, C., Yavuz, M., & Kaya, I. (2010). Fuzzy and grey forecasting techniques and their applications in production systems. *Production Engineering and Management, 252*, 1–24.

Karmakar, S., & Mujumdar, P. P. (2007). A two-phase grey fuzzy optimization approach for water quality management of a river system. *Advances in Water Resources, 30*(5), 1218–1235.

Kaya, Y. (2015). Hidden pattern discovery on epileptic EEG with 1-D local binary patterns and epileptic seizures detection by grey relational analysis. *Australasian Physical & Engineering Sciences in Medicine, 38*(3), 435–446.

Kayacan, E., Ulutas, B., & Kaynak, O. (2010). Grey system theory-based models in time series prediction. *Expert Systems with Applications, 37*(4), 1784–1789.

Khuman, A. S., Yang, Y. J., Liu, S. F. (2016). Grey relational analysis and natural language processing to: Grey language processing. *The Journal of Grey System, 28*(1), 88–97.

Kim, D., Goh, T., Park, M., et al. (2015). Fuzzy sliding mode observer with grey prediction for the estimation of the state-of-charge of a lithium-ion battery. *Energies, 8*(11), 12409–12428.

Ko, A. S., & Chang, N. B. (2008). Optimal planning of co-firing alternative fuels with coal in a power plant by grey nonlinear mixed integer programming model. *Journal of Environmental Management, 88*(1), 11–27.

Ko, J. Z. (1996). Data sequence transformation and GM(1,1) modeling precision. In S. F. Liu & Z. X. Xu (Eds.), *New advances in grey systems research* (pp. 233–235). Press of Huazhong University of Science and Technology.

Kose, E., & Tasci, L. (2015). Prediction of the vertical displacement on the crest of Keban Dam. *The Journal of Grey System, 27*(1), 12–20.

Ku, L. L., & Huang, T. C. (2006). Sequential monitoring of manufacturing processes: An application of grey forecasting models. *International Journal of Advanced Manufacturing Technology, 27*(5), 543–546.

Kumar, S. S., Uthayakumar, M., Kumaran, S. T., et al. (2015). Parametric optimization of wire electrical discharge machining on aluminium based composites through grey relational analysis. *Journal of Manufacturing Processes, 20*, 33–39.

Kung, C. Y., & Wen, K. L. (2007). Applying grey relational analysis and grey decision-making to evaluate the relationship between company attributes and its financial performance: A case study of venture capital enterprises in Taiwan. *Decision Support Systems, 43*(3), 842–852.

Kuo, Y. Y., Yang, T. H., & Huang, G. W. (2008). The use of grey relational analysis in solving multiple attribute decision-making problems. *Computers & Industrial Engineering, 55*(1), 80–93.

Kuzu, A., Bogosyan, S., & Gokasan, M. (2016). Predictive Input Delay Compensation with Grey Predictor for Networked Control System. *International Journal of Computers Communications & Control, 11*(1), 67–76.

Lai, H. H., Lin, Y. C., & Yeh, C. H. (2005). Form design of product image using grey relational analysis and neural network models. *Computers & Operations Research, 32*(10), 2689–2711.

Lai, Y.Q., Chen, X.T., Qin, Q.W., et al. (2004). Corrosion analysis and corrosion rates prediction of $NiFe_2O_4$ cermet inert anodes. *Journal of Center South University, 35*(6), 896–901.

Lee, W. M., & Liao, Y. S. (2003). Self-tuning fuzzy control with a grey prediction for wire rupture prevention in WEDM. *International Journal of Advanced Manufacturing Technology, 22*(7–8), 481–490.

Li, B., & Wei, Y. (2009). A new GM(1,1) model established with optimized grey derivatives. *Systems Engineering: Theory and Practice, 29*(2), 100–105.

Li, B. Z., & Zhu, X. X. (2007). On the RIS evolutionary mechanism based on dissipation and grey incidence entropy. *Studies of Science of Sciences, 25*(6), 1239–1244.

Li, B. L., & Deng, J. L. (1984). A grey model for the biological prevention and treatment system of cotton aphids. *Exploration of Nature, 3*(3), 44–49.

Li, B. J., Liu, S. F., & Zhu, Y. D. (2000). A method to determine the reliability of grey intervals. *Systems Engineering: Theory and Practice, 20*(4), 104–106.

Li, C., Chen, K. J., & Xiang, X. D. (2015). An integrated framework for effective safety management evaluation: Application of an improved grey clustering measurement. *Expert Systems with Applications, 42*(13), 5541–5553.

Li, C., Yang, Y. J., & Liu, S. F. (2019). Comparative analysis of properties of weakening buffer operators in time series prediction models. *Communication in Nonlinear Science Numerical Simulation, 68*, 257–285.

Li, C. H., & Tsai, M. J. (2009). Multi-objective optimization of laser cutting for flash memory modules with special shapes using grey relational analysis. *Optics & Laser Technology, 41*(5), 634–642.

Li, D. C., Chang, C. J., & Chen, W. C. (2011). An extended grey forecasting model for omni-directional forecasting considering data gap difference. *Applied Mathematical Modelling, 35*, 5051–5058.

Li, G. D., Masuda, S., & Nagai, M. (2015). Predictor design using an improved grey model in control systems. *International Journal of Computer Integrated Manufacturing, 28*(3), 297–306.

Li, J. B., Zhao, J. Y., Zheng, X. X., et al. (2008). A new evaluation method of the state of electric transformers based on grey target theory. *Journal of Jilin University (Industry Edition), 38*(1), 201–205.

Li, P. H., Yang, H. L., Sun, L. L., et al. (2011). Application of gray prediction and time series model in spacecraft prognostic. *Computer Measurement & Control, 19*(1), 111–113.

Li, Q., Liu, S. F., & Javed, S. A. (2022). Two-stage multi-level equipment grey state prediction model and application. *Grey Systems: Theory and Application, 12*(2), 462–482.

Li, Q. X. (2009). Grey dynamic input–output analysis. *Journal of Mathematical Analysis and Applications, 359*(2), 514–526.

Li, Q. X., & Liu, S. F. (2008). The foundation of the grey matrix and the grey input–output analysis. *Applied Mathematical Modelling, 32*(3), 267–291.

Li, Q. X. & Liu, S. F., Lin, Y.(2012). Grey enterprise input-output analysis. *Journal of Computational and Applied Mathematics, 236*(7), 1862–1875.

Li, Y., & Li, J. (2020). Study on unbiased interval grey number prediction model with new information priority. *Grey Systems: Theory and Application, 10*(1), 1–11.

Li, X. B., Sun, H. Y., & Wu, Y. X. (2009). Dynamic tracking multi-variable fuzzy predictive functional control of fuel oil feeding temperature. *Computer Engineering and Applications, 45*(9), 200–203.

Li, X. C. (1999). Expansion of the validity range of the GM(1,1) model of the grey systems. *Systems Engineering: Theory and Practice, 19*(1), 97–101.

Li, X. C., Yuan, Z., Zhang, G. B., et al. (2014). Some properties of grey differential equation GM(1,1,β). *Systems Engineering Theory and Practice, 34*(5), 1249–1255.

Li, X. F., Wang, J. J., Ma, J., et al. (2007). Grey incidence analysis for the quality of cotton fiber materials and the resultant yarns and single yarn strength. *Journal of Applied Science, 25*(1), 100–102.

Li, X. H., Wang, H. T., & Jia, J. Q. (2005). Limit displacement discrimination of stability and reliability analysis of surrounding rock in tunnel and underground engineering. *Geotechnical Mechanics, 25*(6), 850–854.

Li, X. M., Dang, Y. G., & Wang, Z. X. (2012). Harmonic buffer operators with variable weight and effect strength comparison. *Systems Engineering Theory and Practice, 32*(11), 2486–2492.

Li, X. Q., Tan, S. L., & Tang, B. G. (2007). The optimal collocation model of missile nuke based on the grey decision theory. *Fire Control and Command Control, 32*(2), 42–47.

Liang, B., Dai, Y. Y., Chen, T. Y., et al. (2014). Grey relational optimization for shale gas exploration and development areas of complicated geological parameter features. *Journal of China Coal Society, 39*(3), 524–530.

Liang, C. Y., Gu, D. X., & Bichindaritz, I. (2012). Integrating gray system theory and logistic regression into case-based reasoning for safety assessment of thermal power plants. *Expert Systems with Applications, 39*(5), 5154–5167.

Liang, Q. W., Song, B. W., & Jia, Y. (2016). The grey Verhulst model of the costs on the research of torpedoes. *Journal of System Simulation, 17*(2), 257–258.

Liem, D. T., Truong, D. Q., & Ahn, K. K. (2015). A torque estimator using online tuning grey fuzzy PID for applications to torque-sensorless control of DC motors. *Mechatronics, 26*, 45–63.

Lin, C. H., Liu, S. F., Fang, Z. G., & Yang, Y. J. (2022). Spectrum analysis of moving average operator and construction of time-frequency hybrid sequence operator. *Grey Systems: Theory and Application, 12*(1), 101–116.

Lin, C. T., & Yang, S. Y. (2003). Forecast of the output value of Taiwan's opto-electronics industry using the Grey forecasting model. *Technological Forecasting and Social Change, 70*(2), 177–186.

Lin, J. J., Ren, H. Q., & Shen, Z. W. (2009). Study on primary influence factors for application of grey system theory to velocity of explosive forming projectile. *Journal of Projectiles, Rockets, Missiles and Guidance, 29*(3), 112–116.

Lin, J. L., & Lin, C. L. (2002). The use of the orthogonal array with grey relational analysis to optimize the electrical discharge machining process with multiple performance characteristics. *International Journal of Machine Tools and Manufacture, 42*(2), 237–244.

Lin, J. L., Wang, K. S., Yan, B. H., et al. (1999). Optimization of parameters in GM(1,1) model by Taguchi method. *Journal of Grey System, 11*(4), 257–277.

Lin, W., & Ho, Y. X. (1996). Research and application of grey multi-target programming. In S. F. Liu & Z. X. Xu (Eds.), *New advances in grey systems research* (pp. 373–375). Press of Huazhong University of Science and Technology.

Lin, Y., & Liu, S. F. (1999a). Regional economic planning based on systemic analysis of small samples (I). *Problems of Nonlinear Analysis in Engineering Systems, 2*(10), 24–33.

Lin, Y., & Liu, S. F. (1999b). Several programming models with unascertained parameters and their application. *Journal of Multi-Criteria Decision Analysis, 8*, 206–220.

Lin, Y., & Liu, S. F. (2000a). Law of exponentiality and exponential curve fitting. *Systems Analysis Modelling Simulation, 38*(1), 621–636.

Lin, Y., & Liu, S. F. (2000b). Regional economic planning based on systemic analysis of small samples (II). *Problems of Nonlinear Analysis in Engineering Systems, 6*(11), 33–49.

Lin, Y., Liu, S. F., & McNeil, D. H. (1998). The past, present, and future of systems science research. *Systems Science and Sustainable Development* (pp. 245–252). Press of Scientific and Technological Literature.

Lin, Y. H., & Lee, P. C. (2007). Novel high-precision grey forecasting model. *Automation in Construction, 16*(6), 771–777.

Lin, Y. H., Lee, P. C., & Chang, T. P. (2009). Adaptive and high-precision grey forecasting model. *Expert Systems with Applications, 36*(6), 9658–9662.

Liu, B., Liu, S. F., & Dang, Y. G. (2003). Time series data mining techniques based on grey systems theory. *Engineering Science of China, 5*(9), 32–35.

Liu, B., Liu, S. F., Cui, Z. J., et al. (2003). Optimization of time response functions of the GM(1,1) model. *Management Science of China, 11*(4), 54–57.

Liu, J. P., Ji, C. S., & Li, H. (2001). Application of fixed weight grey clustering analysis in evaluating coal mining methods. *Journal of Coal Industry, 26*(5), 493–495.

Liu, J., Xiao, X. P., Guo, J. H., et al. (2014). Error and its upper bound estimation between the solutions of GM(1,1) grey forecasting models. *Applied Mathematics and Computation, 246*, 648–660.

Liu, P. L. (2001). Stability of continuous and discrete time-delay grey systems. *International Journal of Systems Science, 32*(7), 29–39.

Liu, S. A., Du, H. T., & Wang, X. L. (2001). Rough set theory and applications. *Systems Engineering: Theory and Practice, 21*(10), 77–82.

Liu, S. F., & Guo, T. B. (1991). *Grey systems theory and applications.* Henan University Press.

Liu, S. F. (1992). Generalized degree of grey incidence. In S. Zhang (Ed.), *Information and systems* (pp. 113–116). DMU Publishing House.

Liu, S. F. (1993). Von Neumann's economic growing turnpike of Henan Province. In W. Zheng (Ed.) *Systems science and systems engineering* (pp. 316–319). International Academic Publishers.

Liu, S. F., & Yang, L. (1994). *Regional economic evaluation, early warning and regulation.* People's Publishing House of Henan.

Liu, S. F., & Zhu, Y. D. (1994). Study on triangular model and indexes in synthetic evaluation of regional economy. In M. Chen (Ed.) Systems control information methodologies & applications (pp. 1274–1279). HUST Publishing House.

Liu, S. F., & Zhu, Y. D. (1996). Grey-econometrics combined model. *The Journal of Grey System, 8*(1), 103–110.

Liu, S. F., Li, B. J., & Dang, Y. G. (2004). The G-C-D model and technical advance. *Kybernetes: The International Journal of Systems & Cybernetics, 33*(2), 303–309.

Liu, S. F. (1994). Grey forecast of drought and inundation in Henan Province. *The Journal of Grey System, 6*(4), 279–288.

Liu, S. F. (1995). On measure of grey information. *The Journal of Grey System, 7*(2), 97–101.

Liu, S. F., & Xu, Z. X. (Eds.). (1996). *New advances in grey systems research.* Press of Huazhong University of Science and Technology.

Liu, S. F., & Forrest, J. (1997). The role and position of grey system theory in science development. *The Journal of Grey System, 9*(4), 351–356.

Liu, S. F., & Dang, Y. G. (1997). The degree of satisfaction of the floating and fixed positional solutions of LPGP. *Journal of Huazhong University of Science and Technology, 25*(1), 24–27.

Liu, S. F. (1998). Methods of grey mathematics and systems analysis for science and technology management. Ph.D. Dissertation of Huazhong University of Science and Technology, Wuhan.

Liu, S. F., Li, B. J., et al. (1998). Evaluation criteria of regional leading industries and mathematical models. *Management Science of China, 6*(2), 8–13.

Liu, S. F., & Lin, Y. (1998). *An introduction to grey systems: Foundations, methodology and applications.* IIGSS Academic publisher.

Liu, S. F., Dang, Y. G., & Fang, Z. G. (2003). *Grey systems theory and its applications (3rd ed.).* Science Press.

Liu, S. F., Guo, T. B., & Dang, Y. G. (1999). *Grey systems theory and applications* (2nd ed.). Science Press.

Liu, S. F., & Wang, Z. Y. (2000). Entropy of grey evaluation coefficient vector. *The Journal of Grey System, 12*(3), 323–326.

Liu, S. F. (2002). Evaluation criteria for comprehensive scientific and technological strength and mathematical models. *Transactions of Nanjing University of Aeronautics and Astronautics, 32*(5), 409–412.

Liu, S. F. (2003). The appearance, development and current state of grey systems theory. *Management Science of China, 16*(4), 14–17.

Liu, S. F., & Wang, R. L. (2003). Grey incidence analysis of the third industries during the ninth fifth period in Nanjing City. *Nanjing University of Science and Technology, 16*(5), 55–58.

Liu, S. F., & Lin, Y. (2004). An axiomatic definition for the degree of greyness of grey numbers. *IEEE System Man and Cybernetics,* 2420–2424.

Liu, S. F., Tang, X. W., Yuan, C. Q., et al. (2004). Order degree of China's industrial structure. *Economic Advances, 5,* 53–56.

Liu, S. F. (2004). Appearance and development of grey systems theory. *Journal of Nanjing University of Aeronautics and Astronautics, 36*(2), 267–272.

Liu, S. F., & Dang, Y. G. (2004). Technical change and the funds for science and technology. *Kybernetes: The International Journal of Systems & Cybernetics, 33*(2), 295–302.

Liu, S. F., Dang, Y. G., & Lin, Y. (2004c). Synthetic utility index method and venturous capital decision-making. *Kybernetes: The International Journal of Systems & Cybernetics, 33*(2), 288–294.

Liu, S. F., Fang, Z. G., & Lin, Y. (2006). Study on a new definition of degree of grey incidence. *Journal of Grey System, 9*(2), 115–122.

Liu, S. F., & Lin, Y. (Ed.). (2010). *Advance in grey systems research.* Springer.

Liu, S. F., & Lin, Y. (2011). *Grey systems theory and applications.* Springer.

Liu, S. F., Fang, Z. G., & Yuan, C. Q. (2012). Research on ACPI system frame for R&D management of complex equipments. *Kybernetes, 41*(5), 750–760.

Liu, S. F., Yang, Y. J., Cao, Y., et al. (2013). A summary on the research of GRA models. *Grey Systems: Theory and Application, 3*(1), 7–15.

Liu, S. F., Xu, B., Forrest, J., et al. (2013). On uniform effect measure functions and a weighted multi-attribute grey target decision model. *The Journal of Grey System, 25*(1), 1–11.

Liu, S. F., Fang, Z. G., & Yang, Y. J. (2014). On the two stages decision model with grey synthetic measure and a betterment of triangular whitenization weight function. *Control and Decision, 29*(7), 1232–1238.

Liu, S. F., Forrest, J., & Yang, Y. J. (2015). Grey system: thinking, methods, and models with applications. In *Contemporary issues in systems science and engineering* (pp. 153–224). Wiley.

Liu, S. F., Zeng, B., Liu, J. F., et al. (2015). Four basic models of GM(1, 1) and their suitable sequences. *Grey Systems: Theory and Application, 5*(2), 141–156.

Liu, S. F., Tao, L. Y., Xie, N. M., et al. (2016). On the new model system and framework of grey system theory. *The Journal of Grey System, 28*(1), 1–15.

Liu, S. F., Yang, Y. J., Xie, N. M., & Forrest, J. (2016). New progress of Grey System Theory in the new millennium. *Grey Systems Theory and Application, 6*(1), 2–31.

Liu, S. F., & Yang, Y. J. (2017a). Explanation of terms of grey forecasting models. *Grey Systems: Theory and Application, 7*(1), 123–128.

Liu, S. F., & Yang, Y. J. (2017b). Explanation of terms of grey clustering evaluation models. *Grey Systems: Theory and Application., 7*(1), 129–135.

Liu, S. F., Yang Y. J., & Forrest, J. (2017). *Grey data analysis.* Springer.

Liu, S. F., Zhang, H. Y., & Yang, Y. J. (2018). On paradox of rule of maximum value and its solution. *Systems Engineering: Theory and Practice, 38*(7), 1830–1835.

Liu, S. F., Fang, Z. G., Xie, N. M., & Yang, Y. J. (2018). Explanation of terms of Grey models for decision-making. *Grey Systems: Theory and Application, 8*(4), 382–387.

Liu, S. F., Fan, Y., Yang, Y. J., Tan, X. R., et al. (2019). Research on index system for disabled elders evaluation and grey clustering model based on end-point mixed possibility functions. *Journal of Grey System, 31*(4), 1–12.

Liu, S. F., Tang, W., Song, D. J., Fang, Z. G., & Yuan, W. F. (2020). A novel GREY-ASMAA model for reliability growth evaluation in the large civil aircraft test flight phase. *Grey Systems: Theory and Application, 10*(1), 46–55.

Liu, S. F., Li, Q., & Yang, Y. J. (2020). A novel synthetic index of two counts and mathematical model for researcher evaluation. *Grey Systems: Theory and Application, 10*(2), 85–95.

Liu, S. F., Liu, T., Yuan, W. F., & Yang, Y. J. (2022). Solving the dilemma in supplier selection by the group of weight vector with kernel. *Grey Systems: Theory and Application, 12*(3), 624–634.

Liu, W. F. (2013). Generalized relational analysis model of interval grey number. *Journal of Zhengzhou University, 45*(2), 41–44, 89.

Liu, Y. A., Chen, S. C., Zhang, M. J., et al. (2006). The application in target tracking of buffer operator and data fusion technology. *Journal of Applied Sciences, 24*(2), 154–158.

Liu, Y., Jian, L. R., et al. (2013). Probabilistic decision method of hybrid grey cluster, variable precision rough sets and fuzzy set with application. *Journal of Management Engineering, 27*(3), 110–115.

Liu, Y. X., Chen, X. T., Zhang, G. R., et al. (2004). Application of mining algorithm based on gray association rule in aluminum electrolysis control. *The Chinese Journal of Nonferrous Metals, 14*(3), 494–498.

Liu, Y. X., Yang, T. H., Li, R. D. (2007). Grey correlation analysis and prediction model of calcium sulphoaluminate formation reaction in high temperature sulfur fixation phase. *Thermal power generation, 6*, 37–40.

Liu, Z. Y. (1995). Grey production functions with n kinds of input factors. *Systems Engineering: Theory and Practice, 15*(9), 6–8.

Lu, X. H., & Wang C. L. (2013). Simulation on ATO speed controller based on grey prediction control. *City Track Traffic Development Research, 2*, 62–65.

Luo, D. (2006). A characteristic vector method for grey decision making. *Systems Engineering: Theory and Practice, 24*(4), 67–71.

Luo, D., & Liu, S. F. (2004a). Combined grey and rough decision making models. *Journal of Xiamen University, 43*(1), 26–30.

Luo, D., & Liu, S. F. (2004b). On grey dynamic programming. *Systems Engineering: Theory and Practice, 24*(4), 56–62.

Luo, D., & Liu, S. F. (2005). Grey incidence decision-making methods for systems with incomplete information. *Journal of Applied Science, 23*(4), 408–412.

Luo, D., Liu, S. F., & Dang, Y. G. (2003). Optimization of grey GM(1,1) model. *Engineering Science of China, 5*(8), 50–53.

Luo, D., Ye, L., Zhai, Y., Zhu, H., & Qian, Q. (2018). Hazard assessment of drought disaster using a grey projection incidence model for the heterogeneous panel data. *Grey Systems: Theory and Application, 8*(4), 509–526.

Luo, M. F. (1994). Fault detection, diagnosis and prognosis using grey system theory. Monash University, A Dissertation Submitted for the Degree of Ph.D in Engineering.

Luo, Q. C. (1989). Design and application of grey optimized input-output models. *Systems Engineering: Theory and Practice, 9*(5), 55–59.

Lv, D. G., Wang, L., & Zhang, P. (2007). A grey incidence method for fuzzy multi-attribute decision-making regarding structural design. *Journal of Harbin University of Technology, 39*(6), 841–844.

Lv, F. (1997). Study on the distinguishing coefficients of grey systems incidence analysis. *Systems Engineering: Theory and Practice, 17*(6), 49–54.

Mahmoudi, A., Feylizadeh, M. R., Darvishi, D., & Liu, S. F. (2018). Grey-fuzzy solution for multi-objective linear programming with interval coefficients. *Grey Systems: Theory and Application, 8*(3), 312–327.

Mahdaviani, S. H., Parvari, M., & Soudbar, D.(2016). Simultaneous multi-objective optimization of a new promoted ethylene dimerization catalyst using grey relational analysis and entropy measurement. *Korean Journal of Chemical Engineering, 33*(2), 423–437.

Mao, S. H., Gao, M. Y., & Xiao, X. P. (2015). Fractional order accumulation time-lag GM(1,N,τ) model and its application. *Systems Engineering—Theory & Practice, 35*(2), 430–436.

Memon, M. S., Lee, Y. H., & Mari, S. I. (2015). Group multi-criteria supplier selection using combined grey systems theory and uncertainty theory. *Expert Systems with Applications, 42*(21), 7951–7959.

Meng, X. F., Wang, C. M., He, F. X., et al. (2012). On life prediction of gun tube based on combined model of grey linear regression. *Journal of Nanjing University of Science and Technology, 36*(4), 635–638.

Mi, C. M., Liu, S. F., Wu, Z. P., et al. (2009). Study on the sequence of strengthening buffer operators of the opposite directional accumulation. *Control and Decision, 24*(3), 352–355.

Miao, C. H., Li, X. C., & Lu, J. H. (2018). Soil pH value grey relation estimation model based on hyper-spectral. *Grey Systems: Theory and Application, 8*(4), 436–447.

Mohammadreza, S., Seyed, H. R. H., & Shide, S. H. (2013). A fuzzy grey goal programming approach for aggregate production planning. *The International Journal of Advanced Manufacturing Technology, 64*(9–12), 1715–1727.

Morán, J., Granada, E., Míguez, J. L., et al. (2006). Use of grey relational analysis to assess and optimize small biomass boilers. *Fuel Processing Technology, 87*(2), 123–127.

Mousavi, S. M., Mirdamadi, S., Siadat, A., et al. (2015). An intuitionistic fuzzy grey model for selection problems with an application to the inspection planning in manufacturing firms. *Engineering Applications of Artificial Intelligence, 39*, 157–167.

Mu, R., & Zhang, J. Q. (2008). Layered comprehensive evaluation based on grey incidence analysis. *Systems Engineering: Theory and Practice, 28*(10), 125–130.

Nagesh, S., Narasimha, M. H. N., Pal, R., et al. (2015). Influence of nanofillers on the quality of CO_2 laser drilling in vinylester/glass using Orthogonal Array Experiments and Grey Relational Analysis. *Optics and Laser Technology., 69*, 23–33.

Olson, D. L., Zhang, J., & Wu, W. (2005).The method of grey related analysis to multiple attribute decision making problems with interval numbers. *Mathematical and Computer Modelling, 42*(9–10), 991–998.

Olson, D. L., & Wu, D. S. (2006). Simulation of fuzzy multi-attribute models for grey relationships. *European Journal of Operational Research, 175*(1), 111–120.

Özdemir, A., & Özdagoglu, G. (2017). Predicting product demand from small-sized data: Grey models. *Grey Systems: Theory and Application, 7*(1), 80–96.

Oztaysi, B. (2014). A decision model for information technology selection using AHP integrated TOPSIS-Grey: The case of content management systems. *Knowledge-Based Systems, 70*, 44–54.

Palanci, O., Alparslan Gok, S. Z., Ergun, S., et al. (2015). Cooperative grey games and the grey Shapley value. *Optimization, 64*(8 Special Issue), 1657–1668.

Pawlak, Z. (1982). Rough sets. *International Journal of Computer & Information Sciences, 11*(5), 341–356.

Pawlak, Z. (1991). *Rough sets: Theoretical aspects of reasoning about data.* Kluwer Academic Publishers.

Peng, F., Wu, G. P., & Fang, M. (2005). Grey programming cluster and application in evaluation ofoil and gas cap layer. *Journal of Hunan University of Science and Technology, 20*(2), 5–10.

Peng, Y., Zhang, X. L., Xu, W., et al. (2018). An optimal algorithm for cascaded reservoir operation by combining the grey forecasting model with DDDP. *Water Science & Technology Water Supply, 18*(1), 142–150.

Pitchipoo, P., Venkumar, P., & Rajakarunakaran, S. (2015). Grey decision model for supplier evaluation and selection in process industry: A comparative perspective. *International Journal of Advanced Manufacturing Technology, 76*(9–12), 2059–2069.

Prasad, K. S., Chalamalasetti, S. R., & Damera, N. R. (2015). Application of grey relational analysis for optimizing weld bead geometry parameters of pulsed current micro plasma arc welded inconel 625 sheets. *International Journal of Advanced Manufacturing Technology, 78*(1–4), 625–632.

Qi, M., Deng, L., Ge, D., et al. (2007). Decomposition method for colored medical figures designed by using the maximum threshold entry value of the grey incidence space. *Journal of Chinese University of Science and Technology, 37*(12), 1543–1545.

Qian, W. Y., & Dang, Y. G. (2009). The GM(1,1) model based on fluctuating sequences. *Systens Engineering: Theory and Practice, 29*(3), 149–154.

Qian, W. Y., Dang, Y. G., & Liu, S. F. (2012). Grey GM(1,1,t) model with time power and application. *Systems Engineering Theory and Practice, 32*(10), 2247–2252.

Qiao, G. L, Zhang, W. M., & Xue, S. (2009). Speed control based on fuzzy PID control with grey prediction in the deep sea stepping system. *Journal of China Coal Sciety, 34*(11), 1550–1553.

Qiu, W., Li, S., Zhao, Q. L., et al. (2007). Grey evaluation and model prediction of the forest coverage in Heilongjiang Province. *Journal of Harbin University of Technology, 39*(10), 1650–1652.

Rajesh, R., & Ravi, V. (2015). Supplier selection in resilient supply chains: A grey relational analysis approach. *Journal of Cleaner Production, 86*, 343–359.

Rajeswari, K., Lavanya, S., & Lakshmi, P. (2015). Grey fuzzy sliding mode controller for vehicle suspension system. *Control Engineering and Applied Informatics, 17*(3), 12–19.

Ramesh, S., Viswanathan, R., & Ambika, S. (2016). Measurement and optimization of surface roughness and tool wear via grey relational analysis. *TOPSIS and RSA Techniques. Measurement, 78*, 63–72.

Rao, C. J., Xiao, X. P., et al. (2006). Generalized accumulated generating operation and its generating space. *Dynamics of Continuous Discrete and Impulsive Systems-Series B-Applications & Algorithms, 13*, 517–521.

Rao, R., & Yadava, V. (2009). Multi-objective optimization of Nd:YAG laser cutting of thin superalloy sheet using grey relational analysis with entropy measurement. *Optics & Laser Technology, 41*(8), 922–993.

Ruan, A. Q., & Liu, S. F. (2007). Grey GERT networks and precision of grey estimates based on customers' demand. *Systems Engineering, 25*(12), 100–104.

Sadeghi, M., Rashidzadeh, M. A., & Soukhakian, M. A. (2012). Using analytic network process in a group decision-making for supplier selection. *Informatica, 23*(4), 621–643.

Sahoo, S., Dhar, A., & Kar, A. (2016). Environmental vulnerability assessment using Grey Analytic Hierarchy Process based model. *Environmental Impact Assessment Review, 56*, 145–154.

Salmeron, J. L., & Gutierrez, E. (2012). Fuzzy Grey Cognitive Maps in reliability engineering. *Applied Soft Computing, 12*, 3818–3824.

Samet, H., & Mojallal, A. (2014). Enhancement of electric arc furnace reactive power compensation using Grey-Markov prediction method. *IET Generation Transmission & Distribution, 8*(9), 1626–1636.

Sarpkaya, C., & Sabir, E. C. (2016). Optimization of the sizing process with grey relational analysis. *Fibres & Textiles in Eastern Eupore, 24*(1), 49–55.

Senthilkumar, N., Sudha, J., & Muthukumar, V. (2015). A grey-fuzzy approach for optimizing machining parameters and the approach angle in turning AISI 1045 steel. *Advances in Production Engineering & Management, 10*(4), 195–208.

Sevastjanov, P., & Dymova, L. (2009). A new method for solving interval and fuzzy equations: Linear case. *Information Sciences, 179*, 925–937.

Shen, V., Chung, Y. F., & Chen, T. S. (2009). A novel application of grey system theory to information security. *Computer Standards & Interfaces, 31*(2), 277–281.

Shui, N. X., & Qing, T. C. (1998). Some theoretical problems about grey systems' GM(1,1) model. *Systems Engineering: Thoery and Practice, 18*(4), 59–63.

Singh, K. J. (2018). Optimization of process parameters of powder mixed EDM for high carbon high chromium alloy steel (D2 steel) through GRA approach. *Grey Systems: Theory and Application, 8*(4), 388–398.

Singh, T., Patnaik, A., & Chauhan, R. (2016). Optimization of tribological properties of cement kiln dust-filled brake pad using grey relation analysis. *Materials & Design, 89*, 1335–1342.

Song, Z. M., Liu, X. Q., & Wang, S. Z. (1997). Curvature simulation method of grey exponential curves. *Systems Engineering: Theorey and Practice, 17*(6), 55–57.

Song, Z. M., Tong, X. J., & Xiao, X. P. (2001). Grey GM(1,1) model of the type of central approximation. *Systems Engineering: Theorey and Practice, 21*(5), 110–113.

Song, Z. Q., & Tan, C. Q. (1997). Analysis criteria, treatments and applications of refined evaluation of petroleum deposits using grey systems theory. *Systems Engineering: Theory and Practice, 17*(3), 74–82.

Su, B. T., & Xie, N. M. (2018). Research on safety evaluation of civil aircraft based on the grey clustering model. *Grey Systems: Theory and Application, 8*(1), 110–120.

Sun, X. D., Jiao, Y., & Hu, J. S. (2005). A decision-making method based on grey incidences and ideal solutions. *Management Science of China, 13*(4), 63–68.

Sutanto, S., Go, A.W., Ismadji, S., et al. (2015). Taguchi Method and grey relational analysis to improve in situ production of FAME from sunflower and jatropha curcas kernels with subcritical solvent mixture. *Journal of the American Oil Chemists Society, 92*(10), 1513–1523.

Tabaszewski, M., & Cempel, C. (2015). Using a set of GM(1,1) models to predict values of diagnostic symptoms. *Mechanical Systems and Signal Processing, 52–53*, 416–425.

Tamura, Y., Zhang, D. P., Umeda, N., & Sakeshita, K. (1992). Load forecasting using grey dynamic model. *The Journal of Grey System, 4*(4), 49–58.

Tan, G. J., Wang, L. L., & Cheng, Y. C. (2011a). Prediction method for cable tension state of cable-stayed bridges based on grey system theory in could areas. *Journal of Jilin University (Engineering and Technology Edition), 41*(2), 170–173.

Tan, J. B., & Lv, Y. J. (2008). Weak uniformity and uniformity of grey judgment matrices and their properties. *Systems Engineering: Theory and Practice, 28*(3), 159–165.

Tan, X. R. (1997). Grey medical incidence theory and applications. Ph.D. Dissertation of Huazhong University of Science and Technology, Wuhan.

Tan, X. R., Deng, J. L., Ren, S. Y., et al. (2011). Study on grey relational methodology of clinical trials. Scientific and technological achievements of China.

Tang, K., Zhou, N., & Fan X. Y. (2012). Analysis on the factors influencing the gas well productivity of S2 gas pool in Permian of Zizhou gas field. *Computing Techniques for Geophysical and Geochemical Exploration, 34*(6), 723–728.

Tang, W. M. (2006). A new prediction model based on grey support vector machine. *Journal of Systems Engineering, 21*(4), 410–413.

Tang, W. X. (1995). A new method for estimating the parameters of the GM(1,1) model and hypothesis testing. *Systems Engineering: Theory and Practice, 15*(3), 20–49.

Talafuse, T. P., & Pohl, E. A. (2018). Small sample discrete reliability growth modeling using a grey systems model. *Grey Systems: Theory and Application, 8*(3), 246–271.

Thananchai, L. (2008). Grey prediction on indoor comfort temperature for HVAC systems. *Expert Systems with Applications, 34*(4), 2284–2289.

Tian, J. Y., & Lu, Y. (2007). Research on grey prediction model of the slab temperature in heating furnace. *Journal of Northeast University, 28*(S1), 6–10.

Tong X. J., Chen M. Y., & Zhou L. (2002). On AGO effect of the grey model. *Systems Engineering-Theory and Prectace, 22*(11), 121–125.

Vallee, R. (2008). Book reviews: Grey information: Theory and Practical applications. *Kybernetes, 37*(1), 89.

Varun, A., & Venkaiah, N. (2015). Grey relational analysis coupled with firefly algorithm for multi-objective optimization of wire electric discharge machining. *Proceedings of the Institution of Mechanical Engineers Part B-Journal of Engineering Manufacture, 229*(8), 1385–1394.

Verma, A., Sarangi, S., & Kolekar, M. H. (2014). Stator winding fault prediction of induction motors using multiscale entropy and grey fuzzy optimization methods. *Computers & Electrical Engineering, 40*(7), 2246–2258.

Wang, H. Y., Liu, L., & Liu, L. (2007). Applied study of BP neural networks based on GRA and PCA. *Management Review, 19*(10), 50–55.

Wang, J. F., & Liu, S. F.(2009). On measuring and sorting for the efficiency index of interval DEA based on interval position and grey incidence model. *Systems Engineering and Elactronics, 31*(6), 2146–2150.

Wang, J. J., & Jing, Y. Y. (2008). Integrated evaluation of distributed triple-generation systems using improved grey incidence approach. *Energy, 33*(9), 1427–1437.

Wang, J. Q., Ren, S. X., & Chen, X. H. (2009a). Preference ordering of grey stochastic multi-criteria decision makings. *Control and Decision Making, 24*(5), 701–705.

Wang, L. J. (2009). Study on cooperation mechanism of one to many in perishable products system. *Journal of Huazhong University of Science and Technology, 37*(8), 12–15.

Wang, W., Wu, M., Cao, W. H., et al. (2010). Fuzzy-expert control based on combination grey prediction model for flue temperature in coke oven. *Control and Decision, 25*(2), 185–190.

Wang, W. P. (1997). Study on grey linear programming. *The Journal of Grey System, 9*(1), 41–46.

Wang, W. P., & Deng, J. L. (1992). Intension measurement of information contained in grey systems propositions. *Grey Systems Theory and Practice, 2*(1), 41–43.

Wang, W. P. (1994). Theory and methods on dealing with grey information. Ph.D. Dissertation of Huazhong University of Science and Technology, Wuhan.

Wang, X. C. (2007). On relationship between rough sets and grey systems. *Fuzzy Systems and Mathematics, 20*(6), 129–135.

Wang, X. L., & Nie, H. (2008). On a method of fatigue life prediction based on grey system model GM(1,1). *Transaction of Nanjing University of Aeronautics and Astronautics, 40*(6), 845–848.

Wang, X. Y., & Yang, X. (1997). Grey systems methods for quality predictive control and diagnosis. *Systems Engineering: Theory and Practice, 17*(5), 105–108.

Wang, X. M. (1993). Grey dynamic models for analyzing economic growths and cycles. *Systems Engineering: Theory and Practice, 13*(1), 42–47.

Wang, X. M., & Luo, J. J. (1986). *Collected programs for grey systems prediction, decision-making, and modeling.* Press of Science Dissemination.

Wang, X. M., & Luo, J. J. (1986). *Software for grey system modeling of forecasting and Decision-making.* Science Popularization Press.

Wang, X. M., Zhang, J. Z., & Wang, R. (2001). *Grey system analysis and practical calculation program.* Wuhan: Press of Huanzhong University of Science and Technology.

Wang, Y. H., & Dang, Y. G. (2009). A comprehensive posterior evaluation method based on the grey fixed weight clustering of the D-S evidence theory. *Systems Engineering: Theory and Practice, 29*(5), 123–128.

Wang, Y. Y., & Cao, Y. H. (2010). Grey neural network model of aviation safety risk. *Journal of Aerospace Power, 25*(5), 1036–1042.

Wang, Z. X., Dang, Y. G., & Liu, S. F. (2008). An optimized GM(1,1) model based on discrete exponential functions. *Systems Engineering: Theory and Practice, 28*(2), 61–67.

Wang, Z. X., Dang, Y. G., & Liu, S. F. (2009). Variable buffer operators and their acting strengths. *Control and Decision, 29*(8), 1218–1222.

Wang, Z. X. (2013). On derivative model of power model GM(1,1). *Systems Engineering Theory and Practice, 33*(11), 2894–2902.

Wang, Z. X., Hipel, K. W., Wang, Q., et al. (2011). An optimized NGBM(1,1) model for forecasting the qualified discharge rate of industrial wastewater in China. *Applied Mathematical Modeling, 35,* 5524–5532.

Wang, Z. L., Liu, Y. L., & Shi, K. Q. (1995). On Kenning grey degree. *The Journal of Grey System, 7*(2), 103–110.

Wang, Z. L. (1998). A theory for grey modeling techniques. Ph.D. Dissertation of Huazhong University of Science and Technology, Wuhan.

Wang, Z. L., & Li, X. Z. (1996). Ordering and inequalities of grey numbers. In S. F. Liu & Z. X. Xu (Eds.), *New Advances of Grey Systems Research* (pp. 364–366). Press of Huazhong University of Science and Technology.

Wei, Y., Kong, X. H., & Hu, D. H. (2011). A kind of universal constructor method for buffer operators. *Grey Systems: Theory and Application, 1*(2), 178–185.

Wei, Y., Zeng, K. F. (2015). The simplified relational axioms and the axiomatic definition of special incidence degrees. *Systems Engineering—Theory & Practice, 35*(6), 1528–1534.

Wiecek-Janka, E., Nowak, M., & Borowiec, A. (2019). Application of the GDM model in the diagnosis of crises in family businesses. *Grey Systems: Theory and Application, 9*(1), 114–127.

Wong, H., Hu, B. Q., & Xia, J. (2006). Change-point analysis of hydrological time series using grey relational method. *Journal of Hydrology., 324*(1), 323–338.

Wu, A. X., Xi, Y., Yang, B. H., Chen, X. S., et al. (2007). Study on grey forecasting model of copper extraction rate with bioleaching of primary sulfide ore. *Acta Metallurgica Sinica, 20*(2), 117–128.

Wu, C. C., & Chang, N. B. (2003). Grey input–output analysis and its application for environmental cost allocation. *European Journal of Operational Research, 145*(1), 175–201.

Wu, C. C., & Chang, N. B. (2004). Corporate optimal production planning with varying environmental costs: A grey compromise programming approach. *European Journal of Operational Research, 155*(1), 68–95.

Wu, D. D., & Olson, D. L. (2010). Fuzzy multiattribute grey related analysis using DEA. *Computers & Mathematics with Applications, 60*(1), 166–174.

Wu, J. H., & Wen, K. L. (2001). Rolling error in GM(1,1) modeling. *Journal of Grey System, 13*(1), 77–80.

Wu, L. F., Liu, S. F., Fang, Z. G., et al. (2015). Properties of the GM(1,1) with fractional order accumulation. *Applied Mathematics and Computation, 252*, 287–293.

Wu, L. F., Liu, S. F., Yao, L. G., et al. (2015). Using fractional order accumulation to reduce errors from inverse accumulated generating operator of grey model. *Soft Computing, 19*(2), 483–488.

Wu, L. F., Liu, S. F., & Yang, Y. J. (2016). A grey model with a time varying weighted generating operator. *IEEE Transactions on Systems Man Cybernetics-Systems, 46*(3), 427–433.

Wu, L. F., Liu, S. F., & Yang, Y. J. (2018). Using the fractional order method to generalize strengthening buffer operator and weakening buffer operator. *IEEE-CAA Journal of Automatica Sinica, 5*(6), 1074–1078.

Wu, L. Y., Wu, Z. P., & Li, M. (2013). Quadratic time- varying parameters discrete grey model. *Systems Engineering Theory and Practice, 33*(11), 2887–2892.

Wu, Q., & Liu, Z. T. (2009). Real formal concept analysis based on grey-rough set theory. *Knowledge-Based Systems, 22*(1), 38–45.

Wu, Y., Yang, S. Z., & Tao, J. H. (1988). Discussions on grey prediction and time series prediction. *Journal of Huazhong University of Science and Technology, 16*(3), 27–33.

Wu, Z. P., Liu, S. F., Cui, L. Z., et al. (2009). Some new weakening buffer operators designed on monotonic functions. *Control and Decision Making, 24*(7), 1055–1058.

Wu, Z. J., & Wang, A. M. (2007). Grey-fuzzy comprehensive evaluations for the planning of reconstructible manufacturing systems. *Mechanic Engineering of China, 18*(19), 2313–2318.

Xia, J., & Zhao, H. Y. (1996). Grey artificial neural network models and applications in short-term prediction of runoffs. *Systems Engineering: Theory and Practice, 16*(11), 82–90.

Xia, T. B., Jin, X. N., Xi, L. F., et al. (2015). Operating load based real-time rolling grey forecasting for machine health prognosis in dynamic maintenance schedule. *Journal of Intelligent Manufacturing, 26*(2), 269–328.

Xiao, J., & Zhang, W. W. (2009). Grey incidence analysis applied to fault diagnosis of drone crash. *Journal of Sichuan Ordnance Engineering, 30*(9), 112–115.

Xiao, X. P., Xie, L. C., & Huang, D. R. (1995). A computational improvement of degrees of grey incidence and applications. *Mathematical Statistics and Management, 14*(5), 27–30.

Xiao, X. P. (2002). On methods of grey systems modeling. Ph.D. Dissertation of Huazhong University of Science and Technology, Wuhan.

Xiao, X. P., Song, Z. M., & Li, F. (2004). *Foundations of grey technology and applications*. Science Press.

Xiao, X. P., Liu, J., Guo, H. (2013). Properties and optimization of generalized accumulation grey model. *Systems Engineering Theory and Practice, 33*(1), 1–9.

Xiao, Y., Shao, D. G., Deng, R., et al. (2007). Comprehensive evaluation supporting systems for the effects of hydraulic engineering projects. *Journal of Wuhan University, 40*(4), 49–52.

Xie, N. M., & Liu, S. F. (2005). Research on discrete grey model and its mechanism. *In IEEE International Conference on Systems, Man and Cybernetics, v 1, IEEE Systems, Man and Cybernetics Society, Proceedings - 2005 International Conference on Systems, Man and Cybernetics* (pp. 606–610).

Xie, N. M., & Liu, S. F. (2006). Generalized discrete grey models and their optimal solutions. *Systems Engineering: Theory and Practice, 26*(6), 108–112.

Xie, N. M., & Liu, S. F. (2007). Research on the multiple and parallel properties of several grey relational models. In *International Conference on Grey Systems and Intelligence Service* (pp. 183–188).

Xie, N. M., & Liu, S. F. (2008a). Discrete grey models of multi-variables and applications. *Systems Engineering: Theory and Practice, 28*(4), 100–107.

Xie, N. M., & Liu, S. F. (2008b). Characteristics of discrete grey models of approximately exponential sequences. *Systems Engineering and Electronics, 30*(5), 863–867.

Xie, N. M., & Liu, S. F. (2008c). Affine characteristics of discrete grey models. *Control and Decision, 23*(2), 200–203.

Xie, N. M., & Liu, S. F. (2009). Discrete grey forecasting model and its optimization. *Applied Mathematical Modelling, 33*(1), 1173–1186.

Xie, N. M., Liu, S. F., & Yuan C. Q. (2014). Grey number sequence forecasting approach for interval analysis: A case of China's gross domestic product prediction. *The Journal of Grey System, 26*(1), 45–58.

Xie, N. M. (2018). Interval grey number based project scheduling model and algorithm. *Grey Systems: Theory and Application, 8*(1), 100–109.

Xie, Y. M., Yu, L. P., & Chen, J. (2007). Application of grey theory in deep drawing robust design. *Chinese Journal of Mechanical Engineering, 43*(3), 54–59.

Xiong, H. J., Chen, M. Y., & Ju, T. (1999). Two classes of grey models of control systems. *Journal of Wuhan Jiaotong Science and Technology, 23*(5), 465–468.

Xiong, H. J., Xiong, P. F., Chen, M. Y., et al. (1999). A generalization of the grey systems SCGM(1,h)_c model. *Journal of Huazhong University of Science and Technology, 27*(1), 1–3.

Xiong, P. P., Zhang, Y., Zeng, B., & Yao, T. X. (2017). MGM(1, m) model based on interval grey number sequence and its applications. *Grey Systems: Theory and Application, 7*(3), 310–319.

Xu, C., Sun, X. Y., & Wang, H. L. (2010). Application grey-economitrics model in traffic volume prediction. *Highway Engineering, 35*(5), 34–38.

Xu, J. P. (1995). On a kind of information grey number. *The Journal of Grey System, 7*(2), 111–130.

Xu, W. X., & Zhang, Q. S. (2001). An integrated computational scheme basedon on grey theory and fuzzy mathematics. *Systems Engineering: Theory and Practice, 21*(4), 114–119.

Xu, Y. D., & Wu, Z. Y. (2003). Grey pricing of stocks based on limited rationality and inefficiency of stock markets. *Journal of Management Engineering, 17*(2), 115–117.

Xu, Z. X., & Wu, G. P. (1993). *Grey Systems theory and grey prediction on mineral deposits*. Press of Chinese University of Geology.

Yamaguchi, D., Li, G. D., & Nagai, M. (2007). A grey-based rough approximation model for interval data processing. *Information Sciences, 177*(21), 4727–4744.

Yan, L. Y., & Mao, L. X. (1998). An application of the GM(1,1) model. *Systems Engineering: Theory and Practice, 18*(10), 104–106.

Yan, S. L., Liu, S. F., Zhu, J. J., et al. (2014). The ranking method of grey numbers based on relative kernel and degree of accuracy. *Control and Decision, 29*(2), 315–319.

Yan, S. L., Liu, S. F., Fang, Z. G., et al. (2014). Method of determining weights of decision makers and attributes for group decision making with interval grey numbers. *Systems Engineering—Theory & Practice, 34*(9), 2372–2378.

Yan, S. L., Liu, S. F., Liu, J. F., et al. (2015). Dynamic grey target decision making method with grey numbers based on existing state and future development trend of alternatives. *Jouranl of Intelligent & Fuzzy Systems, 28*(5), 2159–2168.

Yang, J., & Wong, W. G. (2014). Improved unbiased grey model for prediction of gas supplies. *Journal of Tsinghua University (Science & Technology), 54*(2), 145–148.

Yang, S., Ren, P., & Dang, Y. G. (2009). Opposite directional accumulation generation and optimization of grey GOM(1,1) model. *Systems Engineering: Theory and Practice, 29*(8), 160–164.

Yang, T.S., Yang, P., Dong, X.S., et al. (2008). Method for predicting fault status of satellite based on gray system theory. *Computer Measurement & Control, 16*(9), 1284–1285, 1307

Yang, Y. A., Wei, J., Feng, Z. R., et al. (2008). Application of grey incidence analysis in the selection of the optimal initial orbit of satellites. *Systems Engineering and Electronic Technology, 30*(2), 308–311.

Yang, Y., Wang, S. W., Hao, N. L., Shen, X. B., & Qi, X. H. (2009). On-line noise source identification based on the power spectrum estimation and grey relational analysis. *Applied Acoustics, 70*(3), 493–497.

Yang, Y. J., Liu, S. F., & John, R. (2014). Uncertainty representation of grey numbers and grey sets. *IEEE Transactions on Cybernetics, 44*(9), 1508–1517.

Yao, J. B., & Hu, W. W. (2008). Grey evaluation of operational efficiency of OTH ground-wave radar. *Armament Automation, 27*(4), 12–14.

Yao, T. X., & Liu, S. F. (2009). Characteristics and optimization of discrete GM(1,1) model. *Systems Engineering: Theory and Practice, 29*(3), 142–148.

Yao, T. X., Liu, S. F., & Xie, N. M. (2010). Study on the properties of new information discrete GM(1,1) model. *Journal of System Engineering, 25*(2), 164–170.

Ye, J., Li, B. J., & Liu, F. (2014). Forecasting effect and applicability of weakening buffer operators on GM(1, 1). *Systems Engineering-Theory and Practice, 34*(9), 2364–2371.

Yeh, Y. L., Chen, T. C., & Lin, C. N. (2001). Effects of data characteristics for GM(1,1) modeling. *Journal of Grey System, 13*(3), 121–130.

Yi, C. H., & Gu, P. L. (2003). Predictions on energy using buffer operators of grey sequences. *Journal of Systems Engineering, 18*(2), 189–192.

Yin, J. J., Liang, X. J., Xiao, C. L., et al. (2012). Application of matter-element extension method based on grey clustering theory in the ground water quality evaluation-Example with Taonan City. *Water Saving Irrigation, 6*, 52–55.

Yuan, Z. J., Sun, C. X., Yuan, Z. Y., et al. (2005). Method of grey clustering decision2making to state assessmentof power transformer. *Journal of Zhongqing University, 28*(3), 22–25.

Zhao, G. G., Sun, Y. K., Xu, Y. J., et al. (2007). Grey decision ana lysis of threat estimation in ant-im issile combat of surface warship. *Tactical Missile Technology, 3*, 32–35.

Zeng, B., Liu, S. F., & Meng, W. (2011). Development and application of MSGT 6.0 (modeling system of grey theory 6.0) based on Visual C# and XML. *The Journal of Grey System, 23*(2), 145–154.

Zeng, B., & Liu, S. F. (2014). Prediction model of stochastic oscillation sequence based on amplitude compression. *Systems Engineering Theory and Practice, 34*(8), 2084–2091.

Zeng, X. Y., & Xiao, X. P. (2009). Ways to generalize the GM(1,1) model and applications. *Control and Decision Making, 24*(7), 1092–1096.

Zhang, D. H., Shi, K. Q., & Jiang, S. F. (2001). Grey predicting power load via GM(1,1) modified. *Journal of Grey System, 13*(1), 65–67.

Zhang, J. J., Wu, D. S., & Olson, D. L. (2005). The method of grey related analysis to multiple attribute decision making problems with interval numbers. *Mathematical and Computer Modelling, 42*(9), 991–998.

Zhang, K., & Liu, S. F. (2009). A novel algorithm of image edge detection based on matrix degree of grey incidences. *The Journal of Grey System, 9*(3), 265–276.

Zhang, K. (2014). Multi-variables discrete grey model based on driver control. *Systems Engineering Theory and Practice, 34*(8), 2084–2091.

Zhang, K., Zhong, Q., & Zuo, Y. (2017). Multivariate grey gradient incidence model and its application. *Grey Systems: Theory and Application, 7*(2), 236–246.

Zhang, L., Ren, L.Q., Tong, J., et al. (2004). Study of soil-solid adhesion by grey system theory. *Progress in Natural Science, 14*(2), 119–124.

Zhang, L. P., Zheng, Y. L., & Zhang, X. L. (2015). An overview of the application of grey system theory in modern medicine. *Journal of Mathematical Medicine, 28*(3), 430–434.

Zhang, M. A., Yuan, Y. B, Zhou, J., et al. (2008). Coordinated urban economic responsibility for the aftermath of potential disasters as analyzed using grey systems models. *Systems Engineering: Theory and Practice, 28*(3), 171–176.

Zhang, N. L., Meng, X. Y., Li, Z. H., et al. (2009). Grey two-layered programming problem and its solution. *Systems Engineering: Theory and Practice, 27*(11), 132–138.

Zhang, P., & Sun, Q. (2007). Prediction of the economic development of Baotou city using comparisons of grey theory and statistics. *Mathematical Statistics and Management, 26*(4), 595–601.

Zhang, Q. S. (1996). Deviate information theory of grey hazy sets. Ph.D. Dissertation of Huazhong University of Science and Technology, Wuhan.

Zhang, Q. S. (2001). Difference information entropy in grey theory. *Journal of Grey System, 13*(2), 111–116.

Zhang, Q. S., Deng, J. L., & Fu, G. (1995). On grey clustering in grey hazy set. *The Journal of Grey System, 7*(4), 377–390.

Zhang, Q. S., Han, W. Y., & Deng, J. L. (1994). Information Eentropy of discrete grey number. *The Journal of Grey System, 6*(4), 303–314.

Zhang, X. Y., Wang, Z. L., & Masatake, N. (2006). Research on affective interaction models of robot. *Computer Engineering, 32*(24), 6–12.

Zhao, P. D., & Xia, Q. L. (2009). Achievements and contributions of Chinese scholars in the development of mathematical geology. *Geoscience—Journal of China University of Geosciences, 34*(2), 225–231.

Zheng, D. J., Gu, C. S., & Wu, Z. R. (2005). On time varying prediction model of the deformation of slope with multi-factors. *Journal of Rock Mechanics and Engineering, 24*(17), 3180–3184.

Zheng, Z. N., Wu, Y. Y., & Bao, H. L. (2001). Pathological problems of the GM(1,1) model. *Management Science of China, 9*(5), 38–44.

Zhou, C. S., & Deng, J. L. (1989). Stability analysis of grey discrete-time systems. *IEEE Transactions on Automatic Control, 34*(2), 173–175.

Zhou, C. S., & Deng, J. L. (1986). The stability of grey linear system. *International Journal of Control, 43*(1), 313–320.

Zhou, W. J., Dang, Y. G., & Xiong, P. P. (2013). Grey clustering model for interval grey number with variable and fixed weights. *Systems Engineering Theory and Practice, 33*(10), 2590–2595.

Zhou, Z. J., & Hu, C. H. (2008). An effective hybrid approach based on grey and ARMA for forecasting gyro drift. *Chaos, Solitons & Fractals, 35*(3), 525–529.

Zhu, J. M., Lei, J. T., Huang, Z. W., et al. (2012). Pneumatic position servo control system based on grey correlation compensation control. *Journal of Mechanical Engineering, 48*(20), 159–167.

Zhu, S., Shi, L. Y. (2013). Research on supervision of private equity investment fund based on grey game theory. *Journal of Northeast University, 34*(7), 1057–1060.

Index